The Human Canopy

The Human Canopy

Homo erectus, Homo soloensis,
Homo pekinensis and Homo floresiensis

Valéry Zeitoun

BAR International Series 1937
2009

Published in 2016 by
BAR Publishing, Oxford

BAR International Series 1937

The Human Canopy

ISBN 978 1 4073 0452 6

BAR Publishing is the trading name of British Archaeological Reports (Oxford) Ltd. British Archaeological Reports was first incorporated in 1974 to publish the BAR Series, International and British. In 1992 Hadrian Books Ltd became part of the BAR group. This volume was originally published by John and Erica Hedges Ltd. in conjunction with British Archaeological Reports (Oxford) Ltd / Hadrian Books Ltd, the Series principal publisher, in 2009. This present volume is published by BAR Publishing, 2016.

Printed in England

BAR
PUBLISHING

BAR titles are available from:

BAR Publishing
122 Banbury Rd, Oxford, OX2 7BP, UK
EMAIL info@barpublishing.com
PHONE +44 (0)1865 310431
FAX +44 (0)1865 316916
www.barpublishing.com

The research described in this volume was under-taken in the Laboratorium Bio dan Paleo anthropologi at the University of Gadjah Mada in Yogjakarta, at the Geological Research and Development Centre of Bandung and at Pusbang Arkeologi, Jakarta in Indonesia, at the Forschung Senckenberg Institute of Frankfurt in Germany, at the Rijksmuseum for Natural History at Leiden in the Netherlands, at the Musée de l'Homme, the Galerie de Paléontologie and the Laboratoire d'Anatomie Comparée of the Muséum national of d'Histoire Naturelle in Paris as well as at the Laboratoire de Prehistoire et d'Anthropologie de l'Université de Médecine in Marseille and at the Laboratoire d'Anthroplogie de l'Université de Bordeaux I in France.

CONTENTS

Acknowledgements

This study takes as a starting point the metho-dological basis of research published in 2001 and enhanced by the detailed analysis of several hitherto unseen fossils. The work presented reached its conclusion thanks to the support of Véronique Barriel, Pascal Tassy, Philippe Janvier and Anne-Marie Tillier. The resulting study is in homage to Professor Teuku Jacob who, through almost fifteen years, welcomed me so warmly in his laboratory and permitted me to study the whole collection of fossils available. The study is also in homage to Professor Sastrohamidjojo Sartono who through his boundless enthusiasm enabled me to visit and work in the different Institutes of Bandung. John de Vos, Jens Lorenz Franzen, Fachroel Aziz, Tony Djubi-antono and Harry Widianto allowed me to study original fossils in their care and I am very grateful to them. Thanks to their generosity in welcoming me, Madame Marie-Antoinnette De Lumley and Monsieur Henry De Lumley enabled me to include in this study the ensemble of fossils attributed to *Homo erectus*.

My thanks would be incomplete without mentioning the BLEUSTEIN BLANCHET found-ation for the Vocation that allowed me to make my first research trip in Indonesia and also the FYSSEN foundation for the help to undertake my laboratory work. Finally, I must thank Yves Coppens and Herbert Thomas of the Collège de France for their support.

I am grateful to Laurence Billault for the making off of the book and to Garry Holding for the translation. Finally through the British Archaeological Reports, thanks must be given to good old England for remaining a refuge of freedom of speech ; a freedom which has been forgotten and battered by others in spite of many revolutions.

The major interest of palaeoanthropologists resides in the understanding of evolutional mechanisms which led to the origins of our species: *Homo sapiens* and furthermore, our sub-species: *Homo sapiens sapiens*. In this quest for our origins, research over recent years has focussed on the variability which could exist amongst the different groups of human fossils and, more particularly, that concerning our Neandethalian cousins (Maureille, 2007). A palaeo-populationary approach was rendered possible through the relatively large number of Neandertalian fossils available. The definition of human fossil groups (their taxonomic rank) and their relationship structure linking the different groups (their phylogenetic relationship) have not, however, been clearly established. It will be accepted that it is difficult to talk of the variability of a particular category before the defining limits can be established. Answers to these questions (definition of limits and variability) must be found before research can continue and before evolutionary mechanisms can be understood. Certain authors – for the most part biologists – see a resurgence of typology in this approach. In fact, this is necessarily the case as it is not possible to consider variability before biological limits have not been determined.

This study proposes to examine the case of *Homo erectus* (Dubois, 1893) whose phylogenetic position and taxonomic status remain unclear inspite of considerable research aimed at identifying this taxon from archaic forms of *Homo sapiens*. The analysis put forward here belongs to the domain of alpha-taxonomy and will lead to the taking into consideration of the definition of several human fossil taxons. Particular interest will be taken of South-East Asia – the geographical zone which saw the birth of *Homo erectus* in the history of science. Even if there is no consensus of opinion concerning the *tempo* or any clear definition of his spreading (see Howells, 1980; Tillier and Vander-meersch, 1982; Pope and Cronin, 1984; Hooijer and Kurten, 1984), *Homo erectus sensu lato* is the first acknowledged taxon of the species *Homo* away from its African origins. Studying this fossil is therefore of undeniable interest when considering human evolution. Furthermore, as Howells (1980) underlined, its role in the coming of the species *Homo sapiens* is, without doubt, one of the most important subjects concerning Pleistocene hominoids. We shall see that this point of view continues to be held inspite of different labels applied to African *Homo erectus* specimens: *Homo ergaster* (Wood, 1984; Stringer 1984; Andrews, 1984; Groves, 1989), or others coming just from Africa: *Homo georgicus* (de Lumley & Lordkipanidze, 2006) and even *Homo antecessor* (Bermudez *et al.,* 1997). According to Piveteau (1982), the biological status of *Homo erectus* has evolved in line with discoveries to a far greater extent than for any other human taxonomic line, but even more so in accordance with scientific paradigms. Any study of *Homo erectus* therefore presents an additional epistemological aspect which is interesting to note. The large number of controversies centered on the choice of criteria, methodology to be employed in the recognition of a species and the plurality of interpretations issuing from anatomical and chronological data (divergences concerning observations, disagreement over stratigraphical positions) accord *Homo erectus* the status of group fossil which is both heterogenous and badly defined. The different methods of phlogenetic reconstruction, because they are based on different and even contradictory conceptions as well as on too frequently authoritative arguments, have also contributed in sowing confusion. One of the aims of this work being presented is to present an analysis of the kinship and relationships and to put forward a taxonomic status for fossil specimens brought to the *Homo erectus* species by different authors. Three linking questions will therefore be discussed:

- What is the taxonomic status of *Homo erectus*?
- What is its phylogenetic position?
- What place do Solo Man and Flores Man occupy?

A study of the *calvaria* was undertaken in order to find answers to these questions. This anatomical complex is one of the best preserved and best represented amongst human fossils. It allows us to compare a large number of fossils and consists of a considerable number of anatomical characteristics reputed to be diagnostic. In addition, in the «legal» register of international zoological classification, this anatomical complex is that which originally permitted the identification of *Homo erectus* from a type specimen: the Man of Trinil. This study will also permit the putting forward of a formal answer on this point.

I. Different Acceptances of *Homo erectus*.

I.1. The Age of Discoveries.

Remains of *Homo erectus* were first put to light in Asia towards the end of the 19th Century. In 1890, Dubois discovered a small fragment of a mandible at Kedung Brubus (Java) and in 1891, the crown of a skull of Trinil (the renowned Pithecanthropus) was unearthed from the banks of the Solo River in Central Java. In the 1920's, the Sinanthropes were discovered in China near Pekin. Following this, in the 1930's, with the work of von Koenigswald, the volcanic dome of Sangiran (Java) proved to be very rich in terms of human fossils relating to this group. In Africa, and according to Rightmire (1990), the first peremptory proof of the presence of *Homo erectus* was the discovery of the mandibles of *Atlanthropus mauritanicus* Arambourg, 1954. In 1960 L.S.B. Leakey discovered the specimen « OH9 » at Olduvai, in Tanzania. Since the 1970's and upto today, numerous fossils attributed to *Homo erectus*, have been found in East Africa, China and Indonesia. The most exceptional example, consisting of a whole skeleton was found in the banks of the Nariokotome in Kenya (Brown *et al.*, 1985; Walker and Leakey, 1994). The most impressive series – in terms of polymorphism – is certainly that of Dmanisi (de Lumley *et al.*, 2006; Skinner *et al.*, 2006). A polymorphism, in any case, which begs the legitimate question of the uniqueness of this species (Lee, 2005). According to the summary chart drawn-up by Howell (1991), *Homo erectus* appears some 1.9 million years ago in Eastern Africa following excavations in East Turkana and Olduvai. Following on, *Homo erectus* spreads and develops in Africa and also in Asia. Radiochronological data positions a very ancient *Homo erectus* in Indonesia (Jacob and Curtis, 1971; Swisher *et al.*, 1994). The last representatives of the species would also seem to be Asian. In fact, Indonesian fossils from the Ngandong site – first of all dated as being some 300,000 years old now seem to be between 25,000 and 52,000 years old (Swischer *et al.*, 1996) or between 35,000 and 50,000 years for the Sambungmacan 1 specimen which belongs to the same anatomical series (Falguères *et al.,* 2001). Flores Man could also be a very late representative (Zeitoun *et al.*, 2007).

The presence of *Homo erectus* in Europe is subject to controversy (Tillier and Vandermeersch, 1982; Wood, 1991; Tillier, 1995) even if the discovery (Dzaparidze *et al.*, 1989) of the Georgian site of Dmanisi (eventhough in Asia) and the ancient chronology attributed to the fossils from Atapureca (Carbonell *et al.,* 1995) have re-launched the debate (see especially Tillier, 1995). Since then, authors have re-directed the debate through using new denominations. Thus, in Europe, specimens of *Homo antecessor* (Bermudez *et al.,* 1997) would seem to be common ancestors of Neanderthal and modern man and replace «*Homo erectus*» name whereas in Africa it would seem to be *Homo ergaster* (Wood, 1984) that replaces the name of *Homo erectus*. *De facto*, *Homo erectus* would be limited to Asian specimens uniquely. Before considering this new semantic cutting-up of the ancient world table of *Homo erectus*, it would be useful to reconsider the historical data which demonstrate the succession of phases of multiplying and of regrouping of names given to human fossils.

In this way, efforts at rationalisation by Dobzhansky (1944) and of arbitration by Mayr (1950) for some time, led authors (Campbell, 1965; Howell, 1978) to regroup the taxa *Anthropopithecus erectus* Dubois 1892, *Pithecanthropus erectus* Dubois 1893, *Sinanthropus pekinensis* Black 1927, *Homo (Javanthropus) soloensis* Oppenoorth 1932, *Telanthropus capensis* Broom and Robinson 1949, *Sinanthropus officinalis* von Koenigswald 1953, *Atlanthropus mauritanicus* Arambourg 1954, *Homo leakeyi* Heberer 1963 and *Tchadanthropus uxoris* Coppens 1965 under the name of *Homo erectus*. This denomination was adopted by the symposium of Wartenstein in 1962 for which the papers published in 1963 by Washburn. Campbell (1963) became one of the most successful broadcasters of this decision-making. Ever since, Mayr and Dobzhansky, human fossils, different to *Homo erectus* have, for the most part been attributed to sub-species of *Homo sapiens* (see Campbell 1963 and 1972). Contrary to this, as is underlined by Howells (1980), every Pleistocene fossil which could not be attributed to *Homo habilis* or *Homo sapiens* has generally been considered to be *Homo erectus*. Thus, over the years the species *Homo erectus* has become the refuge of all those human taxa left on the wayside. For some years now, in unaccordance with basic rules of classification, the new anthropological fashion is once again the multiplication of taxon labelling.

I.2. The Characteristics of *Homo erectus*.

In spite of the diagnosis of *Homo erectus* put forward by Howell in 1978 which included African fossils, Howells (1980) underlines that for historical reasons linked to the first discoveries in the field, the hypodigm (the ensemble of specimens attributed to one species according to Simpson's 1961 definition) of *Homo erectus* and the description of supposedly diagnostic characters of this species still rests, for the great majority, on Asian individuals (see also Wood, 1984). A description of the *calvaria* of *Homo erectus* can be resumed by a list of quite general characters used by most authors:

- long, low skull of small capacity (700 << 1250 cm^3);
- cranium vault moderate;
- maximum cranium width situated in the lower part of the temporal bone;
- maximum biparietal width in low position;
- very pinched rear of the skull;
- presence of a sagital carina edged with depressions or parasagittal planes;
- thick skull sides.

Concerning the frontal bone:

- forehead low and slightly convex;
- developped sus-orbitary torus and often separated from the shell by a supra-toral fissure;
- pronounced post-orbital constriction;
- thick eyebrow sockets;
- absence of frontal bulge.

Concerning the parietal bone:

- flattened and rectangular parietal bone;
- parietal sides converging upwards;
- marked temporal crests ending in a « *torus angularis* »;
- absence of parietal bulge.

Concerning the temporal bone:

- low temporal bone squama;
- jointed mastoidian and sus-matoidian crests;
- generally small mastoidal processus, oriented mesially;
- thick tympanic bone;
- wide angle between the typanic bone and the petrosal;
- very large mastoidian incisure with poorly developed juxta-mastodian eminence.

Concerning the occipital bone:

- occipital squama very large and low;
- occipital plane shorter than the nuchal plane;
- presence of a clear occipital torus with supra-toral fissure;
- coinciding of the *inion* and the *opistho-kranion*;
- *inion* well separated from the *ndinion*;
- presence of an occipito-mastoidian crest.

This list represents a simplified table which allows attention to be drawn to the anatomical portrait of the species *Homo erectus*. The analysis put forward in this study re-examines, in detail, this list of characters and also takes on board different studies and monographs dealing with the subject. The choice was made to limit the study to the *calvaria* as this is the anatomical complex which characterises the type specimen of *Homo erectus*. None of the characters concerning the studied specimens was left aside and most of the characters described by the authors were taken into consideration as far as possible. In this way, the results obtained constitute a reliable basis in order to be able to establish the hypodigm of the species *Homo erectus* if the species does indeed exist.

I.3. The uncertain position of *Homo erectus* in classification and its imprecise taxonomic status.

Dubois'earliest publications (in particular Dubois, 1896) affirm that *Pithecanthropus erectus* (a name inspired by *Pithecanthropus alalus* created by Haeckel in 1866) was an intermediairy form between ape and man, in conformity with the Theory of Evolution (Darwin, 1871). Since then, the phylogenetical position of *Pithecanthropus erectus* has changed in the light of the Australopithecines discoveries and of the species *Homo habilis* Leakey, Tobias and Napier, 1964. Several diagnoses, or more exactly, several lists of caracters, have been proposed (Arambourg, 1957; Le Gros Clark, 1964; Howell, 1978; Howells 1966 and 1980) in order to describe *Homo erectus*. Within the framework of a global gradual vision inspired in anthropology by Schwalbe (1904) and developed by Arambourg (1958), Tobias and von Koenigswald (1964) situate *Homo erectus* in a secondary evolutionary stage between *Homo habilis* and *Homo sapiens*. Even if, from 1966, L.S.B Leakey envisaged a *Homo habilis / Homo sapiens* line distinct from *Homo erectus*, the idea of a secondary evolutionary stage is a paradigm which still dominates today. This vision lasted until twenty five years ago, when one could still read a plea in favour of phylogenetic gradualism at the heart of the *Homo* genus (Cronin *et al.*, 1981). As we shall see later, with the species *Homo ergaster*, it is with L.S.B. Leakey's hypothesis of 1966 of a lack of continuity between the *Homo erectus* line and that of *Homo sapiens* which re-surfaces and which, once more is favoured by contemporary authors.

One period more influenced by the model of punctuated equilibrium put forward by Eldredge and Gould (1972), saw the debate oriented towards the status of *Homo erectus* as a species or sub-species of *Homo sapiens*. Certain authors question the very existence of the *Homo erectus* species (Robinson, 1967; Thoma, 1973; Jelinek, 1978 and 1981). A short time after this phase of questioning, and up until recently, a static vision of the species (Rightmire, 1991) opposed a multi-regional evolutionary vision (Wolpoff, 1991). Only as recently as 1980, the cladist school of thought commenced to influence anthropology. Thus, discussion of diagnoses founded on what was actually the « *total morphological pattern* » of the pheneticians (see diagnoses put forward by Le Gros Clark, 1964 and Howell, 1978) was replaced, little by little, by discussions on the value and the polarity of observed characters and on the possibility of constituting a hypodigm.

During a second stage, discussions based on exclusive derivated characters were followed on by presentations of cladograms (*id est* phylogenetic hypotheses). In this way, authors influenced by the cladist approach or their vocabulary (Santa-Luca, 1980; Andrews, 1984; Stringer, 1984; Wood, 1984; Hublin, 1986) suggested lists of characters proper to *Homo erectus*. These authors attempted to clarify both the taxonomic status and the phylogenetic position of the species *Homo erectus* or Man classified under this name tag. During this period Stringer (1991) concluded that *Homo erectus* was rather more an « evolutionary tendency » than a species. This taxon would not be a species but a «morphograde» (Stringer, 1984; Hublin, 1986) or even a «chronograde» (Wolpoff *et al.*, 1994). From this, theoretically, the species *Homo sapiens* would be subordinated (Hublin, 1991). Stringer (1984) and Andrews (1984) distinguished Asian specimens only as constituting a true taxon of a specific order (Braüer held an opposite view in 1990 but Jacob had already done so in 1976). These are opinions which clash but with no illustrations for means of support. Thus, for Rightmire (1984) *Homo erectus* is an authentic species even if there is no proof. *Homo erectus* is truly a palaeo-species rather than an arbitrary grade or a stage in the evolutionary process of a line according to Rightmire (1990). These attempts in no way clarify the question in that cladistics has for the most part been followed as it was fashionable to do so. Cladist vocabulary was more greatly employed to the detriment of its methodological tools (Hublin, 1986). When cladistic analyses were employed, they were so on characters which were only globally identified. Furthermore, when computerised analyses were undertaken, they were only applied to groups of specimens lumped *a priori*. In this way phylogenetic relationships of invented groups were put forward using cladist vocabulary. This kind of study can not answer the question of phylogeny whereas basing research using specimens as a starting point, this question should be resolved. According to Gilbert *et al.,* (2003), cladistic analysis below the generic level is said to be potentially confounded by a variety of factors and species often present complex morphological and genetic clines across their geographic range that pose particular problems for paleontologists. The use of individual specimens as operational taxonomic unit's (OTU's) in phylogenetic studies obviously eliminates variation within each OTU but there is no reason to consider that this approach generates results with little meaning for lineage level systematics as far as the method is to construct a phylogeny and taxonomy without circularity. Indeed, composing multi-specimen OTU's suggests that some information is already known.

I.4. The Origin of Conflicts.

The definition of *Homo erectus* can rest on general characters or on the presence or absence of morphological characters. However, palaeoanthrpologists have for a long time been sensitive to stratigraphical arguments (Murril, 1981; Vandermeersch, 1989; Sartono, 1991a; Pope, 1992). This method of reasoning evidently goes towards Gingerich's (1977 and 1979) stratophenetics. In 1978, Jelinek wrote: « *because evolution is, in principle, a his-*

torical process, exact dating is primordial if we wish to try to understand the way or ways evolution has and is operating ». This author reminded us that the principal preoccupation of anthropologists is to discover the processes of evolution but he does not address the notion of pattern. In this way, for Jelinek, either the preliminary work of reconstitution of the diagram of relationships between fossil Man is considered as given or it is constructed from chronological data reputed to be unreliable. One of the aims of this study is to highlight that chronological criteria is not a priority of phylogenetic construction. It is the diagram of relationship between different fossil Men the pattern constructed from the only morphological and/or morphometrical characters which dictate an understanding of evolutionary mechanisms. Those who, like the majority of biologists, work on the evolutionary process need to adapt to changes in the hypodigm constitution which are inherent in the modification of diagnoses. This modification results in the taking into consideration of either new fossils or the identification of pertinent anatomical characters. Furthermore, the hypodigms are markers which allow the evaluation and orientation of evolutionary mechanisms. It is useful, therefore, that hypodigms be consensual and that the phylogenetic history which links them together be known. In consequence, it is a matter of stating what the phylogeny is and what taxonomy of the entities is for which we seek to understand the evolutionary mechanisms. Even if this is a moveable basis in line with progress following discoveries in the field, the alpha-taxonomy constitutes the basic essential elements required in all reliable statements concerning evolutionary mechanisms. Barriel (1991 and 1994) highlighted three principal contradictions related to reconstruction of phylogeny:

the diversity of methodologies and approaches used;
the choice of species and taxa;
the choice of characters.

It is desirable to discern what is brought in relation to these three sources and not to mix them. Additionally, it is important to apply one irrefutable and explicit method in accordance with criteria chosen precisely for the characters. This method must also be applicable to specimens without neglecting the individual variability once the latter is quantifiable. For this, it is of course a matter of living specimens for which, in theory at least, it is possible to test the biological criteria of interfecundity which are indications of belonging to the same species and sub-species.

I.4.1 Methods and Approaches.

Through the inductive method, it is possible to deduce general laws based on particular events. This is to say that repeated observations allow the stating of predictions which, once confirmed, become theories. Whether a general rule can then be seen to be reliable depends on the number of observations which can confirm it to be so. Counter-examples are then « exceptions which confirm the rule ». But this is a closed method and tends to protect its hypotheses. On the other hand, in the the hypothetico-deductive method, clearly presented by Popper (1973 and 1978) and recommended by Bernard as early as 1865 (Janvier *et al.*, 1980), observation is always undertaken within the framework of an hypothesis. Theories which are proposed can be refuted. The applications and the predictions are only interesting for theoretical reasons. By this, it is meant that they can be used in order to check the theory (Matile *et al.*, 1987). Understanding progresses through the modification of previous understanding, by changing or by being rejected. The counter-example destroys the theory. The hypothetico-deductive method is open and seeks to destroy its hypotheses.

It would seem to be the case in anthropology that phylogenetic understanding has principally evolved through an accumulation of particular events in accordance with the inductive method. Most of the time, fossils are grouped together according to their global similarity. Ressemblances are morphological or the result of calculations reached through metrical data. But ressemblance is not data, it is an interpretation and the three principal schools of Systematics are characterised by their different conceptions of similarity (Tassy, 1986; Matile *et al.*, 1987).

Evolutionary Systematics associated with the work of Huxley (1940, 1957, 1958), Mayr (1950, 1981, 1986) and Simpson (1951, 1961) is based on genealogy, but also on resemblance. The grade, stage of progress and anagenetical unity for Huxley (1958), has an essential role for the establishing of a classification. Morphological divergence, belonging to an adaptive niche, the richness in species and the fact that the classified groups are made of taxons which originate in the same ancestor **without necessarily understanding all of** the descendants make up the rules of the classification.

Phenetic Systematics founded by Cain (1954), Sokal and Sneath (1963), Sneath and Sokal (1973) considered that phylogeny could not be acknowledged as being objective. It considers that the degree of global similarity permitted the establishment of classifications through using mathematical techniques. The evolutionary significance of phenotypical characteristics is not considered.

Phylogenetic Systematics or Cladistics founded by Hennig (1950, 1966) and first of all developped by Brundin (1968), Nelson (1970, 1972, 1978), Nelson and Platnick (1981), Cracraft (1974), Cracraft and Eldredge (1979), Eldredge and Cracraft (1980) and Wiley (1978, 1979, 1981), tends to be based solely on the genealogy of species. It consists of taking into account the evolutionary significance of characteristics in order to identify monophyletic lines or clades. Cladistic analysis is founded on the breaking up of the concept of similarity. Ressemblance can in fact be the result of three distinct processes:

a ressemblance resulting from the phenomenon of homoplasy (convergence and reversion);

a ressemblance due to the sharing of plesiomorphic characters (close ressemblance to the original state);

a ressemblance due to the sharing of amorphic characters (different from those in the original state) which are indicators proper to members of the group.

Characters remote from the initial state, found in one of the taxa considered are then called autapomorphic.

I.4.2 Species and Taxa.

The nature even of fossilised material poses a problem; a problem which is all the more difficult to resolve as the theme studied resides exactly in the determination of the definition of the material taken into account. If the work of restitution of the phylogeny is carried out with species defined beforehand when the primary aim of analysis is actually the recognition of species and their relationships to each other, how can one constitute a phylogeny founded in this way? The definition and the limitations of the species or of the taxon which constitute the object of the study must be correctly understood and two opposing strategies for so doing exist; one is typological and deals with the analysis of characters whereas the other is « populational» and is resolved by statistical approach. Within the framework of an anthropological study which uses cladist concepts, Wood (1984) insists that the detection autapomorphies in *Homo erectus* is sensitive to the slightest modification of the hypodigms of *Homo sapiens* and *Homo erectus.* This observation is trivial and the affirmation simply signifies that the author has applied cladistic analysis to artificial groups – groups supposedly identified *a priori*. The author supposes taxinomic understanding as acquired and the research deals only with preconceived group relationships.

Let us consider two hypodigms A* and B* constituted *a priori*: in such a way that $A^* = (A_1, A_2, A_3, ..., A_i, B_x)$ and $B^* = (B_1, B_2, B_3,..., B_j)$. If the taxa « B » are characterised by an exclusive derivated character « b », this state cannot appear as an autapomorphy as the taxon B_x is arbitrarily included in the hypodigm A*. This example shows that is wrong to want to reconstruct a phylogeny from arbitrary hypodigms. It also underlines that the autapomorphic quality of a character can not be decreed *a priori* - *id est*: before putting a cladistic analysis englobing the ensemble of the characters at our disposition into practice. Leigh's (1992) study on the evolution of the cranial capacity in ancient *Homo erectus* and *Homo sapiens* illustrates the misguided application of hypodigms constituted *a priori*. Leigh takes stratigraphic data as a starting point. He then considers the differences in cranial capacity over time in the *sensu lato* human lineage. The results show that different evolutionary mechanisms of encaphalisation are possible according to the chronology selected for the taxons (differing according to different authors), according to the constitution of the hypodigms (varying according to different authors) but also according to the techniques employed in order to reconstitute the cranial capacity of incomplete fossils. If differences in the composition of hypodigms exist according to different authors, it would seem that, generally speaking, the lack of definition in the criteria for choice for the majority of the approaches leads to an inability to observe evolutionary mechanisms. Following on from Tattersall (1986), osteological studies – and therefore palaeontological studies – tend generally to underestimate the number of species. According to Tattersall, it is not even certain that if evolutionary mechanisms are created, they may not be sufficiently « visible » or « readable » in order to be able to be used in taxonomy. (Tattersall here makes allusion to anagenesis and to necessary arbitrary limitations in order to define species). In accordance with cladistic theory, Tattersall sees these evolutionary bonds (the apparition or loss of characteristics) as the most reliable indicators of speciation. Thus, he states that it is more productive to affirm the existence of several species when several morphs are recognisable. This same research strategy was employed by Wood (1984 and 1992a) who studied *Homo erectus* in his globality without fear of evoking the existence of supplementary taxa between *Homo habilis* and *Homo erectus* on the one hand and, *Homo erectus* and *Homo sapiens* on the other. Rightmire (1990) also affirms a leaning towards this working hypothesis bringing up the necessity of introducing a specific intermediate taxon between *Homo erectus* and *Homo sapiens*. At that time, Rightmire favoured the re-evaluation of the species *Homo heidelbergensis* Schoetensack, 1908. Bonde (1977) also evoke the re-integration of *Homo heidelbergensis* to the rank of species and of intermediary taxon between *Homo erectus* and *Homo sapiens*. Some anthropologists wish to be closer to « biological reality » through adopting a « populational » point of view (in accordance with that of Mayr, but who, it must not be forgotten, researched into neontological populations). In this way, according to Maureille (1994) and in opposition to Tattersall, the absence of any work of synthesis on morphology is generally linked to the methodology employed in studies of prehistoric populations. According to Maureille the notion of morphological variability can only be conceived without a « populational » vision of the sample being studied. He also adds that palaeoanthropologists, for a long time, restricted the variability of fossil specimens in order to bring them closer to a set « type » which had been conceived from an individual held to be representative of the population. Because of the weakness of the fossil sample, the morphology of each specimen led to the creation of a sub-species or even of a new genus. In this way, according to Maureille, typological analysis led to the impossibility of any phylogenetical study and led to an artificial multitude of taxons. Certainly, since the origins of comparative anatomy, and notably – according to Geoffroy de Saint-Hilaire (1818) – the criteria for the recognition of a species is typological, but Maureille (1994) makes no distinction between the « type » in its legal acceptance (holotype) which serves as a historical reference for the classification of bionomial nomenclature and the morphological type. It is clear that a species made up of more or less varied populations cannot be represented by a unique specimen. In palaeontology, the hypodigm is what is mostly used in the definition of a species in its globality. What remains is the recognition of the hypodigm in itself as being the fruit of research. The hypodigm cannot be a postulate as its composition inexorably underpins the definition of the species. The method of analysis is important as the « populational » vision which consists of studying a phenotypical ensemble of characters does not always guarantee the degree of phenogenetic relationships. In Nelson and Platnick's school of thought (1981), it is not possible to escape from a typological classification as, by definition, the analysis is based on on an analysis of characters. In line with this, we shall pursue an approach which is absolutely typological – even if this approach is contested by the Systematics School of thought (see Mayr, 1974 and 1981) which is concerned with current populations composed of species rather than the species themselves. Palaeoanthropologists claim to study the species whereas they are only confronted with specimens. It is only after analysis they are able to categorise or not the specimens into higher groups. However, a way of surmounting the stumbling block, consisting of creating artificial and – *a priori* – hypodigms, it suffices, simply, to work on individuals. Thus, in this study, each specimen will be taken as an operational taxonomic unit (Otu). Its taxonomic rank will be considered as being unknown and this understanding is exactly the initial aim of this study. The domain of investigation of the study being limited to hominids, it is the previous work carried out by different authors which guided our choice of specimens and served in the search for pertinent characters. Characters common to humans, hominids and primates will be excluded in that they do, actually, make up the intrinsic « stock » and they are not discriminating in the establishing of categories amongst the specimens studied.

I.4.3 The Characters Studied.

Systematic characters are those which allow us to differentiate two taxa from each other or those which indicate that two taxa are similar to each other but different from a third. The significance and the nature of the charaters must be sought in interspecific comparison. It is important to determine if the characters are **homologous** or **homoplastic**. According to evolutionary concepts, homology is the similarity inherited from a common ancestor whereas homoplasy is not (Simpson, 1961). The criteria for recognising homology is morhphological. It is based on palaeontological knowledge and especially those which are embryological. Two characters are said to be homologous if they represent different stages in one same series of transformation. For such series, the term **morphocline**, proposed by Maslin (1952) has come into current useage. Two characters are acknowledged as being homoplastic when similarly, when they are not homologous from one taxon to another. The common ancestor did not therefore possess the character. There are different types of homplasy:

- **Convergence** is the appearance of similar characters, independently in several different lineage (the case of adapting to the environment) one talks of parallelism when the convergence concern taxons which are phylogenetically quite close,

- **Reversion** is the return to a morphological state which is similar to the initial one after a series of transformation.

In works which deal with the affinity of taxons of the type *Homo*, invoking the phenomenon of homoplasy often permits the upholding all phylogenetic hypotheses possible. However, only one tree linking different evolutionary units – previous or present exists –: this is the evolutionary tree, which describes the history of the descendency (Tassy, 1991). The origin of conflicting interpretations in the different sources of data arises in the detection and explanation of the pairing of homology/homoplasy. Homoplasy affects the data issuing from all sources (Patterson, 1988) but Cladistics can quantify this. Wood (1991) insisted on the role of the choice of characters. He re-examined the 121 cranial characters put forward by Weidenreich (1943) in order to define *Sinanthropus pekinensis* Black 1927. This list is not a diagnosis as several characters are common to *Homo sapiens* (in the example given by this author, *Homo sapiens* corresponds to both *Homo sapiens sapiens* and *Homo sapiens neanderthalensis*). But this is also the case for diagnoses of *Homo erectus* as stated in the work of Le Gros Clark (1964), Howell (1978) and Howells (1966 and 1980). In Cladistics, the list of autapomorphic characters of one taxon constitutes a robust diagnosis. But this list is the result of an analysis and under no circumstances can be an *a priori* postulate. The characters studied in this work come from anthropological literature which deals more specifically with *Homo erectus*. A revision of these characters and of the control of their presence in specimens was carried out. Hitherto unpublished metrical data were also gathered from the specimens. We do not know – *a priori* – if the ensemble of the morphological and metrical characters taken into account is « valid » characters from the point of view of Systematics. Only the analysis undertaken allows identification, *in fine* those which are pertinent. The « pattern » obtained *a posteriori* - the cladogram – will provide elements of response.

II. Material and Methods of Analysis.

Albeit that neither the status nor the phylogeny of *Homo erectus* gives rise to a consensus, much research is directed towards studying the evolutionary process between *Homo erectus* and *Homo sapiens*. The majority of anthropological studies concerning *Homo erectus* favour process to pattern. This being the case, anthropologists are like travellers trying the read an itinerary from a map. But this desire to read a map assumes that the map exists, that the itinerary exists; *id est* – that the different stages are known and the route between the different stages are indicated on the ground. We here propose to re-centre the anatomical studies undertaken by different researchers in the framework which here is biosymetrics and this, thanks to Cladistics. This method imposes itself as it permits the elimination of taxa which do not correspond to natural groupings from phlogenetic classifications. The groupings accepted by Cladistics are those which contain all the descendents of an ancestral species (= clade). The groups of other living beings which share a general evolutionary level without, for all, needing to be closely related were rejected (= grade). It should be noted that Mayr himself acknowledged Cladistics as the only rational means of construction of phylogeny (Mayr, 1986), even if he did not admit that the classification must be the only and exact reflection of phylogeny.

II.1. Cladistics.

Systematic phylogenetics (or Cladistics) founded by Hennig (1950, 1966) tends to be based uniquely on the genealogy of species. It consists of taking into account the evolutionary significance of characters in order to establish monophyletic groups or clades. It is based on the theory of the modification of characters during descendence albeit that all sexually reproductive organisms have a single genealogical history: one phylogeny. The characters are transmitted from one ancestral species to its daughter-species and therefore, these carry the character signs of their common ascendance. The modification of certain ancestral characters gives the lineage attributes which are proper to them. When one species is divided, its proper characters have been transmitted to its descendents with or without modification. In this way, a closely related group of species presents a mosaic of primitive and evolved states of the characters. This is the mosaic-type evolution put forward by De Beer (1954) or the **heterobathmy** of Hennig (1966). The definitions of different terms belonging to Cladistics or to its concepts are not all listed here. It is suggested that readers consult the work of Wiley (1978), Wiley *et al.* (1991), Darlu and Tassy (1993) and Tassy (1994).

II.1.1 The Principle.

From the point of view of the transformation of the characters during evolution, the character closest to the initial state is a **plesiomorphic** character. The character furthest away from the initial state is an **apomorphic** character. The sharing of a plesiomorphy by two taxons is a **symplesiomorphy**. The sharing of an apomorphy is a **synapomorphy**. When one apomorphy is exclusive to one taxon, one talks of an **autapomorphy**. What is important for a character to be an apomorphy is the precise level at which it appears for the first time in the genealogical sequence and not in the geological sequence. The fundamental principle of Cladistics is that the synapomorphy alone permits the submitting of a hypothesis of close relationship. Symplesiomorphies cannot serve to form monophyletic groups. A group is **monophyletic** if it includes the ancestral species of the group and all of its descendents. The citeria for recognition is the identification of at least one apomorphic character shared with the group and inherited from its ancestral species. For those who hold tenets of phylogenetical systematics, a supraspecific ranked taxon should always be monophyletic. Those groups of which the members only share plesiomorphic characters are said to be **paraphyletic**. One must bear in mind that a character, or more precisely a state of character, only derivates or is primitive when in relation to another. A given character which is expressed according to three ordered states such as: Θ→ O → Φ in this case "O" is primitive in relation to "Φ " and "O" is derivated in relation to "Θ".

II.1.2 The Method.

The analysis of characters consists of reconstituting the series of transformation of homologous characters. Once the clines are recognised, their polarity must be determined. A **cline** is a linear se-quence of characters derived directly from one to another, the **sequence** being the ordered evolutionary series. The aim of the phylogenetic analysis is the construction of a rational, un-arbitrary diagram which must be able to be submitted to being refuted by the introduction of new characters and/or new taxa. The shortest tree is the one which permits the controlling of the possibility of refuting (by the said method of parcimony). The cladistic analysis is the method which claims to make demands on no other previous model on the evolutionary process. The differential evolutionary speeds have no influence on the principle of analysis. Hlusko (2004) and many others object to cladistics, claiming that the evolution is not sufficiently parsimonious. This is certainly the case but the scientific method which describes the evolutionary process has to be so in order to avoid the multiplication of hypotheses presented in *ad hoc* fashion which support a result.

II.1.3 The Criteria.

In Cladistics, relationships of kinship are established on the synapomorphic series corresponding to the degree of ancestry of the common ancestor (the ancestry is not directly stratigraphical). At the specific level, the hypotheses are given as follows:

- two species are more closely related to each other rather than with a third if they have a common ancestor which is not also an ancestor of the third. This type of kinship is expressed in the form of a cladogram. A cladogram is the image of a structure, and not of a process. The bifurcations of a cladogram represent the minimum number of events of speciation necessary in order to be able to report back on the structure observed. The sharing of derivated characters permits the construction of the cladogram. How can these series of transformation be polarised?

The Ontogenetical Criterion.

Haeckel (1866) was the first to interpret the « law » of ontogenetical precedence (first stated by von Baer in 1828) in an evolutionary perspective. What first appears in ontogenesis is primitive. Exceptions to the law of recapitulation of Haeckel were put forward by Beer (1930, 1958) and Gould (1977). But, redefined at the level of characters and not of whole organisms (Nelson, 1973 and 1978), this rule maintains all of its phylogenetical interest. Nelson's formulation is as follows: given that the ontogenetical transformation of a character from a more general state to a less general state, the more general state is primitive and the less general state is evolved. In other words, and to use Tassy's (1994) formulation, the comparison of two species (X and Y) of which one (X) shows the ontogenetical transformation of a character a → b and the other (Y) which does not lead to a transformation a → a. State a, counted twice (the most general) is plesiomorphic, and state b counted once (the least general) is apomorphic. This re-affirmation means that the ontogenesis of the character of a towards b recapitulates the phylogenie as a is plesiomorphic and b apomorphic (**Figure 1**). Nelson's rule is based on continuous ontogenies and must only be retained as a research strategy and not as a biological rule (Tassy, 1994).

The Out-Group Criterion rests on the principle of parsimony. If an homologous character is observed in the group studied and also found outside the group, it is pleisomorphic. If the character is only found in the interior of the group studied it is apomorphic (Wiley, 1978; Wiley *et al.*, 1991; Darlu and Tassy, 1993).

The Criterion of Geological Precedence or **Palaeontological Criterion** can be considered as an accessory criterion. In fact, this criterion is only operative once the monophyly of the group being studied is acquired (Hennig, 1966) and with the condition that the fossil can be classified as being very rich. Janvier (1984), states that this criterion is not a strong argument for phylogenetic construction. The chronological or stratigraphical parameter cannot be an argument which can alone permit the reconstruction of a kinship structure.

For Howells (1980) the gathering together of contemporary material in defining a species seems to be a means which defends itself. If this research strategy seems to stand up in an anagenetical and gradualist vision of evolution, it excludes the diversity and must then be seen as having an inductive approach. In addition, it implies that all analysis of fossils discovered until today in an unknown stratigraphic context must be excluded. To return to Mayr's objection (1974 and 1986)

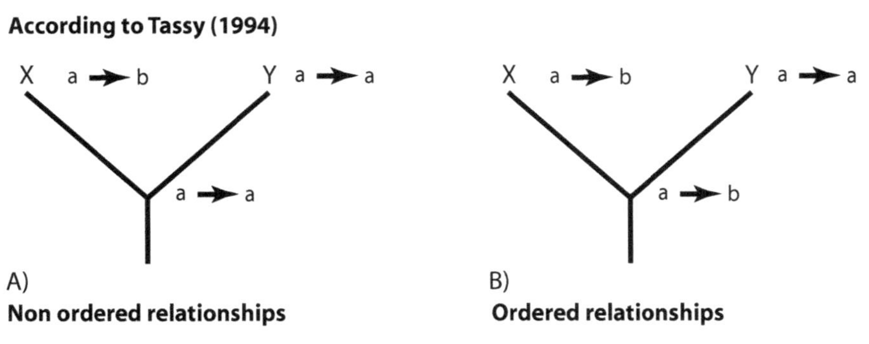

Figure 1 Nelson's rule

A) the ontogenetic transformation a → a observed in X corresponds to one evolutive step from ancestral state a.

B) the ontogenic transformation a → b (one step) is supposed to be present in the ancestor and the absence of transformation observed in Y is considered as a lost (one step).

The hypothesis B) (two step) is less parsimonious than the hypothesis A). The most parsimonious hypothesis A) implies that a is primitive and b derived.

relating to the adoption of a classification based on Cladistics, Habgood (1989a) clarifies that this is shaky as the introduction of a unique character or a new discovery leads to changes in the cladogram. These reservations should rather be taken as an indicator of scientific quality of the approach when faced with rigid and conservative paradigms. One classification should not be seen as an over-full filing cabinet into which all new fossils should be stuffed at all cost. On the contrary, it must be possible to build, adapt or extend the cabinet as and when new discoveries are made.

II.1.4 Objections.

It is not a question of affirming that Evolution is parsimonious but a question of re-stating that scientific discourse concerning Evolution must be (*versus* Hlusko, 2004)**.** Producing endless *ad hoc* hypotheses in line with inductive method diagrams is not an acceptable way of working in Science. In palaeoanthropology, Trinkaus (1990) camped on his anti-cladistics position when this became « egemonic » in the palaeoanthropological press in the 1990's. In this same « sociological » register and inversely, it must be remembered that other authors claimed to be the same in the 1980's when they were only employing the vocabulary without producing any cladistic analysis (see Hublin, 1986). Others have applied cladistic analysis to groups which were supposedly well-defined hypodigms. Objections quickly followed and Chamberlain and Wood (1987) stated that the fiability of results obtained from several cladistic analyses is influenced by the choice of the characters and the choice of the composition of operational taxonomic units. See also Wood (1984) who made clear that the detection of autapomorphies is sensitive to the smallest amount of modification in the hypodigms. In the spirit of alpha-taxonomy (the search for the taxonomic status of the taxa studied) which is at the heart of this study, this type of assertion can only confirm that cladistic analysis which aims to restitute phylogeny has for a long time not been correctly applied: phylogenetic hypotheses resulting from the analysis of artificial hypodigms makes no sense. Even if he justifiably stresses that the coding of characters remains far from being perfect Curnoe (2003) paints a picture of the limitations of Cladistics which is not totally founded. His objections refer to genetic data and cannot be applied when it is a question of working within the framework of fossil species for which biological information is *de facto* restricted. As he does indicate, without suggesting a solution, when the aim is to determine to which group any taxon belongs to before investigating its evolutionary history, it is wrong to work on artificial taxa and wrong to make use of defined and *a priori* oriented characters. In the study we shall put forward a series of solutions. These few lines of text are not anecdotal as they define the scientific direction which is to be followed. The work must be based of the anatomical data of specimens, and continue by describing and coding the latter, producing a phylogenetic analysis from the specimens in order to exclude or seek groupings, defining groups by going from common derivated characters which arise from the analysis and finally naming the groups of specimens which define a hypodigm.

II.2. The Material.

The characters employed in this study are osteological characters proper to the *calvaria*. This anatomic ensemble permits a comparison of a large number of specimens linked to *Homo erectus* but also many other fossil taxa. A series of male dayaks skulls (N=30, Musée de l'Homme in Paris) was taken as a reference sample for *Homo sapiens sapiens*. This restricted choice is linked to the considerable number of parameters retained for each skull and for the time required for their collection. Female growth being sooner attainted during individual development, male skulls therefore present greater « expressiveness » of the variety of morphological characters. The lack of any regional continuity between *Homo erectus* and *Homo sapiens* today is not, however, a *priori* taken for granted. Two other fossil groups were also selected as potential representatives of the *Homo sapiens* species: Neandertalians [*Homo sapiens neanderthalensis* (King, 1864)] and specimens of an uncertain taxonomic position attributed to *Homo sapiens* said to be «archaic». Out-group references were gorillas [*Gorilla gorilla* (Savage and Wyman, 1847)], chimpanzees [*Pan troglodytes* (Blumenbach,1799)], and two complete specimens (Sts 5 and KNMER 406) brought respectively to Australopithecines [*Australopithecus africanus* (Dart ,1925) and *Australopithecus boisei* (Leakey,1959)]. The great Apes (11 *Pan troglodytes* and 11 *Gorilla gorilla*) which were studied came from the Laboratoire d'Anatomie Comparée du Muséum national d'Histoire Naturelle in Paris. To the extent that the aim of this study was to distinguish different categories within the genus *Homo*, the taxons selected for Out-group was appropriate. A study based on the genus *Australopithecus* or *Paranthropus* would no doubt have required considering another choice. The aim was to classify the human fossil specimens within the framework of Systematics. Taxonomic indicators shown in the description of specimens were not taken into account. Each specimen was considered in isolation. In all, there were 67 operational taxonomic units: 64 fossil specimens and 30 Modern Man specimens (1 Otu), 11 adult gorillas (1 Otu) and 11 adult chimpanzees (1 Otu). Following morphological analysis, there exists no other distinctive criteria in the Modern Man series. Each individual appeared as identical to the next, bearing in mind the definition of characters selected. The same observation was true within the gorilla series and the chimpanzee series. In this way, these tree series appear as three operational taxonomic units. The list of the specimens studied is presented according to geographical zones.

The ontogenic criterion being strongest from the point of view of Cladistics, immature individuals (Modern Man, gorillas and chimpanzees) were studied through the search for polarisation of character states. Whenever it was possible to follow the evolution of one homologous character during growth within a given taxon (Modern Man or Modern chimpanzees) this additional information was included. Thus, if polarisation of characters was identified, it was taken into account in the organisation of the coding of the states of characters during the analysis. Immature fossil specimens are indicated by the symbol: « ° » in the description of specimens.

II.2.1. List of Specimen Fossils.

The different level of importance of studies put forward by different authors, and even the nature of the fossils do not permit for a homogeneous description of information obtained from the specimens studied. In fact, certain fossils aroused more interest than others. Several of them had a taxonomic status accepted by all of the anthropologists and therefore did not give rise to much discussion. Others, on the other hand, gave rise to bitter discussions and many points of view clashed. This was also the case when the exact stratigraphic position of the fossils was not known and that several dating methods (for the most part unknown at the time of discovery) have been put into practice since. In the case of the Indonesian specimens, the stratigraphy of the sites was particularly stressed upon as these fossils are reputedly poorly dated. However, many studies – certainly contradictory – exist and point out what the relative chronological position is relating to the Indonesian fossils. Stratigraphical data which permitted the chronological situating of the Chinese fossils were also noted down

Summary of Information concerning the specimens studied.

Sterkfontein 5 (South Africa): This *calvarium* is that of an *Australopithecus africanus* (Dart, 1925) discovered in the Sterkfontein valley (Transvaal) at level 4 of «Type Site» by Broom in 1947. Geomorphological data gives an age of 3.3 Ma (Partridge, 1973), with the faunic data situating the age between 2.7 and 2.2 Ma (Johanson and Edey, 1983).

Saldanha 1 (South Africa): This specimen is represented by a *calva* discovered in the province of the Cape by Singer in 1953. Undated, it is morphologically close to Kabwe 1, of the Ngandong series and specimens from Zhoukoudian according to Singer (1958). It has been assigned to *Homo sapiens rhodesiensis* (Woodward, 1921) by Campbell (1963).

Kabwe 1 (Zambia): Discovered in 1921 by T. Zwigelaar, this *calvarium* is situated chronologically in a period defined as a « Middle Stone Age » post-Acheulaen by Oakley (1960). For Vrba (1982) the specimen belongs to at least the isotopic stage 9, whereas Partridge (1982) finds isotopic stage 7 or the beginning of isotopic stage 6. Day (1986) situates it at isotopic stage 6 or even older. This fossil is described as being *Homo sapiens rhodesiensis* (Woodward, 1921) by Howell (1980). It had been rebaptised *Cyphanthropus rhodesiensis* by Pycraft in 1928. Close to specimens of Ngandong by Weidenreich (1943), it is considered by Rightmire (1986a) as a *Homo sapiens*. In 1990, Rightmire went further and proposed that with Arago, Petralona and Ndutu, it was part of an ensemble of archaic *Homo sapiens* which could be called *Homo heidelbergensis* Schoetensack, 1908.

OH 9 (Tanzania): This specimen was discovered on the ground of the «Member I» of the « Louis Leakey Korongo » site (Leakey, 1971). This incomplete *calvaria* is supposed to belong to the upper part of « Member II » of which the crown is dated between 1.2 and 1.15 Ma. by Leakey and Hay, (1982). Described as *Homo erectus* by Howell (1978) and Rightmire (1979), it corresponds to the type *Homo leakeyi* Heberer, 1963. According to Maier and Nkini (1985) this specimen is really *Homo erectus*.

Ndutu 1 (Tanzania): The indirect dating of a tuff, similar to the upper layers of Masek, to which this very incomplete *calvaria* supposedly belongs to, is around 400,000 years old according to Clarke (1990). For Mturi (1976) the more recent age of 500,000 to 600,000 years is put forward. According to Rightmire (1986a) it belongs to Middle Pleistocene and is not as old as the crown of «Masek Bed». According to Bräuer (1991) this fossil is contemporary with Bodo and of isotopic stage 11. It was first of all classified as being *Homo erectus* by Clarke (1976). It is seen as being a bridge between *Homo habilis* Leakey, Tobias and Napier, 1964 and *Homo sapiens* by this author in 1990 whereas Rightmire (1983 and 1986a), considers it to be *Homo sapiens*.

LH 18 (Tanzania): Discovered on the Laetoli site, this *calvaria* was dated by the U-Th method as being of between 90,000 and 129,000 years old (Bräuer, 1989) in line with its stratigraphical position relative to the « Ndutu Beds » of « the Olduvai Gorge ». The specimen is decribed as being *Homo sapiens* having affinities with *Homo erectus* by Day *et al.* (1980). It is deccribed as being an archaic *Homo sapiens* by Bräuer and Leakey (1986), and Bräuer and Rimbach (1990).

KNMER 406 (Kenya): This *calvarium* is that of an *Australopithecus boisei* discovered by M. G. and R.E.F Leakey in 1978 in area 10 of the Ileret zone in the « KBS Member » of the « Koobi Fora Formation », under the Okote complex. Dated as being between 1.65-1.7 Ma. (Wood, 1991). Chamberlain and Wood (1987) and Wood (1992b) whilst preserving membership of the gender *Paranthropus*.

KNMER 1470 (Kenya): This *calvarium* originates from area 131 of Koobi Fora, 35.5 metres under the tuff KBS of « Lower Member » which was dated 2.6 Ma. by the radio-isotopes (Fitch and Miller, 1970) and of 2.7 to 3 Ma. Palaeomagnetic data (Brock and Isaac, 1974). For Wood (1991) this specimen belongs to « Upper Burgi Member » of the « Koobi Fora Formation » under the tuff KBS dated only between 1.88 and 1.9 Ma. Fitch *et al.* (1996) also date this specimen as being of between 1.9 and 2.5 Ma. For Leakey (1973a), Day *et al.* (1975) and Wood (1991) the specimen belongs to an indeterminate species of the type *Homo*. For Howell (1978) it is an *Homo habilis* and also for White and Kimbel (1981), Tobias (1983) and Johanson *et al.* (1987). It is seen to be different than specimens OH 24 and KNMER 1813, considered to be much more archaic forms of *Homo habilis* by Leakey *et al.* (1978). For Groves (1989) it should be attributed to *Homo rudolfensis* Alexeev 1986.

KNMER 1813 (Kenya): The small cranial capacity of this specimen relegates it to a primitive stage. Initially described as being an *Australopithecus* (Leakey, 1974), this fossil is close to *Australopithecus africanus* according to Walker and Leakey (1978). This incomplete *calvarium* was attributed to *Homo habilis* by Howell (1978), White *et al.* (1981), Tobias (1983). For Groves and Mazak (1975) it's either an *Homo habilis* or an *Homo ergaster* Groves and Mazak, 1975. In 1989 Groves made *Homo ergaster* into a paratype. An eventual connection with *Homo erectus* was suggested by Hublin (1983), Wood (1984) and Stringer (1984). Finally it has been classified as being an indeterminate species of type *Homo* by Wood (1991). A study carried out by Liebermann *et al.* (1996) saw this specimen as being an *Homo habilis sensus stricto*. The fossil comes from « Upper Burgi Member » of the « Koobi Fora Formation » from under the tuff KBS dated between 1.88 and 1.9 Ma. By Feibel *et al.* (1989). Complementary dating in accordance were put forward by Isaac *et al.* (1971), Maglio (1971), Vondra *et al.* (1971), Brock et Isaac (1974) and Fitch *et al.* (1974).

KNMER 3733 (Kenya): This *calvarium* was discovered in area 104 of Koobi Fora, between the tuffs KBS and Koobi Fora/BBS. Tuff KBS, underlying, supposes a date of 1.78 Ma. (Wood, 1991). This specimen is presented as being *Homo erectus* Leakey and Walker (1976) and confirmed as such by Walker and Leakey (1978) as well as by Walker (1981). Coppens (1983) and Tobias (1978 and 1991) also adopt this idea. According to Howell (1978), Rightmire (1979 and 1984) and Bilsborough (1983), it is completely compatible with the sinanthropes whereas Wood (1984), Stringer (1984), Andrews (1984) and Groves (1989), see in this fossil a clear anatomical distinction with the fossils from Asia. Bilsborough and Wood (1986) and Grooves (1989) even accord it as having an affinity with *Homo ergaster*. Wood (1991) classifies it as being from the species *Homo* having affinities with *Homo erectus*. Clarke (1990) and Wood (1992a) connect it to *Homo ergaster* Groves and Mazak, 1975.

KNMER 3883 (Kenya): *Calvarium* discovered in the area 1/1A of Ileret belonging to « Okote Member » of the « Koobi Fora Formation » *id est*: dated between 1.5 and 1.65 Ma. according to Wood (1991). Closely linked to KNMER 3733 for Picq (1983) and Rightmire (1984). Considered as belonging to a species of the genus *Homo* with affinities to *Homo erectus* for Wood (1991). Clarke (1990) and Wood (1992a) link it to *Homo ergaster* Groves and Mazak, 1975.

KNMWT 15000° (Kenya): Discovered to the West of Lake Turkana, on the Eastern bank of a small stream south of Nariokotome, this almost complete skeleton is situated at the lowest part of « Natoo Member » which is dated at 1.53 Ma. for Walker and Leakey (1994), or an estimation of 1.6 Ma. for Brown *et al.* (1985). It is supposedly a juvenile form of KNMER 3733, considered as being a *Homo erectus* for Leakey and Walker (1976 and 1985). Wood (1992a) sees a close link with KNMER 3733 and KNMER 3883, of *Homo ergaster* Groves and Mazak, 1975.

Eliye Springs 1 (Kenya): According to Bräuer (1989), no stratigraphical data exists concerning this *calvarium* which had a part of the front facial part previously eroded. For Bräuer and Leakey (1986) it is a specimen of « recent » archaic *Homo sapiens.*

Bodo 1 (Ethiopia): This specimen consists only of a frontal part and a back part of the *calvaria.* The stratigraphical position of the specimen corresponds with the limit of Middle and Upper Middle Pleistocene according to Bräuer (1984a). In so doing he refers to the isotopic stage 11 (Bräuer, 1989). Rightmire (1986b) sees it as being *Homo sapiens.*

Omo Kibish 2 (Ethiopia): This *calvaria* was found out of its stratigraphical context in the South-West of Ethiopia near the river Omo. For Stringer (1978) Omo 2 is morphologicaly close to Saccopastore, Ngandong and Kabwe. Howell (1978) insists on its distinct difference with *Homo erectus* and *Homo rhodesiensis* and concludes that it is an *Homo sapiens* close to Modern Man. For Day and Stringer (1982 and 1991), it is not a modern *Homo sapiens* and must be a transitional specimen between *Homo erectus* and *Homo sapiens*. It is considered as being an archaic form of *Homo sapiens* by Bräuer and Leakey (1986). For Rightmire (1983) it is an early *Homo sapiens* as certain characters relate it to Ndutu and Kabwe.

Salé 1 (Morocco): This *calvarium* which was illegally exported from Morocco is described as being that of an evolved Pithecanthropus (Jaeger, 1973). It was discovered in levels attributed to the Téniftien (Jaeger, 1975). An E.S.R. dating, of the dental enamel at the same level as the *calvarium*, gives a result of around 420,000 years (Hublin, 1991). Related to evolved *Homo erectus* for Jaeger (1975), Hublin (1985) more readily relates it to *Homo sapiens* and stresses the existence of the pathological mark on the occipital shell.

Jebel Irhoud 1 (Morocco): Discovered in a byrites mine in West Morocco, near Safi, this *calvarium* fossil was first of all described as being that of a Neanderthal Man by Ennouchi (1962). It is a *Homo sapiens* which does not figure in the Neanderthalian line for Howells (1973a) and Hublin (1989). According to Hublin (1991) it would be contemporary with isotopic stage 6.

Devil's Tower° (Gibraltar): In a cave in Gibraltar, several anatomical elements of a *calvarium* were discovered at a Mousterian level and acknowledged as belonging to the same individual by Garrod (1928). According to Dean *et al.* (1986) by correlation with a neighbouring cave, the Mousterian occupation would be contemporary with isotopic stage 5. This immature specimen is a Neadertalian (Tillier, 1988).

Arago (France): Discovered in the Eastern Pyrenees in the Arago cave on the Tautavel site, the two specimens studied **Arago XXI** and **Arago XLVII** are associated with the Adenian faunic complex corresponding to isotopic stages 9 and 10 (Chaline, 1985) *id est*: aged more than 350,000 years. For de Lumley *et al.* (1984) the estimated age of the base of the ensemble III from where these fossils originated is around 450,000 years, *id est*: isotopic stage 12.

Arago XXI is a frontal bone and one side associated to *Homo erectus* (Lumley and Lumley, 1971), to the ante-Neandertalians (Lumley, 1976) and finally to the pre-Neandertalians (Vandermeersch, 1982b). It has « erectoidal traits » for Vlcek (1986). For Spitery (1982) it is a *Homo erectus*.

Arago XLVII is a parietal done attributed to *Homo erectus* by Grimaud (1982). The two anatomical parts are most often described as being an ensemble. Thus, this ensemble is closer to early *Homo sapiens* than to *Homo erectus* from China for Wu and Wu (1982). In (1985) de Lumley and Sonakia talked of evolved *Homo erectus*. **These two anatomical elements were associated together for the analysis.**

La Chapelle-aux-Saints (France): Discovered in 1908, in the Corrèze region in a small cave – de la Bouffia Bonneval – in the Brive area. The *cranium* studied is that of a complete *Homo sapiens neanderthalensis* skeleton associated with Mousterian civilization of the type Quina according to Vandermeersch (1965). E.S.R dating suggests between 44,000 and 60,000 years (Mellars, 1986; Raynal, 1990).

La Ferrassie 1 (France): Discovered in the Dordogne in the large shelter of La Ferrassie, this skeleton is defined as being that of a Neanderthalian by (1976). One attempt at dating by thermoluminescence on flints originating from levels C-D as defined by Peyrony, indicates the date of 72,000 years (Mellars, 1986).

La Quina H5 (France): Discovered in the Charentes, this *cranium* is that of a Neanderthalian specimen assigned to Würm I-II (Vandermeersch, 1965). In 1976, it was attributed to Würm II by this author. A series of measures by thermoluminescence indicates a date in the region of 65,000 years for the flints associated with this individual (Mellars, 1986).

Swanscombe (Great Britain): Discovered in Kent south of the Thames, this *calvaria* is dated as being interglacial Holstein following stratigraphy reported by Dines (1964). Successively named *Homo marstoni* by Paterson, 1940, *Homo swanscombensis* by Kennard, 1942, it is the holotype of *Homo sapiens proto-sapiens* Montandon, 1943 as the nomenclatures which preceded are not in conformity. It is held to be at the basis of the divergence of *Homo neanderthalensis* King, 1864 and *Homo sapiens* by Weiner and Campbell (1964). According to Vandermeersch (1982b), Hublin and Tillier (1991), it is a preNeandertalian.

Spy 1 (Belgium): This *cranium* was discovered in a cave at a level containing evidence of lithic industry, described as being Quina-type Mousterian (Bordes,1959). Fauna from ancient Würm is associated with the stratigraphical level of this Neanderthalian specimen (Stringer *et al.*,1984).

Engis 2° (Belgium): Discovered in 1829 in a cave in the Province of Liège, this immature *calvarium* is the most ancient discovery of a human fossil. The archaeological infilling of the cave is unknown but this juvenile individual is related to Mousterian levels by Stringer *et al.* (1984). The fossil is a representative of the Neanderthalian lineage according to Tillier (1983).

Monte Circéo 1 (Italy): Discovered on the floor of a cave on the Guattari site, this Neandertalian *calvarium* is linked to a Pontinian facies Mousterian industry, *id est*: a period corresponding to the end of Würm I or the beginning of Würm II (Blanc, 1939). Dating carried out by E.S.R. on the calcite or the fauna gives a minimum age of between 50,000 and 60,000 years (Grün et Stringer, 1991).

Velika Pecina° (Croatia): This immature individual is represented by a frontal discovered by the town of Gornec on the Ravna Gora towards Varazdin. It comes from a protoaurignacian level whose age is greater than the underlying level dated at 38,500 years (Smith, 1976). For Mann and Trinkaus (1974) it is Neandertalian, but for Smith (1976) it is even closer to Upper Palaeolithic Modern Man.

Petralona 1 (Greece): Discovered in Macedonia, this *calvarium* is described as being Neandertalian (Kokkoros and Kanellis, 1960; Poulianos, 1967; Bonis and Melentis 1982, 1991). On the other hand, for Hemmer (1975) and Stringer (1983), it is an *Homo erectus*. According to Kurten and Poulianos (1977) the fauna associated with the specimen belongs to Holsteinien (Mindel-Riss). Datings obtained through E.S.R. on the calcite covering the crown gave rise to a minimum age of between 160,000 and 240,000 years (Hennig *et al.*, 1981).

Zuttiyeh (Middle East): Discovered in the cave of Mugharet el Zuttiyeh, the estimated age of this fossil is situated between 200,000 and 250,000 years old according to Bar Yosef and Vandermeersch (1993). According to Vandermeersch (1981) this is a *Homo sapiens*, and more precisely, *Homo sapiens sapiens* in 1982 (Vandermeersch, 1982a p.298). Sohn and Wolpoff (1993) put forward ressemblances with the Sinanthropes. For these authors this specimen is not Neandertalian but, all the same, it is not Modern.

Skhul V (Near East): This *cranium* is that of an individual discovered on the Mugharet-es-Skhul site. Chronologically, it belongs to the rainy period preceding the humid period of « Tabun B ». Level B of the site where this specimen comes from showed evidence of lithic industry of Levalloisian facies (Boutié, 1979). Attempts at dating by E.S.R. give a bracket of between 66,000 and 113,000 years (Stringer *et al.*, 1989). Using the techñique of thermoluminescence, Mercier *et al.* (1993) situate level B as being between 101,000 and 137, 000 years old whilst a previous study gave a larger margin (30,000 to 210,000 years old according to Mercier, 1992). Indicated as being a possible hybrid of Neanderthal and Modern Man, this specimen has been linked to Modern Man by Howell (1956) and Vandermeersch (1981). According to the study by Bräuer and Rimbach (1990), this *cranium* shows certain affinities with Neandertalian Man.

Amud 1 (Near East): Discovered in a cave, this individual's grave is situated at the summit of Member B1 of the site's stratigraphy (Chinzei, 1970). This level is attributed to a period including between ancient Würm and recent Würm. E.S.R. dated between 38,000 and 54,000 years old (Grün and Stringer, 1991). Dates obtained by thermoluminescence (Valadas in Rak *et al.*, 1994) indicate an age including between 50,000 and 60,000 years for the lower levels of the grave. For Suzuki (1970) this individual is Neandertalian.

Shanidar 5 (Irak): Composed of a temporal bone, a frament of a parietal bone and the face, this Irakian speciement is associated with level D of the site by Trinkaus (1983, 1991) working on the previous studies of Stewart (1959) and Solecki (1960). This Neandertalian specimen is associated with istopic stage 4.

Teshik Tash° (Ouzbekistan): According to Bordes (1955), the levels of the cave where the skeleton of this child was discovered are contemporary with the interclacial periods of Würm I / Würm II. For Weidenreich (1945) this specimen was an intermediate form between Neandertalian and Modern Man. For Gremiatskij (1949) and for Tillier (1988) this specimen is without doubt Neandertalian.

Narmada (India): Discovered in the middle region of the Narmada valley on the site of Hathnora (Madhya Pradesh), this half-*calvarium* is of the Upper Pleistocene Period for Biswas and Dasarma (in de Lumley and Sonokia, 1985) or of the Middle Pleistocene according to these authors. But according to Kennedy *et al.* (1991) because of the fact that the specimen had been moved previously it could be older than the level where it was found. According to Lumley and Sonakia (1985) this specimen is an evolved *Homo erectus*. According to Sonakia (1985) the name of this fossil is *Homo erectus narmadaensis* Sonakia, 1984 but according to Kennedy *et al.* (1991) it is an archaic form of *Homo sapiens*.

Sinanthropus (Chine): Searches in the Zhoukoudian site, in the Province of Hopei, to the North of Pekin, has given rise to the discovery of several dozens of human vestiges. The specimens studied were the most complete *calvariae* described by Black (1930 and 1934) and Weidenreich (1943). Dating of the different levels was rendered possible for the different levels from which the specimens originated. Using the U-Th Method, Liu (1983) indicated that levels 1 and 2 could be dated as being between 230,000 and 250,000 years old, level 7 between 370,000 and 400,000 years, levels 8-9 around 420,000 years. Using the Fission tracks method, level 10 was dated at 460,000 years old. Finally, level 12 is older than 500,000 years but older than 730,000 years according to palaeomagnetic indicators.

Sinanthropus III was discovered in the locus E, in an innermost recess of the lower cave to the East of the « Lower Fissure ». This specimen is a *calvarium* - the most ancient on the site according to Pope (1992). According to Wu and Dong (1983) level 10 is presumed to have contained the oldest specimen, some 460,000 years old.

Sinanthropus X (*calvaria*) and **Sinanthropus XI** (*calvaria*) were found in levels 8-9, **Sinanthropus**

XII (*calvaria*) in level 9 according to Pope (1992). The Zoukhoudian specimens are considered to be *Sinanthropus pekinensis* Black, 1927. This labelling was proposed by Black because of a single tooth with the labelling being extended to *calvaria* of specimen Sinanthropus III.

Dali (China): Discovered in Shaanxi Province on the third terrace on the banks of the River Luo, the Dali *calvarium* originates from the base of Recent Middle Pleistocene, just on the discordance which defines the beginning of Recent Lower Pleistocenede (Li, 1983). The level containing the specimen is contemporary with isotropic stages 6-7, *id est*: around 300 000 years old for Wang (1985). For Wu (1991) it is more recent (190 to 230 000 years old). The specimen is a *Homo sapiens daliensis* Wu, 1981 for Zhou and Wang (1982). It is a transitional form between *Homo erectus* and *Homo sapiens* according to Wu (1983). Considered to be an evolved form of *Homo erectus* for de Lumley and Sonakia (1985) it is also considered to be a pre-Modern form of *Homo sapiens* by Pope (1992).

Maba (China): Discovered in the Province of Guangdong, this incomplete *calvarium* is associated with another fauna dated around 130,000 years old (Wu, 1991). For Pope (1992) this fossil more likely belongs to mid isotopic stage 5 and is considered to be a pre-Modern *Homo sapiens*. For Zhou and Wang (1982) and Wu (1983) it is an early *Homo sapiens*.

Sangiran (Indonesia): The site of Sangiran is an ancient volcanic mound on which numerous human fossile have been found. With the exception of the Sangiran 4 specimen which comes from the Pucangan formation, the Sangiran specimens studied in this work belong to the Kabuh formation. Following Sémah *et al.* (1992) the base of Motopuro, which corresponds to the summit of the Kabuh formation is some 250,000 years old. The base of the Kabuh formation corresponds to the summit of Grenzbank, between 700,000 and 800,000 years old as according to Sémah's (1986) palaeomagnetic studies, the underlying Pucangan formation is situated beteen the magnetic occurrences of Olduvai and Matuyama/Brunhes (0.73 Ma.). According to Hyodo *et al.* (1993) the Kabuh formation can be dated between 0.97 Ma. and 0.73 Ma. The volcanic breach at the base of the Pucangan formation is dated between 1.6 and 1.7 Ma. by Sémah *et al.* (1992). These pieces of data are in line with Ithara *et al.'s* (1985a et b) results who see Grenzbank as being dated between 0.9 and 0.73 Ma. and the datings by Suzuki *et al's*.(1985) fission track dates the fossiliferous level of Pucangan at about 1.16 Ma.

Sangiran 2: This *calvaria* was discovered in 1937 by von Koenigswald. According to the stratigraphy established by De Terra in Van Heekeren (1972), this specimen originates in the Trinil levels of the Kabuh formation. For Sartono (1982) the Sangiran fossils of the Kabuh formation are more recent than 0.73 Ma. For de Vos *et al.* (1982) this specimen comes from Grenzbank and is synchronous with the Trinil biozone. A *Pithecanthropus erectus* for von Koenigswald (1940) and a *Pithecanthropus erectus sensu stricto* for Weidenreich (1945). According to Jacob (1980) this specimen is a *Pithecanthropus erectus* whereas according to Sartono (1976) it is an *Homo palaeojavanicus sangiransis*. In 1982 Sartono considered it as being an archaic form of *Homo erectus*.

Sangiran 3: This incomplete *calva* was discovered in the Kabuh levels and is therefore not older than 0.73 (Sartono, 1982). For Jacob (1980) it should come from the highest level of the Kabuh formation but, on the other hand, according to Vos *et al.* (1982), this specimen is contemporary with the Kedung Brudus biozone. In other words, and according to Leinders *et al.* (1985), it is 0.8 Ma old. It is a *Pithecanthropus erectus* for von Koenigswald (1940) and a *Pithecanthropus erectus sensu stricto* for Weidenreich (1945). According to Sartono (1982), it is an archaic form of *Homo erectus*.

Sangiran 4: Found during the 1938-39 dig of de Von Koenigswald in the Djetis level of the Pucangan formation, this back piece of a *calvaria* is very probably dated between 0.73 Ma. and 1.6 Ma. according to Sartono (1982). It is close to 0.9 Ma. for Matsu'ura *et al.* (1992). This specimen was designated as being *Pithecanthropus modjokertensis* by von Koenigswald in 1940, and as *Pithecanthropus robustus* by Weidenreich in 1945. Jacob (1980) agrees with von Koenigswald whereas Sartono (1982) proposes the trinome *Homo palaeojavanicus robustus*.

Sangiran 10°: Discovered in 1963 by Jacob close to a tributary of the river Tjemoro, this *calva* of a young adult (Jacob, 1967a) comes from the same levels as Sangiran 2 (Sartono, 1964). It is a *Pithecanthropus erectus* for von Koenigswald (1968, 1973 and 1975) and an archaic form of *Homo erectus* for Sartono (1982), Grimaud and Jacob (1983).

Sangiran 12: Discovered in 1965 by Sartono, this *calvaria* eroded previously comes from the Trinil level of the Kabuh formation (Sartono, 1968). According to de Vos *et al.* (1982), this specimen is contemporary with the biozone of Kedung Brudus, which would correspond to an age of 0.8 Ma. according to the work of de Leinders *et al.* (1985). Seen to be *Pithecanthropus erectus* by Jacob (1980) and *Homo palaeojavanicus robustus* for Sartono (1982).

Sangiran 17: Found on the Southern bank of the Putjung stream, a tributary of the River Tjémoro, this almost complete *calvarium* is attributed to the Trinil level of the Kabuh formation by Sartono (1971). It is contemporary with the Kedung Brudus biozone according to de Vos *et al.* (1982) which would correspond to an age of 0.8 Ma. according to Leinders *et al.* (1985). Jacob (1980) situates it in the middle part of the Kabuh formation and would therefore be dated at least 830,000 years old according to the age attributed to the middle part of the Kabuh formation by Jacob (1977). Dated as being around 0.6-0.7 Ma. by Sartono (1982) and closer to 0.78 Ma. according to Mastu'ura *et al.* (1992). This is a *Pithecanthropus erectus* for von Koenigswald and a late *Homo erectus* for Sartono (1982). Jacob (1980) considers it to be a *Pithecanthropus soloensis*.

Sangiran 26: This temporal bone comes from the Kabuh formation in the higher reaches of Sangiran. According to Widianto (1993) it is related to Sangiran 17. According to Jacob (1982) these two specimens belong to the same stratigraphic level, in the middle of the Kabuh formation.

Sangiran 38: This *calva* originated in the Kabuh formation of Kabuh on the Sangiran site. According to Widianto (1993), this specimen is related to specimen Sangiran 12, or to a specimen from the « Trinil-Sangiran » group, as defined by this author, but which, however, would be closer to more robust forms such as Sangiran 4 .

Trinil 2 (Java): Discovered at Trinil on the banks of the River Solo in a meander cutting through Kendeng'stratigraphic series, this *calva* is that of *Pithecanthropus erectus* Dubois 1893. Its interpretation was been very controversial at the time up until Weidenreich in 1922. The deposits from where this specimen originates must be more recent than 0.73 Ma. according to Sartono (1982) who classifies it as being archaic *Homo erectus*.

Modjokerto° (Java): Discovered in the village of Sumber Tengah North of Perning and North-east of Modjokerto, this *calvaria* belongs to the Pucangan formation for Duyfjes (1936) and von Koenigswald (1940). Indirect dating indicates that this individual is very old. Jacob and Curtis (1971) suggest 1.9 Ma. More recently, Swisher *et al.* (1994) suggest 1.81 Ma. According to Sartono (1982), at Perning, the black clays of Pucangan belong to the palaeozone of Matuyama and consequently the fossil is only 0.73 Ma. old. According to de Vos *et al.* (1982) the fauna associated with this specimen belongs to the biozone of Kedung Brudus and is not as old as that of the Trinil site. The geological work (Huffman, 2001) and the relocation of the site by Huffman *et al.,* (2005 and 2006) bring evidence that this specimen should belonged to the oldest period suggested by former authors. This immature specimen of less than 4 years old (Balzeau *et al.,* 2005) was named *Homo modjokertensis* by von Koenigswald in 1936. It was considered to be a more archaic form than *Pithecanthropus erectus* by von Koenigswald (1940).

Sambungmachan 1 (Java): Discovered at the East of Sragen in sedimentary deposits of the River Solo, this *calvaria* comes from the lower levels of Kabuh (Jacob, 1984). According to this author this *calvaria* is 0.9 Ma old. On the other hand, for Sémah *et al.* (1980) its age precedes the Matuyama/Brunhes limit. The stratigraphical level at which this specimen was found does not therefore correspond with the oldest Jaramillo sequence (Sémah, 1986) but it is now clear for everybody that the use of paleomagnetism is a non sense for the dating of continental geological formations. First datings obtained by Gamma Spectronomy (Yokoyama *in* Widianto, 1993) gives a date greater than 300,000 years but this date was later revised and this specimen was then seen to be between 35,000 and 50,000 years old (Falguères *et al.,* 2001). Jacob (1976) considers this individual as a possible intermediate specimen between *Pithecanthropus erectus* and *Pithecanthropus soloensis*. It is also an intermediate form between that of Sangiran and those of Ngandong for Wolpoff (1991). For de Lumley and Sonakia (1985) this specimen is an evolved *Homo erectus*.

Ngandong (Java): Several individual fossils were found in the sediment situated some 20 metres above the present level of the banks of the River Solo (Oppenoorth, 1932). A stratigraphy of the site was put forward in Sartono (1976). Datings carried out on the fauna found in the same alluvial series were between 80,000 and 100,000 years of age for Barstra *et al.* (1988). For these authors the fossilised crowns of Ngandong are « in place ». On the other hand, for de Terra (1943) it is very probable that the bone remains had been disturbed and originated from older levels. According to de Vos *et al.* (1993) Ngandong fauna would have disappeared in Middle Pleistocene. For Jacob (1984) the Ngandong specimens are 250,000 years old and according to Sémah *et al.* (1993), in the region of 300,000 years – obtained by Gamma Spectronomy. Later work by (Swisher *et al.*, 1996) give a considerably more recent age: between 25,000 and 52,000 years.

Ngandong 7 and **Ngandong 12** are almost complete *calvariae*. **Ngandong 1, Ngandong 3, Ngandong 6, Ngandong 10** and **Ngandong 11** are very incomplete *calvariae*.

Ngandong 5 is a *calva* and **Ngandong 2°** is a frontal squama of an immature individual.

These specimens make up the holotype of *Homo (Javanthropus) soloensis* Openoorth, 1932. In a communication which denies the attribution of the discoveries by de von Koenigswald of *Pithecanthropus erectus*, Dubois (1940) employs the trinôme *Homo sapiens soloensis* for the Ngandong specimens. They are seen as being *Homo erectus* by Santa Luca (1980), late *Homo erectus* for Sartono (1982) and evolved *Homo erectus* for Lumley and Sonakia (1985).

Ngawi (Java): Found in the alluvial deposits of the River Solo, this complete *calvarium* has no precise stratigraphic position (Sartono, 1991b). According to Widianto (1993) this specimen is morphologically related to Sambungmacan and Ngandong specimens. It has been shown (Widianto and Zeitoun, 2003) that this is a specific taxon which could be labelled *Homo soloensis*.

LB1 (Flores): Discovered in 2003 in the cave of Luang Bua at Flores (Brown *et al.,* 2004; Morwood *et al.,* 2004), this cranium has a very small capacity (417 cm^3) and is damaged giving rise to arguments concerning its status.

WHL50 (Australia): This *calva* was discovered in the region of the Willandra Lakes. For Habgood (1989b) and Wolpoff (1991) it is an early form of *Homo sapiens*. Dated at around 50,000 years by Thorne and Wolpoff (1992), Bräuer (1989) dates it at 25,000 years.

The study here put forward is not exhaustive. The smaller number of specimens taken into account is due to the inaccessibility of other specimens. Furthermore, craniums deformed because of taphonomic reasons – such as those of Steinheim or Hexian – could not be observed or measured in such a way as to present anatomic reality. Fossils such as that of Bilzingsleben which was insufficiently complete also did not allow the taking of measurements or the making of sufficiently accurate observations (see *infra*, the problem of managing missing data).

II.2.2. Assessment of the Synonymies.

If one refers to the hypodigm proposed by Howell in 1978 in order to define the *Homo erectus* species (Dubois, 1893) and the most currently employed taxonomic attribution for fossils discovered or described since 1978, the synonymy is as follows:

Anthropopithecus erectus Dubois, 1892*, * see below
Pithecanthropus erectus Dubois, 1894,
Sinanthropus pekinensis Black, 1927,
Homo (Javanthropus) soloensis Oppenoorth, 1932,
Homo modjokertensis von Koenigswald, 1936,
Pithecanthropus modjokertensis von Koenigswald, 1940,
Pithecanthropus robustus Weidenreich, 1945
Telanthropus capensis Broom and Robinson, 1949,
Sinanthropus officinalis von Koenigswald, 1953,
Atlanthropus mauritanicus Arambourg, 1954,
Homo leakeyi Heberer, 1963,

Sinanthropus lantianensis Woo, 1964
Tchadanthropus uxoris Coppens, 1965
Homo ergaster Groves and Mazak, 1975.

The ensemble of these labellings is not accepted by all authors for the moment, and, according to Liptak (1969), it was during the symposium of Wartenstein in 1962 (published in 1963: « classification and human evolution », Washburn S. L.(ed.) Chicago Aldin Publishing Company) that the sub-family of the Homininae was seen as consisting of only one genus. Thus the species *Telanthropus, Pithecanthropus*, *Sinanthropus* and *Atlanthropus* were dropped in favour of *Homo*. This symposium supported Mayr's (1650) recommandations which suggested the reduction to two species: *Homo erectus* and *Homo sapiens* as being the most recent hominids positioned «after» *Australopithecus*. During this symposium, several sub-species were put forward to replace named species for which the genus had been dropped. Thus, according to a **chronological** division suggested by Campbell (1963), the synonymy of *Homo erectus* would be as follows:

Homo erectus erectus (Dubois, 1892)*, * see below
Homo erectus pekinensis (Black and Zdansky, 1927)
Homo erectus mauritanicus (Arambourg, 1954)

According to Campbell (*op. cit.*) the creation of the sub-species *Homo erectus heidelbergensis* (Schoetensack, 1908) is not justified as, despite its very particular morphology, this specimen is a unique mandible. According to the Code, Campbell's refusal concerning Mauer's mandible (type of *Homo heidelbergensis* Schoetensack, 1908) cannot be accepted. Furthermore, it should be noted that the trinome *Homo sapiens soloensis* (Openoorth, 1932) retained by Campbell excludes the Ngandong fossils of the *Homo erectus* taxon. This fact is criticised by Jacob (1976) and Santa Luca (1980). In any case, the authors did not pursue Cambell's recommendations. However, Stringer (1987) casts aside the Ngandong series of the operational taxonomic unity corresponding to Asiatic *Homo erectus* of his analysis.

Other terms have been put forward in anthropological literature. Thus, Sartono has re-named fossils already described in mutiplying the trinomes.

Homo erectus ngandongensis Sartono, 1975
Homo palaeojavanicus sangiransis Sartono, 1976
Homo palaeojavanicus robustus Sartono, 1982

The trinome *Homo sapiens erectus* was suggested by Jelinek in 1978, considering that *Homo erectus* was not a species.

In 1980, Howells proposed a review of the trinomes which were the most currenly used.

Homo erectus erectus (Dubois, 1894)
Homo erectus pekinensis (Black, 1927)
Homo erectus modjokertensis (von Koenigswald, 1936),
Homo erectus mauritanicus (Arambourg, 1954),
Homo erectus leakeyi (Heberer, 1963),
Homo erectus ngandongensis Sartono, 1975
Homo erectus yanoumensis Li, Chien, Ma, Pu, Hsing and Chu, 1977
Homo erectus bilzingslebensis Vlcek, 1978
Homo erectus tautavelensis de Lumley and de Lumley, 1979

The discovery of the specimen of the Narmada, affililiated to *Homo erectus*, was labelled according to the Coding: *Homo erectus narmadaensis* Sonakia, 1984. However, this taxonomic belonging was later put into question by (Kennedy *et al.*, 1991). In 1990, Clarke stated that *Homo mauritanicus* (Arambourg, 1954) was a species for which the label was not in accordance with article 13 of the Code.

*** The original publication where the skullcap found in Java was described, with no possibility of confusion regarding the specimen, dates from the third term of 1892 but was only published in 1893 (J. De Vos, personal communication). In conformity with the recommendations of the International Code of Classification which applies to descriptions previous to the 20th Century, the binome should therefore be *Homo erectus* (Dubois, 1893).**

II.3. Computerised Analysis.

II.3.1 The Parsimony Program.

In order to analyse the date, the cladistic cnalysis program of parsimony PAUP version 3.1s (Phylogenetic Analysis Using Parsimony) was employed (Swofford, 1993). This program permits the comparison of several phylogenetical hypotheses and to rapidly select the most parsomonious in terms of number of steps (length of the cladogram) *id est*: the number of synapomorphies. The solidity of the tree is estimated by the consistence index (CI) and the retention index (RI) (Farris, 1989 and 1991). The consistence index measures the relationship between the ideal adjustment of the data and the real adjustment. Thus, for a character **k**, its amplitude **r** is the difference between the largest and the smallest number of character states. The amplitude **R** of the ensemble of the characters of the matrix is $\mathbf{R} = \sum \mathbf{k}\ \mathbf{r(k)}$

The consistence index is then the relationship of **R** along the effective **L** of the tree [**IC** = **R** / **L**]. An consistence index of 1 signifies that the tree under consideration does not present any homoplasy. The consistence index is all the better if the automorphies are more numerous. The retention index takes into account the maximum number of required transformations by the data (**G** = number of steps necessary if all the changes of state only happen along the terminal branches) [**IR = (G - L) / (G -R)**]. In the computerised phylogenetic processing of data, the parsimony programs estimate the state of the characters at the nodes (synapomorphies/ homoplasies), and express the length of the branches according to the number of derivated characters proper to these branches. In this way, the analysis of parsimony corresponds exactly with the phylogenetic analysis by Hennig according to Tassy (1991) and Barriel (1994).

A few authors have put into practice a computerised cladistic analysis with reference to hominids. Chamberlain and Wood (1987) state that the reliability of results obtained in several cladistic analyses is influenced by the choice of characters and the choice of the composition of the operational taxonomic units. As has been previously stated, this goes without saying. It must be clearly stated then that cladistic analysis is nothing other than an analysis of **n** caracteristics and of **N** taxons and has nothing to say concerning characters or taxa which have not been taken into consideration. Reservations indicated by Chamberlain and Wood can be overcome in taking for an operational taxonomic unit the fossil specimen and also taking into account all the information available; *id est*: without prejudging the more or less significant role or discriminating against one characteristic or another from a taxonomic point of view.

II.3.2 The Coding of Characters.

In the case where a character is composed of two states of transformation: **a** and **b**, a binary code suffices (0.1). The rooting of the tree aided by Out-group indicates the sense of the morphocline 0 → 1 or 1 → 0. The absence of information in a taxon is coded by a question mark « ? ». The program PAUP analyses directly the multiple state characters (0, 1, 2, 3). In the study here presented, we only came across these two cases of simple figures. Reliance on computerisation, with its alpha-numeric coding of characters, necessitated a prior rigorous analysis of the characters.

II.3.3 Processing Options.

Matrix computerised processing options to be selected for analysing characters are here briefly presented. More detailed information concerning the parsimony analyses can be consulted in Darlu and Tassy's work (1993).

Ordered and Non-ordered Processing of Multiple State Characters.

One option enables the whole of or part of the multiple state characters in an ordered or non-ordered way. When ordered, the number of steps between two states is equal to the absolute value of their arithmetical difference. For an ordered linear series (**a** - **b** - **c** - **d**) three steps are needed (three evolutive changes) in order to pass from one state « **a** » to the state « **d** » and inversely. In this case, one introduces at the beginning of the analysis, a phlogenetic analysis beforehand. This is the most common approach within the framework of analysis of morphological data (**Figure 2**).

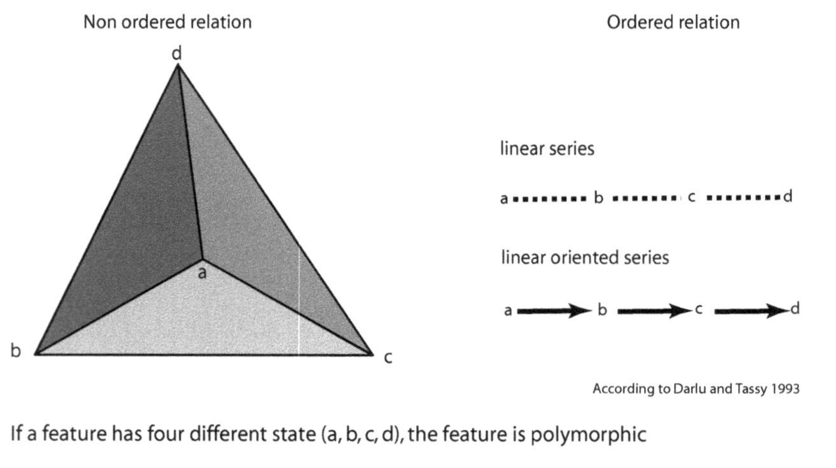

Figure 2: Relation of the states of characters.

When a character is expressed by four different states: a, b, c and d, this character is a multiple states character. If all state can directly be transformed in another one, each transformation cost one step. The serie is non-ordered. When the serie is linear the transformation is ordered. In this case the transformation can be oriented or not.

Certain authors are uncomfortable with this aspect (Hauser and Presch, 1991). On the other hand, when the character is handled in disorderly fashion, any state is considered as being separate from the others by a single step, which does not impose an arbitrary intermediary step (parsimony said of Fitch) (Fitch, 1971), but this implies that the states are not linked hierarchically. In some ways this is a sort of «nullified phylogenetic analysis». If one anatomical or ontogenetic piece of information exists which could permit choosing between these two methods of processing, this must be taken into account.

Optimisation of Transformations.

For the same topology, different distributions of characters can be compatible. In this case certain options maximise the reversions (option ACCTRAN = « accelerated transformation » of the PAUP) whereas others maximise the convergences (option DELTRAN = « delayed transformation » of the PAUP) (Swofford, 1993). These options have no effect on the length or on the topology of the tree but have an effect on the on the state of the nodes of the characters. These can therefore be discussed *a posteriori*. It is also possible to use the option MINF (Swofford, 1993) which exerts changes to the base Out-group / in-group node and directs the transformations towards the terminal branches. This option maximises the autapomorphies in such a way as not to augment the number of synapomorphies to the internal knots of the «in-group».

Weighting.

The option by default results in the same weight for all of the characters, but it is possible to affect the weighting of certain characters and transformations. Weighting leads to the introduction of additional hypotheses. Considering that there is no criterion which favours any one character above another, the option of «default» was adopted.

The Processing of Missing Data.

In the matrix of missing data of characters, missing data is indicated by question marks « ? ». The optimisation of these question mark « ? » is carried out *a posteriori* in relation to the cladogram obtained from the ensemble of the characters. In certain cases, it is not possible to express information as the latter depends on a missing character. Thus, the continuity of two structures is subordinate to the existence of the two structures. If only one structure is present on the fossil, a question mark « ? » indicates that the state of the character cannot be expressed. The use of question mark « ? » is also a strategy for encoding which was applied according to the principle stated in (Barriel and Tassy, 1994) see for example the statement for characters 94 to 97 in the following pages.

The Taking Root.

In the analyses put forward in this study, the rooting of the trees makes use of the Out-group criterion.

II.4. Morphological Data.

Morphological data has two aspects. Morphological characters can be relative descriptions of either the position or the size of an element when compared to another. The characters can still give rise to a more particular and complex definition to a precise combination of elements. Depending on the case, it is possible to affirm if this definition of a character is kept to or not. One can state if the character is present or absent. On the other hand, for example, when authors indicate that the upper part of the occipital squama is more extended than the lower, in order to evaluate this criterion it is preferable to use the metrical data. Indications calculated from craniometrical points which put into relation the upper and lower parts of the occipital bone squama will be used. This type of character is then described in the section «metrical data». When homologous points can be identified, one measure is more appropriate than an observation in relation to the size or the relative position of the objects. With one measure, rather than a description, one thereby introduces a quantification of the information, which is more satisfactory

II.4.1. Description of Morphological Characters.

Globally speaking, data processed under the heading «morphological data» espresses a parameter of shape whereas characters grouped under «metrical data» give rise to greater clarification of formats of different anatomical structures. In this way, an anatomical structure whose « triangular » shape evolves towards one which is « rectangular » and then towards one « circular » in shape will be described in the section « morphological data ». On the other hand, for this same structure, size could be indicated by an index calculated from homologous points (primary homology = anatomical homology). An assessment of anatomical observation and information gathered from literature leads to the proposition of a list of states of characters susceptible of giving account of phylogenetical relationships between specimens affiliated to *Homo erectus*. A total of 123 morphological characters were analysed which were divided into 25 characters for the frontal bone, 16 for the parietal bone, 20 for the occipital bone and 62 for the temporal bone.

II.4.1.1 The Frontal Bone (characters 1 to 25).

The sub-orbital torus is a complex originally defined by Schwalbe (1901). Two elements: an *arcus supraciliaris* and an *arcus supraorbitalis* separated by a third element: the *sulcus supraorbitalis* permitting the definition of the presence of the sub-orbital torus. Cunningham (1908) added the *trigonum supraorbitale* to these three elements. Santa Luca (1980) introduced the term of *torus glabellaris* in order to describe the median part of the orbital arch. Suzuki (1970) and Nara (1994), suggested a typology in order to describe the evolution of the different anatomical complexes of the sub-orbital region which differs to that employed by Spitery (1982 and 1985). This latter typology is more adapted to describing relationships between *sulcus supraorbitalis* and *sulcus postorbitalis*. The anatomical elements used can be described according to one viewpoint taken from the horizontal plane and then the frontal plane. In any case, Smith and Raynard (1980) determined three distances which permitted the ascertaining of the variation of thickness of the sub-orbital bone.

The collective studies of these authors were used in order to describe the orbital arch and to put forward a coding.

- *character 1*: fusion of the superciliary arches (*arcus superciliaris*) at the glabellary level (Figure 3).

<u>Coding</u>: 2 states

* 0: the two *arcus superciliaris* are not fused.

* 1: fusion of the two *arcus superciliaris* at the medial portion.

Figure 3: Left orbite of Ngandong VI.

a) Character: 1 state 1: medial fusion of the *arcus superciliaris*

b) Character: 3 state 1: relief of the upper edge of the *arcus superciliaris* marked, projecting upwards.

- *character 2*: degree of closure of the *sulcus supraorbitalis* (Figure 4).

<u>Coding</u>: 3 states

* 0: *sulcus supraorbitalis* absent as a result of the total fusion between the *arcus superciliaris* and the *arcus supraorbitalis*.

* 1: *sulcus supraorbitalis* present but with incomplete fusion of the *arcus superciliaris* and the *arcus supraorbitalis*.

* 2: *sulcus supraorbitalis* present with complete absence of fusion between the *arcus superciliaris* and the *arcus supraorbitalis*.

- *character 3*: importance of the relief of the upper edge of the superciliary arch (*arcus superciliaris*).

<u>Coding</u>: 2 states

* 0: relief is weak or absent.

* 1: relief marked, projecting upwards.

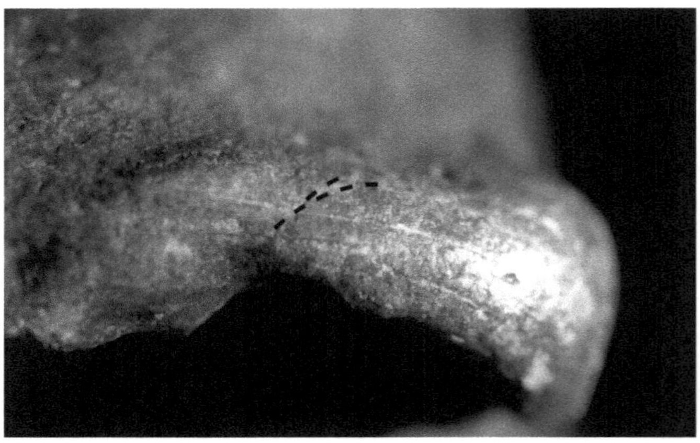

Figure 4: Left orbite of Ngandong XII.

Character 2: state 1: incomplete fusion of the *arcus superciliaris* and the *arcus supraorbitalis*.

- *character 4*: continuity of the post-orbital groove (*sulcus postorbitalis*).

Weidenreich (1943) described a supra-toral groove to be found on Ngandong specimens, attenuated in the glabellary region. This groove does not exist – according to this author – in Pithecanthropes. In 1951, Weidenreich no longer mentions attenuation in the glabellary region but talks of a median triangular depression which marks the separation of a torus which had been originally uniform in the Ngandong series. It is preferable to speak of a post-orbital groove rather than a post-toral groove. In fact, there is still no post-orbital torus.

Coding: 3 states

* 0: *sulcus postorbitalis* absent due to the continuity between the frontal squama and the supra-orbital complex.
* 1: *sulcus postorbitalis* present but incomplete as a result of the interruption of the medial zone.
* 2: *sulcus postorbitalis* present and complete from one edge to the other.

- *character 5*: existence of a lateral post-orbitory depression (Figure 5).

Figure 5: Frontal edge of Ngandong XII.

Character 5 state 1: depression at the lateral part of the *sulcus postorbitalis*.

Coding: 2 states

* 0: absence of a depression at the lateral part of the *sulcus postorbitalis*.
* 1: existence of a depression at the lateral part of the *sulcus postorbitalis*.

- *character 6*: shape of the ensemble of the frontal edge in *norma facialis*.

Howells (1980) insists on the fact that a linear shape of the ensemble of the frontal edge in *norma facialis* and in *norma verticalis* characterise *Homo erectus* whereas the two conditions do not exist in the Ngandong series of fossils. Hublin (1986) insists on the variability of the rectitude of the frontal edge at the heart of the series of human fossils. Distinct studies of the variations of this element according to two anatomical planes were therefore carried out.

Coding: 2 states

* 0: rectilinear contour at the frontal edge in *norma facialis*.
* 1: convex contour at the top of the frontal edge in *norma facialis*.

- *character 7*: existence of a glabellary depression in *norma facialis* (this depression is not linked to the existence of a *torus supraorbitalis*).

Coding: 2 states

* 0: absence of a glabellary depression on the frontal edge in *norma facialis*.
* 1: presence of a glabellary depression on the frontal edge in *norma facialis*.

- *character 8*: shape of the ensemble of the upper frontal edge in *norma facialis*.

Weidenreich (1951) demonstrates an evolutionary gradient of the upper frontal edge since the time of great Apes upto Modern Man. From a rectilinear shape in *norma verticalis*, the frontal edge became more and more curved. But, Stringer (1984) has reservations concerning the phylogenetical significance of the rectitude of the frontal edge when viewed from the back and from above. Hublin (1986) adds that the shape of the frontal edge is variable with *Homo erectus*. Two states were adopted in order to describe the frontal edge in *norma verticalis*.

Coding: 2 states

* 0: rectilinear contour of the frontal edge in *norma facialis*.
* 1: convex contour at the top of the frontal edge in *norma facialis*.

- *character 9*: position of the glabellary zone in relation to the frontal edge in *norma verticalis* (Figure 6).

For Weidenreich (1943) the glabellary projection does not exist in all Sinanthropes. The glabellary zone can vary in the line of the frontal edge. This zone can then be further forward or further back than the frontal edge. It can also be that there is no modification when compared to the line of the frontal edge.

Coding: 3 states

* 0: gabellary zone further back.
* 1: gabellary zone neither projected nor in retreat.
* 2: gabellary zone further forward than the frontal edge.

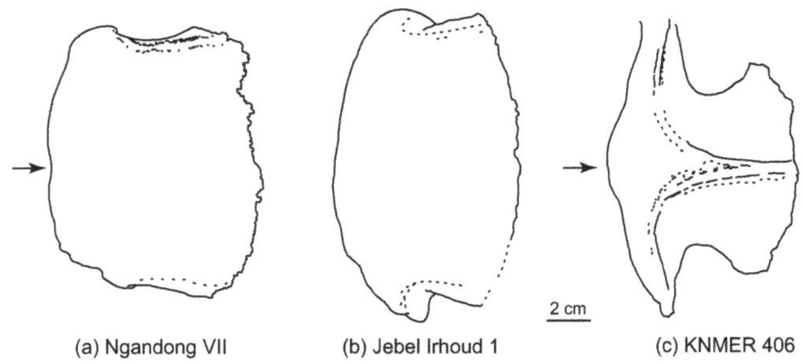

Figure 6: Character 9: position of the glabellary zone in relation to the frontal edge in *norma verticalis*.

a) state 0: gabellary zone further back.

b) state 1: gabellary zone neither projected nor in retreat.

c) state 2: gabellary zone further forward than the frontal edge.

- *character 10*: existence of a «microdepression» of the gabellary zone in *norma verticalis* (Figure 7).

The term microdepression, describes the existence of a small glabellary depression which, for most of the time, is associated with a large projection forward of the ensemble of the glabellary region.

Coding: 2 states

* 0: absence of a microdepression in the glabellary zone.
* 1: presence of a microdepression in the glabellary zone.
* ?: indeterminate state linked to state «0» of character 9.

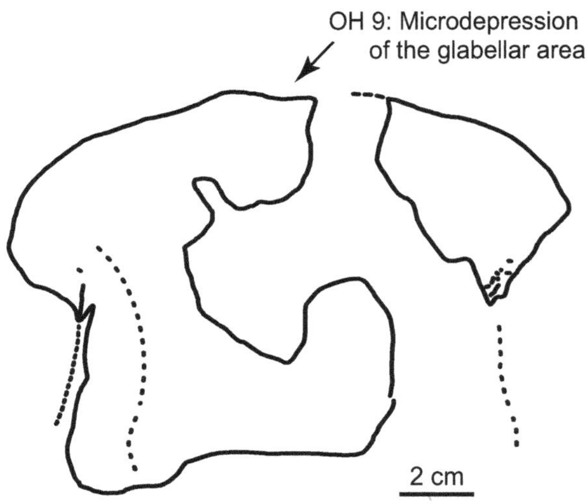

Figure 7: Frontal edge of OH 9

Character 10: state 1: presence of a microdepression in the glabellary zone.

- *character 11*: continuity of the *torus supraorbitalis* (Figure 8 and 9).

MacIntosch and Larnach (1972) and Stringer (1984) are all insistant on the lack of division of the torus. Weidenreich (1943) describes some supraorbital folds linked by a robust glabellary torus or of a single continuous frontal torus.

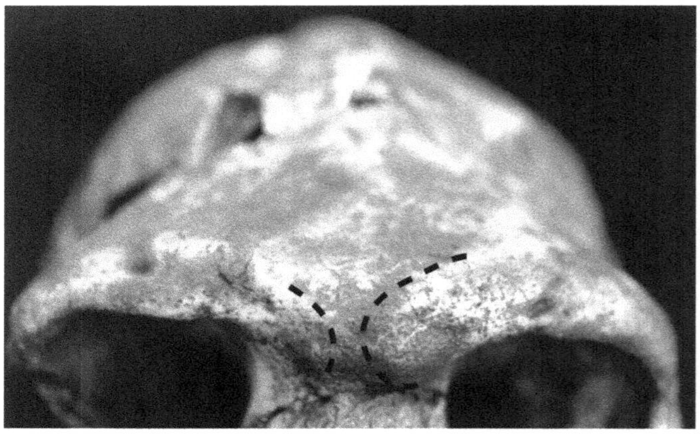

Figure 8: Edge of the frontal bone of Bodo.

Character 11: state 1: presence of two distinct « mono-orbital » toruses.

Coding: 3 states

* 0: *torus supraorbitalis* absent.
* 1: *torus supraorbitalis* incomplete, interrupted in the median zone. With two distinct «mono-orbital» toruses.
* 2: *torus supraorbitalis* present and complete from one edge to the other. Presence of a single « bi-orbital » torus.

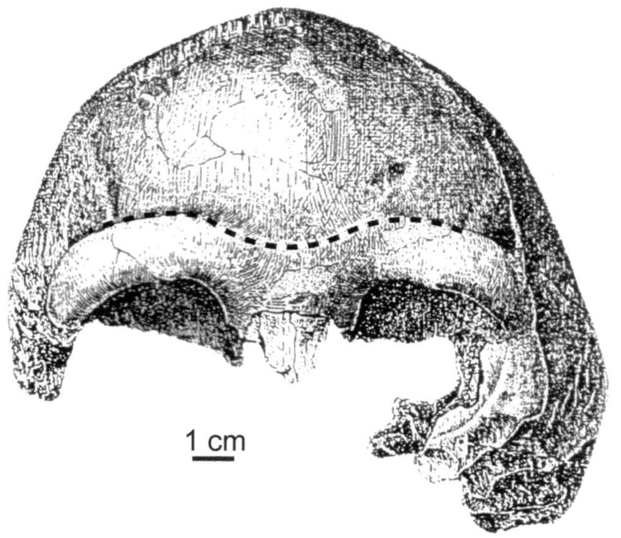

According to Weidenreich (1943)

Figure 9: Frontal view of *Sinanthropus pekinensis* XII.

Character 11: state 2: Presence of a single «bi-orbital» torus.

- *character 12*: type of orbital arch (Figure 10).

Smith and Ranyard (1980) described three segments (medial, middle of the orbit and lateral) in order to define the thickness of the orbital arch. Weidenreich states that the medial segment is more developed in Sinathropes. Either (a) the medial segment, (b) the segment at the centre of the orbital arch and (c) the lateral segment exists for the specimens studied; four states exist according to the relative size of these segments.

<u>Coding</u> : 4 states

* 0 : type G : where (a) > (b), (b) < (c) and (a) < (c).
* 1 : type P : where (a) > (b), (b) < (c) and (a) > (c).
* 2 : type A : where (a) < (b), (b) > (c) and (a) > (c).
* 3 : type M : where (a) > (b), (b) > (c) and (a) > (c).

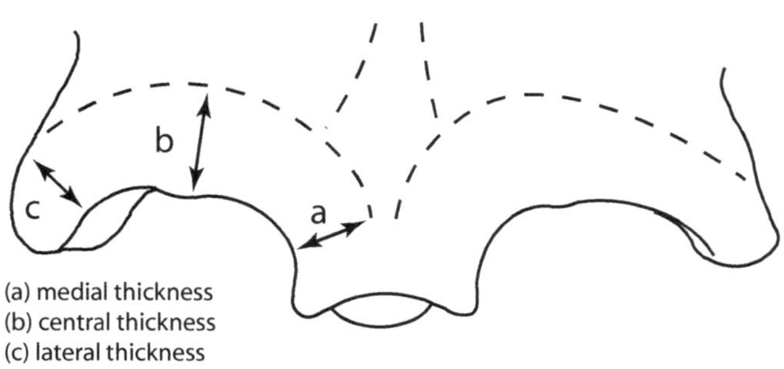

Figure 10

Thickness of the segment of the orbital arch.

- *character 13*: width of the temporal band (the space between the *linea temporalis superior* and the *linea temporalis inferior*; Grimaud, 1982) on the frontal bone.

The development of the temporal crown (the back part of the temporal band) on the frontal bone is taken into account by Bräuer and Leakey (1986) with reference to archaic African *Homo sapiens*.

<u>Coding</u>: 2 states.

* 0: the temporal band is narrow on the frontal bone.
* 1: the temporal band is wide on the frontal bone.

- *character 14*: projection of the temporal band on the frontal bone.

The temporal band is considered as being a characteristic of which the variation in shape is of great interest by Rightmire (1984)

<u>Coding</u>: 2 states.

* 0: the temporal band projects only slightly.
* 1: the temporal band projects considerably.

- *character 15*: existence of a tuberculum at the *frontotemporale* point (Figure 11).

The occasional existence of a small tuberculum was observed during the study of the fossilised skulls, situated on the lateral crown of the frontal bone at the *frontotemporale* point.

<u>Coding</u>: 2 states.

* 0: absence of a tuberculum at the *frontotemporale* point.
* 1: existence of a tuberculum at the *frontotemporale* point.

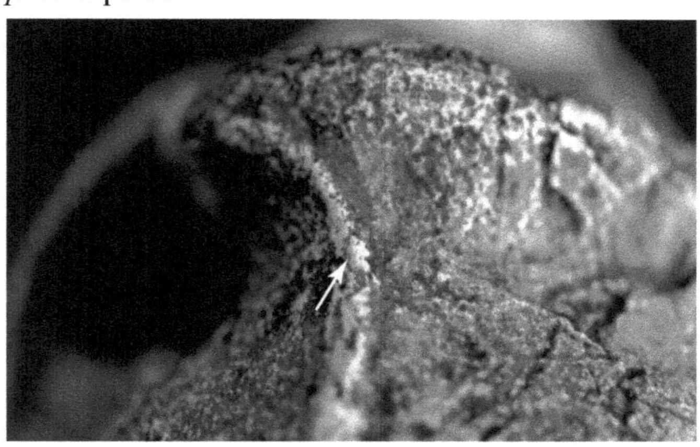

Figure 11: Left orbital trigone of KNMER 3733.

Character 15: state1: presence of a tuberculum at the frontotemporale point.

- *character 16*: existence of a metopic crest (frontale portion of the sagittal crest).

This structure can be seen on the Great Apes and on certain specimens attributed to the *Australopithecus* species.

Coding: 2 states.

* 0: absence of a metopic crest created by the fusion of the temporal lines.
* 1: existence of a metopic crest created by the fusion of the temporal lines.

- *character 17*: existence of a medio-sagittal supraglabellary tuberculum at the junction of the postorbitory groove (*sulcus postorbitale*) and the frontal squama (*squama frontalis*).

During the study of the Neanderthalian *calvariae* specimens, it was discovered that nearly all of them had a spherical projection at the medio-sagittal junction of the *sulcus postorbitale* and the *squama frontalis*. This tuberculum can be felt very easily but it was not possible to photograph this anatomical element.

Coding: 2 states.

* 0: absence of a supraglabellary tuberculum.
* 1: existence of a supraglabellary tuberculum.

- *character 18*: type of metopic keel.

The existence of a metopic keel is considered to be a character proper to *Homo erectus* by Andrews (1984). However, Bräuer and Mbua (1992) submit the hypothesis that the beginnings of a metopic keel is present in the bregmatic portion of the squama of Sterkfontein 5 and a smaller development exists on the KNMER 1813 specimen. Santa Luca (1980) describes a metopic keel which could be contiguous with a bregmatic protuberance in *Homo erectus*. According to Weidenreich (1951) a gradual diminishing in height followed by the total disappearance of the metopic keel towards the bregmatic protrudence exists in the Ngandong fossils.The metopic keel has therefore been defined by its presence, its intensity and also by the position it occupies on the frontal squama. The frontal squama has therefore been divided into four equal parts from the *supraglabellare* to the *bregma*. The study of juvenile fossil specimens shows a gradiant defined according to four different states when the metopic keel is present. The supposed individual age of the juvenile fossils was used in order to establish the gradiant of the states of characters from 0 to 4 without this order being used to constrain later analysis.

Coding: 5 states

* 0: absence of a metopic keel.
* 1: a keel is present but only on the rear half.
* 2: a keel is present on the whole of the frontal squama.
* 3: a keel is only present on the back three quarters of the frontal squama.
* 4: a keel is only present on the second and third quarters of the frontal squama.

- *character 19*: degree of the relief of the metopic keel.

The existence of a metopic protuberance is considered as being a unique character to *Homo erectus* by Hublin (1986).

Coding: 2 states.

* 0: metopic keel only slightly developed.
* 1: metopic keel of sizeable relief.
* ?: indeterminate linked to state «0» of character 18.

- *character 20*: antero-posterior development of the metopic keel.

Observation of the fossilised material revealed the existence of a distinction regarding the evolution of the size of the metopic keel.

Coding: 2 states.

* 0: the lateral edges of the metopic keel are *quasi* parallel.
* 1: the metopic keel is enlarged and flattened-out towards the rear.
* ?: indetermination linked to state «0» of character 18.

- *character 21*: existence of a bregmatic protuberance (Figure 12).

The existence of a bregmatic protuberance is seen to be a characteristic belonging to *Homo erectus* by Hublin

(1986). Stringer (1984) associates the bregmatic protuberance, the coronal reinforcement, the metopic keel and the sagittal keel within an ensemble characteristic of the morphology of *Homo erectus*. Different shapes of bregmatic proturbence, linked to adjacent structures have been described by Santa Luca (1980).

Coding : 2 states.

* 0: absence of bregmatic protrudence.
* 1: presence of a bregmatic protrudence.

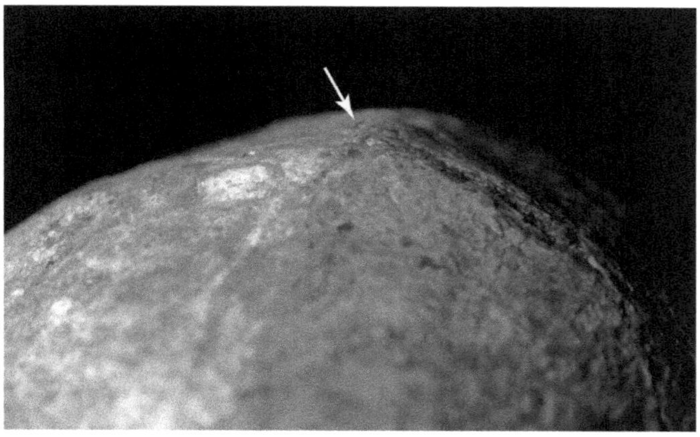

Figure 12: Upper view of Ngandong XII.

Character 21: state 1: bregmatic protrudence.

- *character 22*: existence of an upper coronal reinforcement.

According to Santa Luca (1980) the coronal reinforcement, the thickness of the bone at the level of the coronal suture, is more or less laterally spread-out. This characteristic is particular to *Homo erectus* according to Stringer (1984) and Hublin (1986). This structure comes close to the existence of a bregmatic protruberance in the shape of a cross (Santa Luca, 1980).

Coding: 3 states.

* 0: absence of coronal reinforcement.
* 1: existence of a coronal reinforcement at the upper part of the squama.
* 2: existence of a coronal reinforcement at the upper part which continues to the lateral edge of the squama.

- *character 23*: existence of a precoronal depression.

Coding: 3 states.

* 0: absence of perturbation in the convexity of the frontal squama.
* 1: existence of a precoronal plane.
* 2: existence of a precoronal depression.

- *character 24*: existence of a frontal lump (*tuber frontale*).

Coding: 4 states.

* 0: absence of a distinctive frontal lump.
* 1: existence of a distinctive frontal lump medially situated.
* 2: existence of a distinctive frontal lump centred on the lateral half of the squama.
* 3: existence of a distinctive frontal lump, laterally situated.

- *character 25*: existence of a supratrigonal depression (Figure 13 and 14).

The existence if very small depressions on the medial edges of the rear part of the temporal lines was put forward by Day *et al.*, (1976) in his description of the frontal bone of KNMER 1813. A study of the fossilised materiel indicates that this morphology also exists on other specimens.

Coding: 2 states.

* 0: absence of a supratrigonal depression.
* 1: existence of a supratrigonal depression.

II.4.1.2 The Parietal Bone (characters 26 to 41).

- *character 26*: existence of a sagittal crest.

This structure corresponds to the relief of the temporal lines on the parietal. It is visible on the

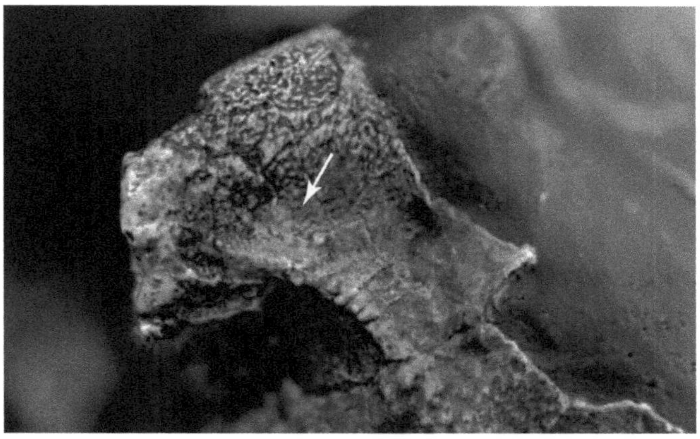

Figure 13: Upper view of the Left orbital trigone of KNMER 1813.

Character 25: state 1: existence of a supratrigonal depression.

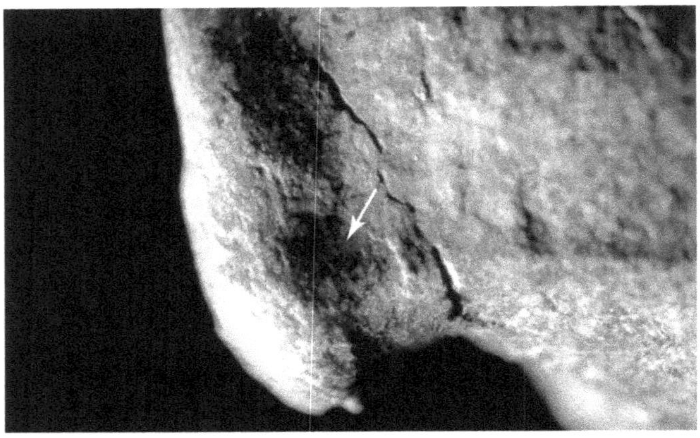

Figure 14: Upper view of the Left orbital trigone of Ngandong VI. Character 25: state 1: existence of a supratrigonal depression.

Great Apes and on certain speciments attributed to the genus *Australopithecus*.

Coding: 2 states.

* 0: absence of a sagittal crest.
* 1: presence of a sagittal crest.

- *character 27*: existence of a sagittal keel on the rear half of the *bregma -lambda* arch.

The development of the sagittal keel is variable (Santa Luca, 1980). The keel can be confined to the bregmatic area (Bräuer and Mbua, 1992). It is held to be a particularity of *Homo erectus* by Andrews (1984). This anatomical element has already been quoted by Weidenreich (1943) as being a specificity of Sinanthropes.

Coding: 4 states.

* 0: absence of a sagittal keel.
* 1: existence of a sagittal keel only on the front part of the front half of the *bregma –lambda* arch.
* 2: existence of a sagittal keel only on the back part of the front half of the *bregma –lambda* arch.
* 3: existence of a sagittal keel on the ensemble of the front half of the *bregma –lambda* arch
* ?: indetermination linked to the state «0» of character 26.

- *character 28*: existence of an obeliac depression (on the third quarter of the *bregma –lambda* arch).

Weidenreich (1943) does not distinguish between obeliac depression and prelambdatic depression, but it would seem to be interesting to pursue the the separate evolutionary process of these two structures. Furthermore, it is conceivable that the expression of the sagittal keel on the third quarter of the parietal bones prevent an observation of a depression. Inversely, the existence of a depression can obliterate the expression of a sagittal keel on the third part of the parietals. A definition of the three states relative to the anatomic structures which could exist on the third quarter of the *bregma- lambda* arch follows.

Coding: 3 states.

* 0: existence of a sagittal keel on the third quarter of the *bregma-lambda* arch.
* 1: absence of a sagittal keel on the third quarter of the *bregma-lambda* arch.
* 2: existence of an obeliac depression on the third quarter of the *bregma-lambda* arch.

- *character 29*: existence of a prelambdatic depression (on the last quarter of the *bregma-lambda* arch).

An absence of keel is described on the last quarter *bregma-lambda* arch of the Sangiran 12 and Sangiran 17 specimens (Sartono and Grimaud, 1983). The plane which has been sometimes described (Grimaud, 1982; Grimaud and Jacob, 1983) in this anatomical part very certainly corresponds to a total absence of keel locally.

Coding: 3 states.

* 0: existence of a sagittal keel on the last quarter of the *bregma-lambda* arch.
* 1: absence of the sagittal keel on the last quarter of the *bregma-lambda* arch.
* 2: existence of a prelambdatic depression on the last sagittal quarter of the *bregma-lambda* arch.

- *character 30*: size of the sagittale keel.

Coding: 2 states.

* 0: existence of a sagittal keel hardly developed.
* 1: existence of a strong sagittal keel.
* ?: indetermination linked to state «1» of character 26.
* ?: indetermination linked to state «0» of character 27.
* ?: indetermination linked to states «1» or «2» of character 28.

- *character 31*: existence of a postcoronal depression.

Observed by Weidenreich (1943), the variations in expression of the postcoronal depression are described amongst the different specimens in the human line studied by Grimaud (1982). A plane and not a depression is decribed on certain fossils (Grimaud and Jacob, 1983; Sartono and Grimaud, 1983).

Coding: 3 states.

* 0: absence of perturbation in the convexity of the parietal squama to the rear of coronal suture.
* 1: existence of a postcoronal plane.
* 2: existence of a postcoronal depression.

- *character 32*: projection of the temporal band (the space between the *linea temporalis superior* and the *linea temporalis inferior*) (Figure 15).

The temporal band corresponds to the space contained between the *linea temporale inferior* and the *linea temporale superior.* It corresponds to the zone of insertion of the deep face of the temporal apenevrose. The variations of position, of relief and also of the existence of a *stephanion* disconnection of these elements within hominids, have been described by Grimaud (1982). Weidenreich (1951) stresses the difference between Sinanthropes, where the temporal line reaches the *tuber parietalis*, and the Ngandong specimens where it is slightly below. For Grimaud (1982) the temporal lines are protuberant in *Homo erectus* (see also Grimaud, 1986).

Coding: 3 states.

* 0: temporal lines hardly or not visible.

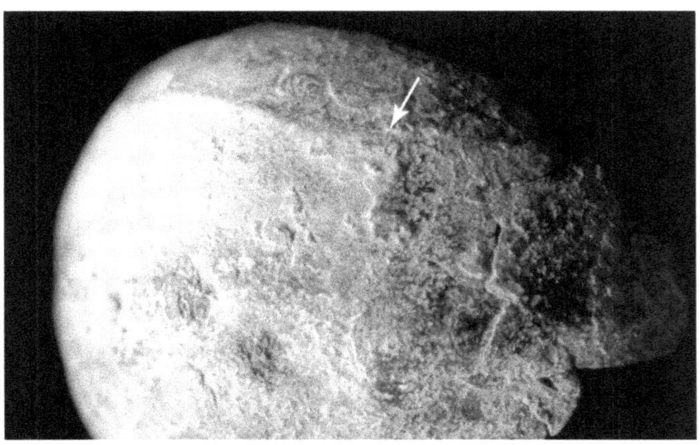

Figure 15: Upper view of Sangiran 38.

Character 32: state 2: temporal lines protubing.
Character 33: state 1: narrow temporal band on the parietal.

* 1: temporal lines clearly visible but not protruding.
* 2: temporal lines protubing.

- *character 33*: width of the temporal band on the parietal bone (Figure 15 and 16).

Coding: 2 states.

* 0: wide temporal band on the parietal bone.
* 1: narrow temporal band on the parietal.

- *character 34*: existence of a disconnection of the temporal band level with the *stephanion* (Figure 16).

Coding: 2 states.

* 0: absence of a disconnection between the frontal and parietal.
* 1: evidence of a disconnection between the frontal and the parietal.

- *character 35*: height of the temporal line on the parietal bone.

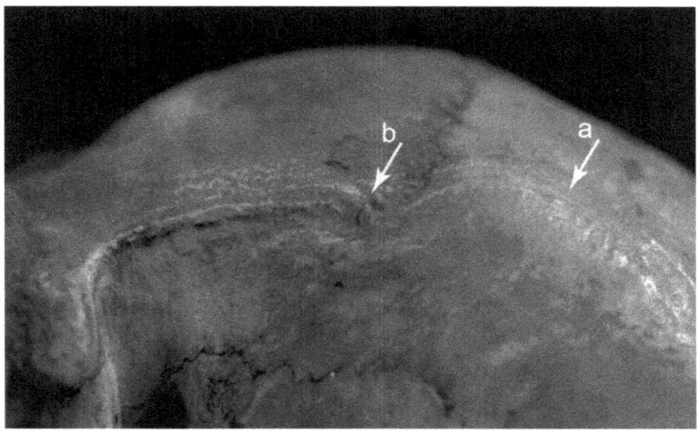

Figure 16: Lateral view of Kabwe.

a) Character 33: state 0: wide temporal band on the parietal bone.
b) Character 34: state 1: disconnection between the frontal and the parietal.

Coding: 2 states.

* 0: the temporal line is in a high position on the patietal in *norma lateralis*.
* 1: the temporal line is in a low position on the parietal in *norma lateralis*.

- *character 36*: existence of a *processus asteriacus* (Figure 17).

The bone can be thicker and constitute a fully closed projection on the lower rear segment of the *sutura lambdoidea,* which would correspond to the *processus asteriacus* of Haferland (Hublin, 1978a). Jacob (1967b) mentions a processus asteriacus

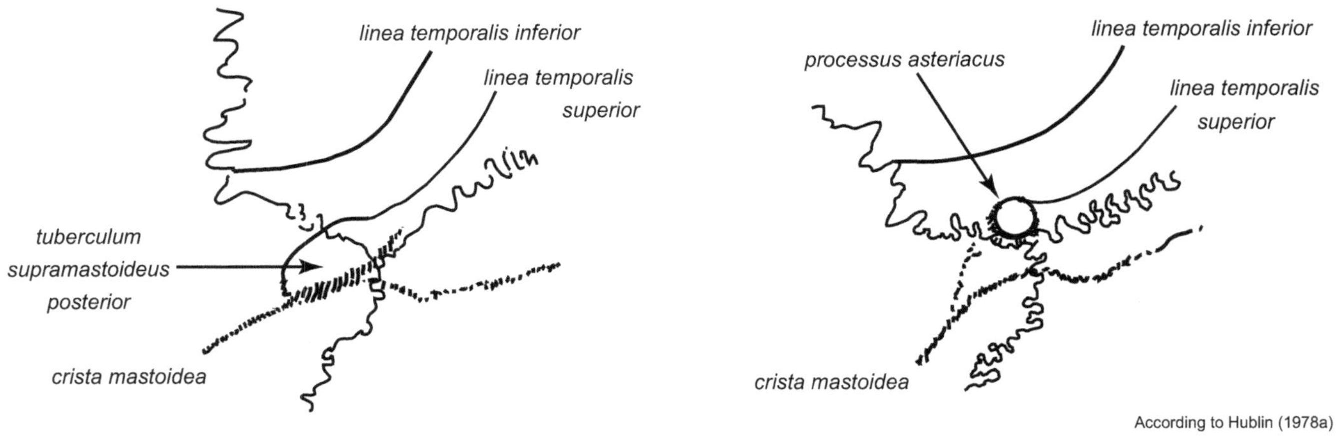

Figure 17

a) Character 36: state 1: presence of a processus asteriacus.
b) Character 75: state 1: presence of a rear sus-mastoïdian tuberculum.

The processus asteriacus is an elevation limited to the inferior segment of the upper temporal line of the parietal bone.
The rear sus-mastoïdian tuberculum is situated on the temporal bone or on the parieto-mastoidian suture.

described by Weidenreich, but this is a *torus angularis* as defined by this author.

Coding: 2 states.

* 0: absence of a *processus asteriacus*.
* 1: presence of a *processus asteriacus*.

- *character 37*: existence of an asteriac depression (Figure 18).

In order for there to be an asteriac depression, there has to be a depression centred on the *asterion* which overlaps the parietal, the occipital and the temporal bones.

Coding: 2 states.

* 0: absence of an asteriac depression.
* 1: presence of an asteriac depression.

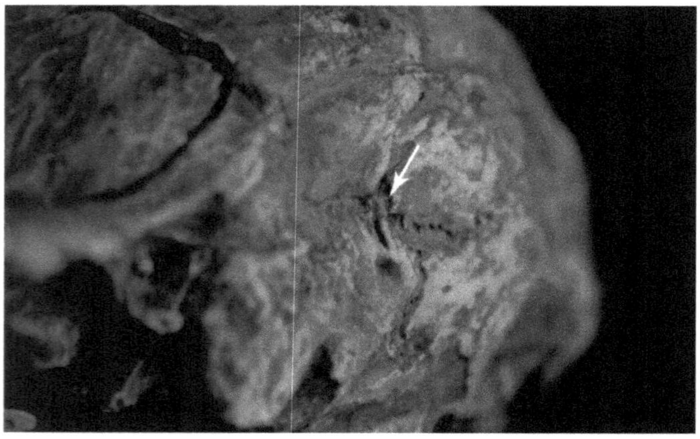

Figure 18: Lateral view of La Chapelle-aux-Saints.

Character 37: state 1: presence of an asteriac depression.

- *character 38*: existence of a parietal bump (*tuber parietale*).

Weidenreich (1943) mentions that there is a difference of expression of the *tuber parietale* between Neanderthalians and Sinanthropes. An analysis of this structure carried out by Grimaud (1982) showed a variation of the expression and a position which was more or less further to the side, laterally of the *tuber parietale*. In line with the view of this author, these results show there is no *tuber parietale* in Australopithecus or in *Homo habilis*. For Grimaud (1986) the *tuber parietale* of *Homo erectus* is flattened.

Coding: 3 states.

* 0: absence of a marked parietal bump.
* 1: existence of a well marked and centred parietal bump.
* 2: existence of a well marked parietal bump to the side, laterally speaking.

- *character 39*: existence of a «*tuber angularis*» (and not «*torus angularis*») (Figure 19).

Several acceptations of the anatomical complex named «*torus angularis*» exist. As described by Weidenreich (1943), the *torus angularis parietalis* is a rounded bump which emerges on the external surface of the parietal, at the angle between the lambdoidal and parieto-mastoidal sutures. For Santa Luca (1980) it is a flat tumescence on the upper part of the upper temporal line. According to Grimaud (1986) this structure is constantly found on fossils

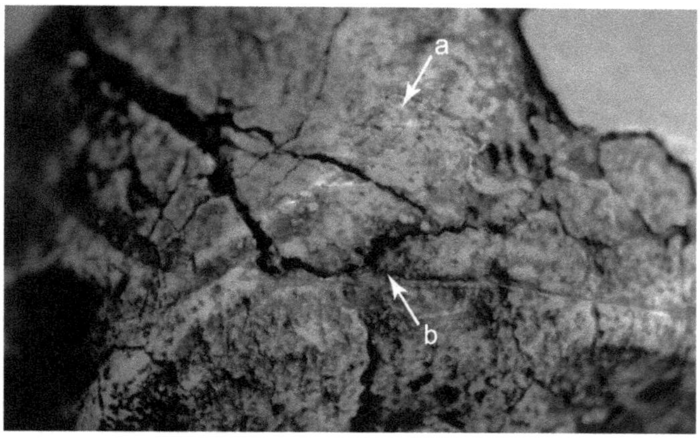

Figure 19: Lateral view of OH 9.

a) Character 39: state 1: existence of a «tuber angularis».
b) Character 37: state 1: presence of an asteriac depression.

from Java. For Andrews (1984) and for Hublin (1986) the *torus angularis* is a typical characteristic of *Homo erectus*. This anatomical complex was decomposed into elements which were possible to observe in isolation on the specimens.The «*tuber angularis*» is a thickening of the bone which is situated at the lower posterior angle of the parietal. Several structures were here recognised which reinforces or otherwise the tuberosity. Not all were associated on the different fossils studied.

<u>Coding</u>: 2 states.

* 0: absence of tuberosity in the posterior zone of the parietal, at the point where the two temporal lines join together earlier after curving towards the front.
* 1: presence of a tuberosity in the posterior zone of the parietal, where the two temporal lines join together beforecurving towards the front.

- *character 40*: existence of a back depression underlining the upper temporal line in the lower back part of the parietal.

Described by Grimaud (1982) for Ngandong 1 and by Sartono and Grimaud (1983) for Sangiran 17, such a depression could correspond to the particular structure given by Bräuer and Mbua (1992) on the ensemble of the back part of the parietal.

<u>Coding</u>: 2 states.

* 0: absence of a depression underlining the upper temporal line.
* 1: presence of a depression underlining the upper temporal line.

- *character 41*: existence of a depression between the two temporal lines in the infero-posterior part of the parietal.

<u>Coding</u>: 2 states.

* 0: absence of a depression between the lower and upper temporal lines.
* 1: presence of a depression between the upper and lower temporal lines.

II.4.1.3 The Occipital Bone (characters 42 to 61)

Several anatomical elements concerning the occipital have been identified. In fact a strong transversal *torus occipitalis* edged by a continuous and medially more developed *sulcus supratoralis*, joining together the *torus angularis*, the supramastoïdal crests as well as the occipito-mastoïdal crests, has been seen as an apomorphy of *Homo erectus* by Hublin (1986). But, several of these structures can exist separately on the different specimens. The anatomical complex, therefore, was decomposed into distinct constituative elements for the different specimens.

- *character 42*: curvature of the upper part of the occipital squama (*planum occipitale*) in *norma lateralis*.

In order to define the limit between the upper and lower parts of the occipital squama (*squama occipitalis*), it is a question of agreeing on the definition of the *inion*. We have chosen to use the definition put forward by Hublin (1978a and c). The *inion* is the point where the two *lineae nuchae superiores* meet in the medio-sagittal plane. It is on the *tuberculum linearum*: either on a rough patch situated on the tuberculum or at the junction of the *proturentia occipitalis externa* and the tuberculum. In the case where the *lineae nuchae superiores* are so weakly developed that they do not go as far as the medio-sagittal plane, they need to be artificially lengthened following the same direction with the *inion* placed at their meeting point. The *lineae nuchae supremae* and the *protuberantia occipitalis externa* which they shape are only taken into consideration to determine the *inion* in the case of fusion with the *tuberculum linearum*. In determining the *inion* it is not recommended to use the *protuberantia occipitalis interna*, or the *endinion*. If the surface between the *lineae nuchae superiores* and the *supremae* form a transversal fold which is more or less pronounced, the *inion* will then always be found – in line with the above definition – at the lower edge of this fold, *id est*: at the level of the *lineae nuchae superiores*. This point must never be placed at the lower extremity of a *protuberantia occipitalis externa* in the shape of a stick.

Coding: 2 states.

* 0: *planum occipitale* flat or slightly concave towards the rear.
* 1: *planum occipitale* convex shape towards the rear.

- *character 43*: shape of the upper part of the occipital squama in *norma occipitalis*.

Coding: 2 states.

* 0: upper shell of the occipital pentagonally-shaped to rounded in *norma occipitalis*.
* 1: upper shell of the occipital triangular-shaped in *norma occipitalis*.

- *character 44*: curvature of the lower part of the lower shell in *norma lateralis*.

Coding: 2 states.

* 0: lower shell of the occipital convex towards the rear in *norma lateralis*.
* 1: lower shell of the occipital flat or slightly concave towards the rear in *norma lateralis*.

- *character 45*: shape of the *torus occipitalis* in *norma occipitalis*.

According to Weidenreich (1940) the *torus occipitalis* is composed of a central section and of two arms resulting in a bifurcation at the proximity of the lambdoidal suture. Santa Luca (1980) takes up the same descriptive elements. According to Weidenreich (1951), the torus occipital of the fossils of Ngandong differ to those of *Sinanthropus* and of *Pithecanthropus* where it is transversal and is separated from the squama by a supratoral groove. Hublin (1978a) was particularly interested in the nature of this anatomical structure and in its variation, and insisted on its phylogenetical importance.

Coding: 4 states.

* 0: absence of *torus occipitalis*.
* 1: existence of a convex *torus occipitalis* high up in *norma occipitalis*.
* 2: existence of a rectilinear *torus occipitalis* in *norma occipitalis*.
*3: existence of a concave *torus occipitalis* towards the top in *norma occipitalis*.

- *character 46*: importance of the *torus occipitalis*.

Coding: 2 states.

* 0: *torus occipitalis* unprojecting or only slightly so.
* 1: *torus occipitalis* projecting strongly.
* ?: indetermination linked to state «0» of character 45.

- *character 47*: lateral spreading-out of the *torus occipitalis* (Figure 20 and 21).

For MacIntosch and Larnach (1972) the *torus occipitalis* has the shape of a median monticule or of a laterally extended wrinkle as far as the mastoidian crest or the occipito-mastoïdian wrinkle.

Coding: 2 states.

* 0: *torus occipitalis* median.
* 1: *torus occipitalis* transversal.
* ?: indetermination linked to state «0» of character 45.

- *character 49*: continuity of the *torus occipitalis* with the *crista mastoidea*.

Figure 20: Posterior view of Sangiran 38.

Character 47: state 0: torus occipitalis laterally extended.

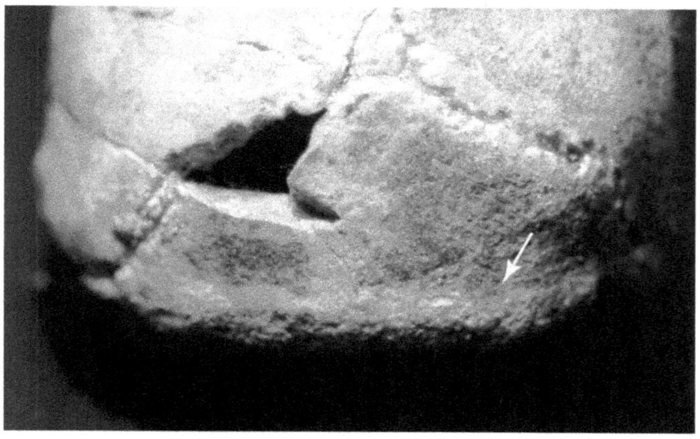

Figure 21: Posterior view of Sangiran 10.

Character 47: state 1: torus occipitalis transversal.

The link between *torus occipitalis* and *crista mastoidea* was indicated by Weidenreich (1940 and 1943). It concerns the Great Apes and *Homo erectus* known at that time. According to Spitery (1985) the absence of a link between *torus occipitalis* and *crista mastoidea* distinguishes the *Autralopithecus africanus* from the pongides.

Coding: 2 states.

* 0: *torus occipitalis* with no direct link with *crista mastoidea.*
* 1: *torus occipitalis* continuous with *crista mastoidea.*
* ?: indetermination linked to the state «0» of character 45.
* ?: indetermination linked to state «0» of character 47.

- *character 50*: continuity of the *torus occipitalis* with the upper temporal line.

Coding: 2 states.

* 0: *torus occipitalis* with no direct line with the upper temporal line.
* 1: *torus occipitalis* continuous with the upper temporal line.
* ?: indetermination linked to state «0» of character 45.
* ?: indetermination linked to state «0» of character 47.

- *character 51*: continuity of the *sulcus occipitalis* on the whole of the upper occipital squama (Figure 22).

A *sulcus supratoralis* can be seen in *Homo erectus* (MacIntosch and Larnach, 1972). *Sulcus occipitalis* will be mentioned here in order not to presume the existence of an occipital torus.

Coding: 3 states.

* 0: absence of *sulcus occipitalis.*
* 1: a discontinued *sulcus occipitalis* is present which effaces itself latero-medially.
* 2: a *sulcus occipitalis* is present, transversally continuous.

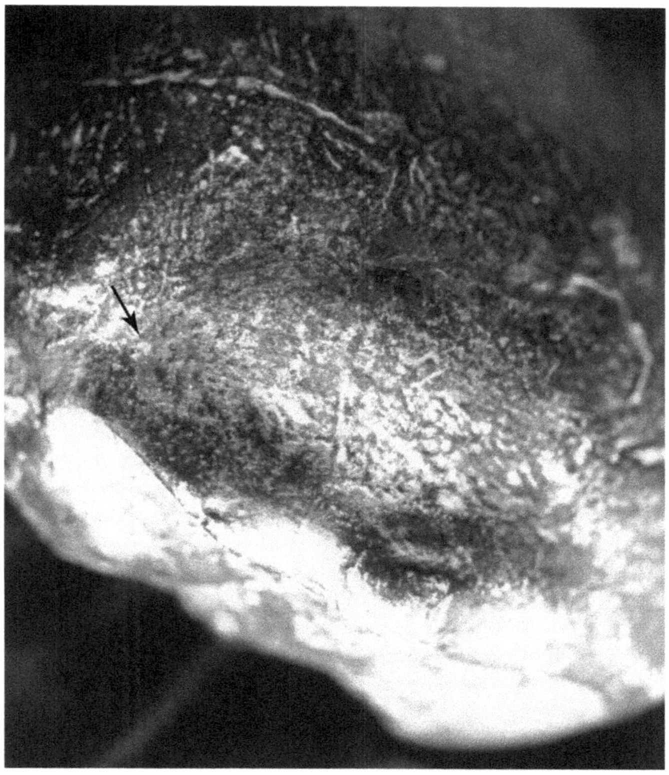

Figure 22: Posterior view of Sangiran 12.

Character 51: state 2: a sulcus occipitalis is present and transversally continuous on the occipitale squama.

- *character 52*: existence of lateral occipital depressions on the *sulcus occipitalis.*

Lateral depressions exist above the torus occipital on certain specimens of Ngandong (Weidenreich, 1951).

Coding: 2 states.

* 0: lateral depression absent.
* 1: lateral depression present.

- *character 53*: existence of an occipital bun (Figure 22 and 23).

The occipital bun is a term used by Boule (1911-13) and redefined by Ducros (1967). It is quantified by

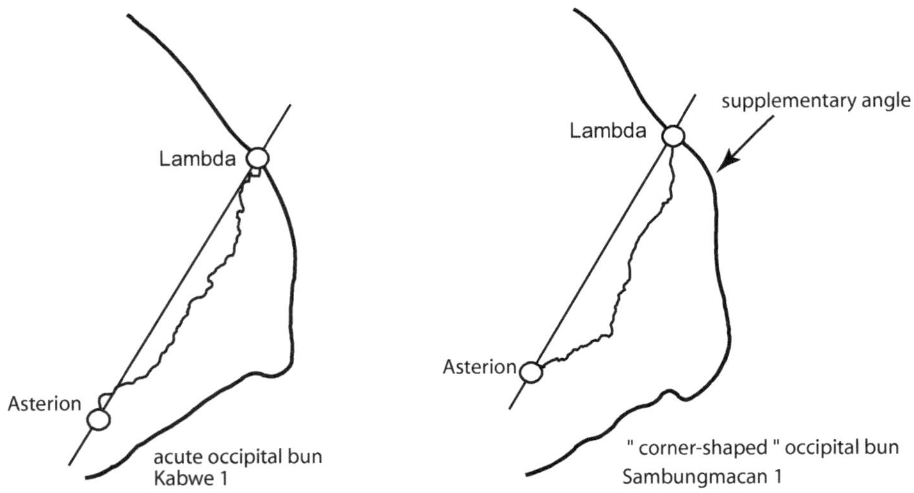

Figure 23: Typology of the occipital bun relatively to the asterion-lambda asterion plane.

Character 53: state 1: presence of an acute occipital bun.
Character 53: state 2: presence or a « corner-shaped » occipital bun.

measures of which the utilisation is difficult. A morphological and typological approach was used, thus: the bun exists when there is a change in curvature between the ensemble of the vault and the occipital independently of the existence of a prelambdatic depression affecting the parietals. An acute occipital buon can be defined by the existence, in *norma lateralis*, of an acute angle between the upper edge of the occipital shell and the occipito-parietal suture. On the other hand, the bun is said to be « cornered » when a supplementary angle affecting the upper edge of the occipital shell is affected.

Coding: 3 states.

* 0: absence of an occipital bun.
* 1: presence of an acute occipital bun.
* 2: presence or a «corner-shaped» occipital bun.

- *character 54*: development of an external occipital bump (*protuberantia occipitalis externa*) (Figure 24).

According to Hublin's acceptance (1978a), the *protuberantia occipitalis externa* is made up by the joining of the *linea nuchae suprema*. The absence of *protuberantia occipitalis externa* is seen to be a characteristic trait of *Homo erectus* by MacIntosch and Larnach (1972) but not by Hublin (1986).

Codage: 2 states.

* 0: absence of an external occipital bump.
* 1: presence of an external occipital bump.

- *character 55*: development of the *tuberculum linearum* (Figure 25).

The *tuberculum linearum* is the relief situated at the junction of the *linea nuchae superior* and the *crista occipitalis* (Hublin, 1978a). A strong *tuberculum linearum* is not seen as being a diagnostic trait of *Homo erectus* by Hublin (1986).

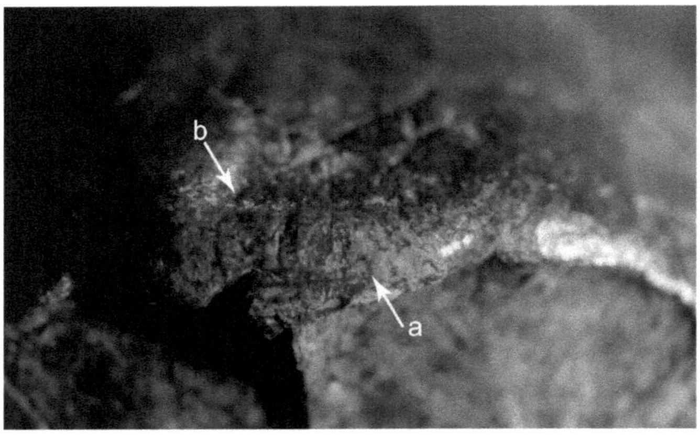

Figure 24: Posterior view of Ngandong VII.

a) Character 54: state 1: presence of an external occipital bump.
b) Character 56: state 1: presence of a depression above the tuberculum linearum.

Coding: 3 states.

* 0: absence of a *tuberculum linearum*.
* 1: presence of a *tuberculum linearum* with a moderate relief.
* 2: presence of a *tuberculum linearum* with a strong relief.

- *character 56*: existence of a medial cavity of the overlying occipital fold of the *tuberculum linearum* in *norma occipitalis* (Figure 24).

A variation of this structure was briefly mentioned by Weidenreich (1951).

Coding: 2 states.

* 0: absence of a depression above the *tuberculum linearum*.
* 1: presence of a depression above the *tuberculum linearum*.
* ?: indetermination linked to state «0» of character 55.

- *character 57*: existence of an external occipital crest (*crista occipitalis externa*) (Figure 26).

The *crista occipitalis externa* is a relief relative to the depth of the insertions of the major complexus and minor right muscles (Hublin, 1978a). This structure was retained as a quality of *Homo erectus* by MacIntosch and Larnach (1972).

Figure 25: Posterior view of Ngandong VII.

Character 55: state 2: presence of a tuberculum linearum with a strong relief.

<u>Coding</u>: 2 states.

* 0: absence of an external occipital crest.
* 1: presence of an external occipital crest.

- *character 58*: shape of the sus-iniac fossa.

The existence of a sus-iniac fossa, described by Fraipont and Lohest in 1887, was defined by Hublin (1978a) as a depression situated above the *inion*, above the highest occipital curved lines. Nara (1994) adopted typological scale in order to describe its evolution in the Neanderthalians. The present observations led to the description of sus-inaic depressions according to which they were spread laterally (roughly rectangular-shaped) or shorter (roughly triangular-shaped) or with the convergence of the lateral edges of these fossas directed upwards or downwards.

<u>Coding</u>: 4 states.

* 0: absence of a sus-iniac fossa.
* 1: presence of a sus-iniac fossa, as broad as high.
* 2: presence of a sus-iniac fossa wider than high.
* 3: presence of a sus-iniac fossa very wide in comparison to its height.

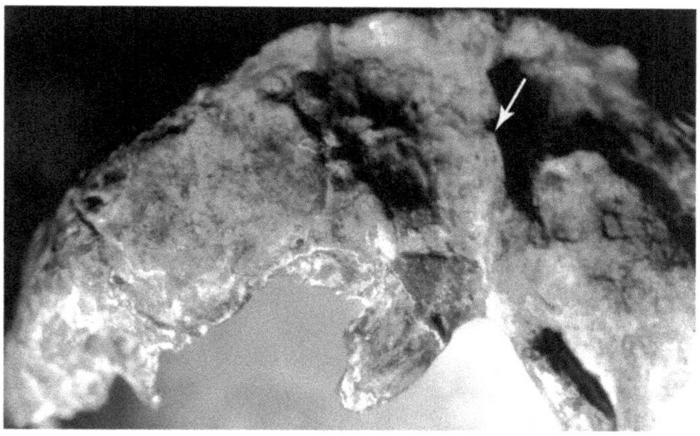

Figure 26: Inferior view of Ngandong I.

Character 57: state 1: presence of an external occipital crest.

- *character 59*: shape of the lateral sides of the sus-inaic fossa.

<u>Coding</u>: 3 states.

* 0: the lateral sides of the fossa converge upwards.
* 1: the lateral sides of the fossa are parallel or arched.
* 2: the lateral sides of the fossa diverge upwards.
* ?: indetermination llinked to state «0» of character 58.

- *character 60*: existence of a retromastoidian process (*processus retromastoideus* by Waldeyer 1909) (Figure 27).

Situated half way between the *inion* and the mastoidal process, this tuberosity was described by Weidenreich (1951) with relation to Ngandong. Hublin (1978a) confirmed that this relief was situated at the meeting point of the *linea nuchae superior* and of the upper rear secondary branch of the *linea nuchae inferior*.

<u>Coding</u>: 2 states.

* 0 : absence of retromastoïdian process.
* 1: presence of a strong retromastoidian process.

- *character 61*: existence of a *crista occipitomastoidea*.

Figure 27: Inferior view of KNMER 3733.

Character 60: state 1: retromastoïdian process.

The *crista occipitomastoidea* first defined by Weidenreich (1940) is in fact, according to Hublin (1978a) a fold created from the junction of the *linea m. obliquus capitis superior* and the medial loccipital and lateral lips of the juxtamoitidian bump. This structure is considered to have a characteristic belonging to *Homo erectus* by MacIntosch and Larnach (1972) as well as by Hublin (1986).

Coding: 2 states.

* 0: absence of *crista occipitomastoidea.*
* 1: presence of *crista occipitomastoidea.*

II.4.1.4 The temporal bone (characters 62 to 123)

- *character 62*: height of the temporal squama on the cranium vault.

The variations in height of the temporal squama were described in studies by Grimaud (1982) and Stringer (1984). *Homo erectus* is generally (Stringer, 1984; Hublin, 1986) considered as having a low temporal shell, but this is also the case for Great Apes.

Coding: 2 states.

* 0: temporal squama in high position.
* 1: temporal squama in low position.

- *character 63*: shape of the temporal squama.

For Weidenreich (1943) the shape of the squama is triangular in the Sinathropes. This triangular shape is indeed also often to be found in *Homo erectus* specimens in the study by Elyaqtine (1995).

Coding: 2 states.

* 0: temporal squama of rounded or polygonal shape.
* 1: temporal squama triangular shaped.

- *character 64*: shape of front edge of the temporal squama.

According to Weidenreich (1951), the rectitude of the front edge of the temporal squama converges in marking the triangular shape of the temporal squama. Observation shows that this character is not necessarily linked to a triangular shape.

Coding: 2 states.

* 0: front edge of the temporal squama is rounded in shape or sinuous.
* 1: front edge of the temporal squama is of a rectilinear shape.

- *character 65*: shape of the upper edge of the temporal squama.

With regards to the Ngandong series, Santa Luca (1980) criticises Weidenreich's findings (1951) which state that the upper edge of the temporal squama is rectilinear and not curved as in Modern Man. For Hublin (1986) the rectitude of the upper edge of the temporal shell is not unique to *Homo erectus.*

Coding: 2 states.

* 0: upper edge of the temporal shell is round or sinuous.
* 1: upper edge of the temporal shell is rectilinear.

- *character 66* : size of the *crista supramastoidea* at the level of the *porion.*

During the time of this study, it appeared that the *crista supramastoidea* was quite large at the level of the *porion* on certain fossil specimens whereas in Modern Man this is not the case.

Coding: 2 states.

* 0: *crista supramastoidea* is weak at the level of the *porion.*
* 1: *crista supramastoidea* is marked at the level of the *porion.*

- *character 67*: continuity and degree of angulation of the *crista supramastoidea* with the root of the zygomatic process (*processus zygomaticus ossis temporalis*)(Figure 28 and 29).

Since it's first description by Boule (1911-1913), and in-line with a more recent study (Stringer *et al.*,1984), the position of the external auditory meatus with reference to the zygomatic arch is held to be a Neanderthalian autmorphy. Nara (1994) showed that variations of this characteristic existed in *Homo sapiens* – including in Neanderthal Man. This study reveals the existence of a discrepancy

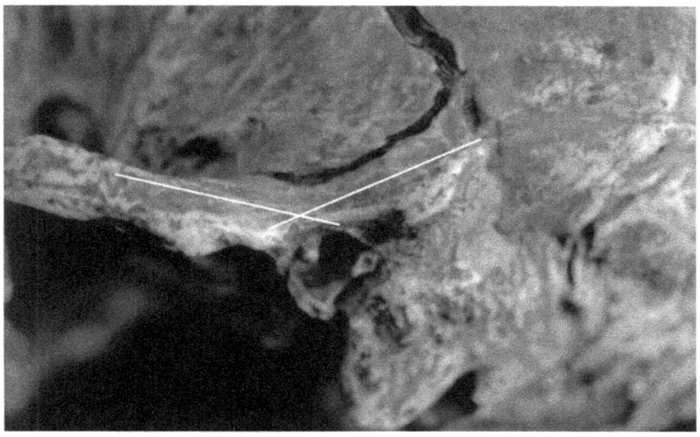

Figure 28: Lateral view of the left temporal of La Chapelle-aux-Saints.

Character 67: state 0: discrepency between the crista supramastoidea and the zygomatic process.

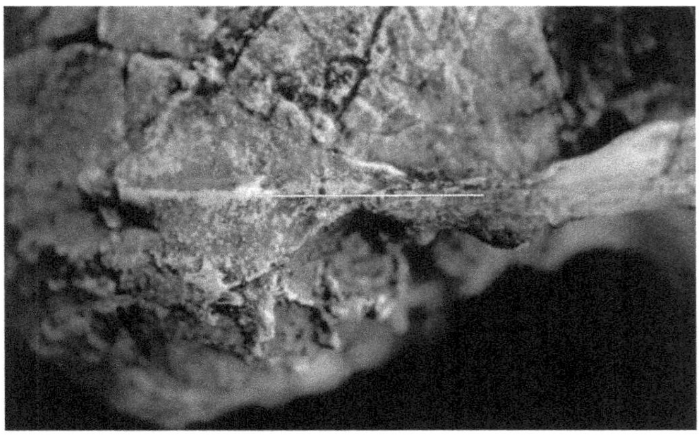

Figure 29: Right lateral view of KNMER 3733.

Character 67: state 1: continuity between the crista supramastoidea and the zygomatic process.

(from the point of view of an angle) between the segment carried by the *crista supramastoidea* and the principal generating axis of the root of the *processus zygomaticus* of the temporal bone. For Santa Luca (1980) the continuity of these two structures, albeit variable, is often to be seen in Ngandong specimens.

Coding: 2 states.

* 0: there is a discrepency between the *crista supramastoidea* and the zygomatic process.
* 1: there is continuity between the *crista supramastoidea* and the zygomatic process.

- *character 68*: continuity of the *crista supramastoidea* with the lower temporal line (*linea temporalis inferior*).

Weidenreich (1951) indicates that in the Ngandong series as well as in the Sinathropes, the *crista supramastoidea* ends abruptly; that it has no direct link with the occipital torus or the *torus angularis* of the parietal. The variation of this characteristic was also raised by Grimaud (1982). Santa Luca (1980) indicated that the *crista supramastoidea* could be continuous with the *linea temporalis inferior*.

Coding: 2 states.

* 0: the *crista supramastoide* has no direct link with the lower temporal line.
* 1: the *crista supramastoidea* is in continuity with the lower temporal line.

- *character 69*: existence of a forward sus-mastoidian tuberculum (*tuberculum supramastoideum anterius*) (Figure 30).

Hublin (1978a) put forward that the *crista supramastoidea* could present a tuberculum which stopped suddenly at the *sutura squamosa*.

Coding: 2 states.

* 0: absence of an anterior sus-mastoïdian tuberculum.
* 1: presence of an anterior sus-mastoïdian tuberculum.

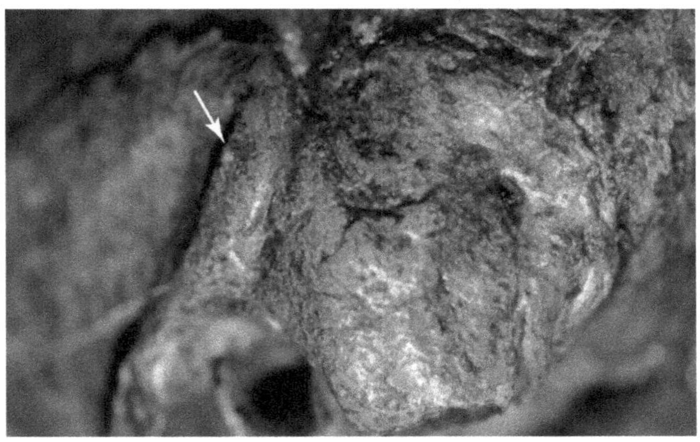

Figure 30: Infero-lateral view of Ngandong XII.

Character 69: state 1: anterior sus-mastoïdian tuberculum.

- *character 70*: size of the *crista mastoidea.*

In his last personal study Hublin (1978a) showed that the *crista mastoidea* can project be very salient. In 1986, he however stated that this trait was not proper to *Homo erectus*.

Coding: 2 states.

* 0: *crista mastoidea* , slightly salient is present.
* 1: a strong *crista mastoidea* is present.

- *character 71*: continuity of the *crista mastoidea* with the upper temporal line (*linea temporalis inferior*).

Coding: 2 states.

* 0: the *crista mastoidea* has no direct link with the upper temporal line.
* 1: the *crista mastoidea* is in continuity with the upper temporal line.

- *character 72*: rear opening of the sus-mastoïdian space between the *crista mastoidea* and the *crista supramastoidea* (Figure 31).

Stringer (1984) considered that a forward union of the *crista mastoidea* and the *crista supramastoidea* existed in *Homo erectus*. This supposes a rear opening and a forward closing of the *sulcus supramastoideus*.

Coding : 2 states.

* 0: rear contact between the *crista mastoidea* and the *crista supramastoidea* leading to the absence of an open sus-mastoïdien groove.
* 1: presence between the *crista mastoidea* and the *crista supramastoidea* of a continuos space open towards the rear.

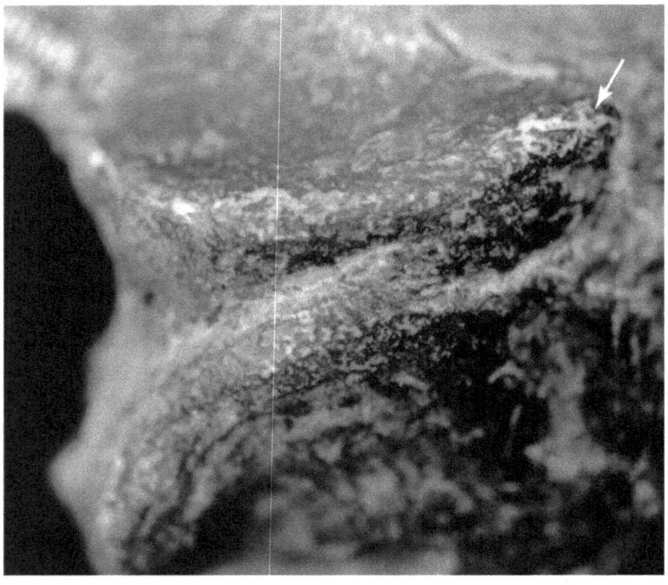

Figure 31: Lateral view of Sangiran 12.

Character 72: state 0: Posterior closure of the sus-mastoidian space.

- *character 73*: size of the sus-mastoïdien space between the *crista mastoidea* and the *crista supramastoidea* (Figure 32 and 33).

Santa Luca (1980) evoked variability in the width of the sulcus *supramastoidea* between the fossils of Ngandong and of Sangiran. For Weidenreich (1943) the *crista mastoidea*, the *crista supramastoidea* and the sus-mastoidian space are developed in the Sinanthropes whereas the groove does not exist in the Ngandong or in the Sinanthropes according to what he wrote in 1951. Hublin (1986) reckons that the more or less large reduction of the *sulcus supramastoidea* cannot be retained as a diagnostic trait of *Homo erectus*.

Coding: 3 states.

* 0: the *crista mastoidea* and the *crista supramastoidea* are side by side.
* 1: presence of a narrow sus-mastoïdien space between the *crista mastoidea* and the *crista supramastoidea*.
* 2: presence of a large sus-mastoïdian space between the *crista mastoidea* and the *crista supramastoidea*.

- *character 74*: forward convergence of the *crista mastoidea* and the *crista supramastoidea.*

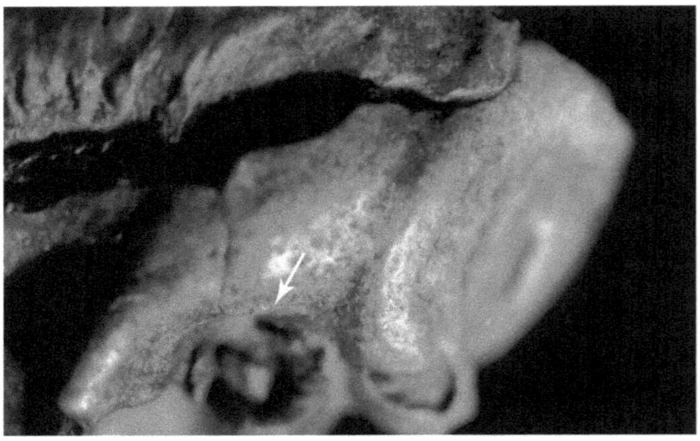

Figure 32: Lateral view of Sangiran 26.

Character 73: state 1: narrow sus-mastoïdian space between the crista mastoidea and the crista supramastoidea.

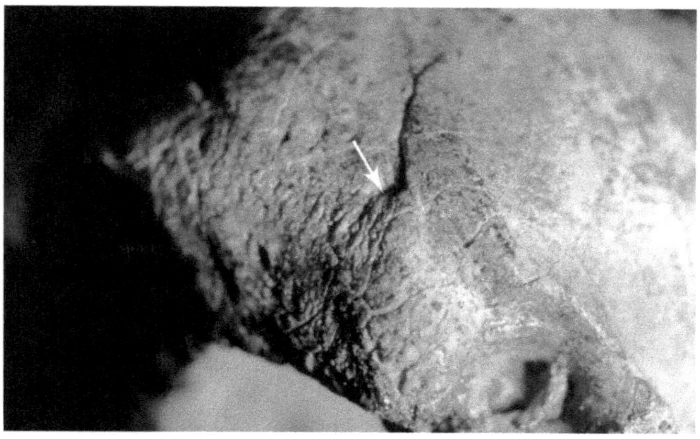

Figure 33: Right lateral view of Sambungmacan 1.

Character 73: state 2: large sus-mastoïdian space between the crista mastoidea and the crista supramastoidea.

Considering more particularly the *cranium* of La Quina H5, Hublin (1978b) demonstrated that the Neanderthalians generally present a convergence toward the front of the mastoidal and supramastoidal crests.

Coding: 3 states.

* 0: the *crista mastoidea* and the *crista supramastoidea* diverge towards the front.
* 1: the *crista mastoidea* and the *crista supramastoidea* are parallel.
* 2: the *crista mastoidea* and the *crista supramastoidea* converge towards the front.

- *character 75*: existence of a rear sus-mastoidian tuberculum (*tuberculum supramastoideus posterior*) (Figure 17).

The *tuberculum supramastoideus posterior* tuberculum situated lower down than the *processus asteriacus* described by Hublin (1978a). It is situated on the temporal bone or straddling the parieto-mastoidian suture.

Coding: 2 states.

* 0: absence of a rear sus-mastoïdian tuberculum.
* 1: presence of a rear sus-mastoïdian tuberculum.

- *character 76*: existence of a premastoïdal tuberculum (*tuberculum mastoideum anterius*).

Hublin (1978b) describes a tuberculum (*tuberculum mastoideum anterius*) in Neanderthal Man situated behind the external auditory meatus at the point where the mastoidal crest branches downwards. The existence of this characteristic is also mentioned by Santa Luca (1978). Elyaqtine (1995) confirmed that this structure could be quite frequently observed in Neanderthals.

Coding: 2 states.

* 0: absence of a pre-mastoidal tuberculum.
* 1: presence of a pre-mastoidal tuberculum.

- *character 77*: existence of a suprameatal spine (*spina suprameatum*) (Figure 34).

The suprameatal spine is not to be found in Sinathropes. However, on certain skulls it is possible to observe a vestigial stage (Weidenreich, 1943). It should be noted that Weidenreich omits to say on which ones. When such a stage exists, it is here considered that the *spina suprameatum* exists. MacIntosch and Larnach (1972) make mention of a "suprameatal tegmen", characteristic of *Homo erectus*. This characteristic was not accepted as being an apomorphy of *Homo erectus* by Hublin (1986) whereas it does figure in the diagnosis of the species according to Howells (1980) and that for Jacob (1984) it is another criterium in recognising *Homo erectus*.

Coding: 2 states.

* 0: absence of a suprameatal spine
* 1: presence of a suprameatal spine.

- *character 78*: shape of the external auditory meatus in *norma lateralis* (Figure 35).

The elliptic shapes of the external auditory meatus horizontally orientated, are prevalent in Sinathropes according to Weidenreich (1943). The elliptic shape is also present in Sangiran 4 whereas a circular shape exists in Sangiran 2 (Weidenreich, 1951). Elyaqtine

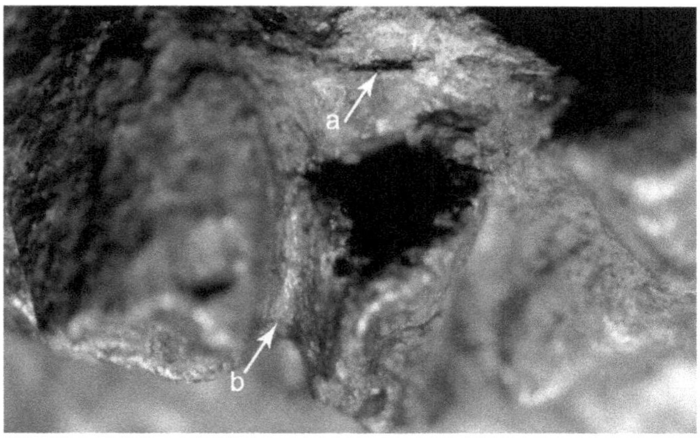

Figure 34: Inferior view of the temporal of Ngandong XII.

a) Character 77: state 1: existence of a suprameatal spine.
b) Character 89: state 1: existence of a cleft between the rear part of the tympanal and the front part of the mastoidal process.

(1995) also took into account the variability of this characteristic in *Homo sapiens* and *Homo erectus*. Nevertheless according to this author, it is not a very pertinent criterion.

Coding: 2 states.

* 0: rounded tympanal section.
* 1: ellipsoidal or ovaoidal tynpanal section.

- *character 79*: orientation of the large axis of the lateral section of the external auditory meatus in *norma lateralis*.

According to Stringer (1984), the large axis of the lateral section of the tympanal is usually orientated vertically in *norma lateralis* in *Homo erectus*.

Coding: 3 states.

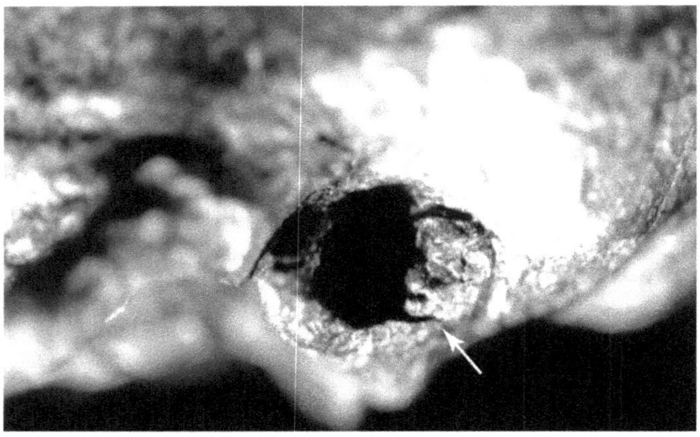

Figure 35: Lateral view of Sangiran 2.

Character 78: state 0: rounded tympanal section.

* 0: large axis orientated obliquely from top to bottom towards the front in *norma lateralis*.
* 1: large axis orientated vertically in *norma lateralis*.
* 2: large axis orientated obliquely from top to bottom towards the rear in *norma lateralis*.

- *character 80*: thickness of the tympanal in *norma lateralis*.

The tympanal is described as being very different in shape, thickness and positioning than in Modern Man. It is more horizontal in the Sinathropes according to Weidenreich (1943). The thickness of the tympanal is measured on the bone perimeter around the external auditory meatus. Classically, measurements are taken vertically on the lower part and another at the front in a vertical plane. Here, we only carry out estimates such as that the general thickness observed in Modern Man is less pronounced and that the thickess in gorillas is greater.

Coding: 2 states.

* 0: tympanal not very thick in *norma lateralis*.
* 1: tympanal quite thick in *norma lateralis*.

- *character 81*: contribution of the tympanal to the rear side of the mandible fossa (Figure 36).

Coding: 2 states.

* 0: the *processus postglenoidalis* contributes significantly to the rear wall of the mandible fossa.
* 1: the tympanal contributes, almost exclusively, to the rear wall of the mandible fossa.
* ?: indetermination linked to state «0» of character 119.

- *character 82*: convexity of the rear wall surface of the mandibular fossa.

The part of the tympanic plate which contributes to the back wall of the glenoidal cavity of the temporal is either flat or more or less convex in Sinathropes whereas it is concave in Modern Man, according to Weidenreich (1943).

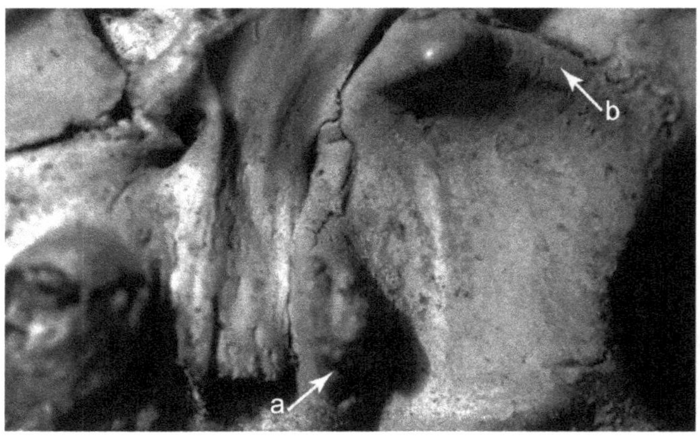

Figure 36: Inferior view of the *fossa mandibularis* of a gorilla.

a) Character 81: state 0: significant contribution of the *processus postglenoidalis* to the rear wall of the mandible fossa.
b) Character 111: state 0: the entoglenoidal formation is a distinct relief of the *tuberculum articulare*.
Character 120: state 2: large transversal stretching of the *processus postglenoidalis*.

Coding: 2 states.

* 0: the surface of the back wall of the mandibular fossa is flat to concave towards the front in *norma lateralis*.
* 1: the surface of the back wall of the mandibular fossa is convex towards the front in *norma lateralis*.

- *character 83*: mastoidal process detached in relation to the base of the skull.

The mastoidal process detached in relation to the base of the skull more so than the size of the mastoidal process itself, is regarded as being a characteristic proper to Neanderthalians. Observation of the material allows one to see that this characteristic also exists elsewhere.

Coding: 2 states.

* 0: in *norma lateralis*, the mastoidal process is not larger than the volume making up the base of the skull.
* 1: in *norma lateralis*, the mastoidal process is clearly detached from the base of the skull.

- *character 84*: size of the mastoidal process.

For Santa Luca (1980) the Sangiran fossils have a somewhat undeveloped mastoidal process. In *Homo erectus*, the mastoidal process is small or even very small for MacIntosch and Larnach (1972). Delson *et al.* (1977) state this character is an autapomorphy of *Homo erectus*.

Coding: 2 states.

* 0: in *norma lateralis*, the mastoidal process has developed little in volume.
* 1: in *norma lateralis*, the mastoidal process is massive.

- *character 85*: development of the *tuberculum articulare*.

According to Weidenreich (1943) Sinathropes do not have a real articular tuberculum but a preglenoidal plane, whereas the Ngandong series has a tuberculum and a deep and narrow fossa. For Weidenreich, the absence of a tuberculum is not primitive and the disappearance of the tuberculum is a trait deriving from gorillas and chimpanzees.

Coding: 2 states.

* 0: in *norma basilaris,* the *tuberculum articulare* is hardly developed on the surface.
* 1: in *norma basilaris*, the *tuberculum articulare* occupies a large surface area.

- *character 86*: projection of the *tuberculum articulare.*

To follow Picq (1983), the projection of the *tuberculum articulare* follows a gradiant. In the Great Apes, the Australopithecus, and the *Homo habilis* it is quite gentle but it is steeper in *Homo erectus* and Modern Man.

Coding : 2 states.

* 0: in *norma basilaris*, the *tuberculum articulare* is not or only slightly projecting.
* 1: in *norma basilaris*, the *tuberculum articulare* is projecting.

- *characteristic 87* : existence of a stretched subtemporal plane.

Coding: 2 states.

* 0: absence of a stretched sub-temporal plane preceding the articulary surface of the glenoidal cavity of the temporal.
* 1: existence of a sub-temporal plane which prolongs the glenoidal cavity of the temporal.

- *character 88*: existence of a pre-glenoidal tuberculum (*tuberculum anterius fossae mandibularis*) (Figure 37).

On certain Asiatic and African human fossils, at the front of the *fossa mandibularis,* a small tuberculum (*tuberculum anterius fossae mandibularis.*) exists at the forward limit of the ascendency of the glenoidal capsule of the temporo-mandibule, on the *tuber articulare.*

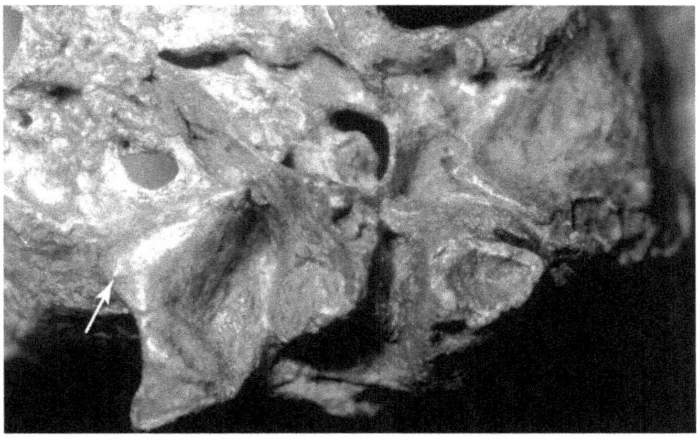

Figure 37: Inferior view of the *fossa mandibularis* of Ngandong XII.

Character 88: state 1: existence of a *tuberculum anterius fossae mandibularis*.

Coding: 2 states.

* 0: there is no *tuberculum articulare* in the ascendency of the articulary of the glenoidal cavity of the temporal.
* 1: there is a tuberculum on the *tuberculum articulare* in the ascendency of the glenoidal cavity of the temporal on the forward part.

- *character 89*: size of the space existing between the tympanal and the front part of the mastoidal process (Figure 34).

As in Modern Man, and contrary to *Homo erectus*, a fusion of the typanic plate and the front part of the mastoidal process does exist on the Kabwe specimen (Santa Luca, 1980). Several authors (Andrews, 1984; Stringer, 1984; Hublin, 1986; Rightmire, 1986; Turner and Chamberlain, 1989) considered that a deep fissure characterised *Homo erectus*. For Kennedy (1991) this characteristic is not unique to *Homo erectus* as it also exists in the Great Apes and in certain Neanderthalians – which this study also confirms in part.

Coding: 3 states.

* 0: the rear part of the tympanal alongside the front part of the mastoidal process.
* 1: there is a cleft between the rear part of the tympanal and the front part of the mastoidal process.
* 2: there is quite a large gap between the rear part of the typanal and the front part of the mastoidal process.

- *character 90*: sagittal length of the mandibular fossa.

Weidenreich (1943) drew up a list of parameters permitting the description of length and depth of the *fossa mandibularis*. Picq (1983) demonstated that several anatomical elements of the temporo-mandibulary joint varied between Graet Apes and Hominids. This study initially directed towards a tentative morphofunctional explanation, proposes the existence of an evolutionary gradiant. In fact, starting from the presumed phylogenetical position of African and Asian specimens and of their respective strati-graphic positions, Picq described a gradual evolutionary mechanism for certain anatomical structures of the temporo-mandibulary articulation. The different anatomical elements described are re-taken on board individually in isolation in order to be tested.

Coding: 2 states.

* 0: short *fossa mandibularis.*
* 1: long *fossa mandibularis.*

- *character 91*: vertical depth of the mandibulary fossa.

Black (1931) established that the *fossa mandibularis* was very deep in Sinanthropus III. Weidenreich (1951) extended this observation to the Ngandong series.

Coding: 2 states.

* 0: shallow *fossa mandibularis.*
* 1: deep *fossa mandibularis.*

- *character 92*: degree of the rear slope of the *tuberculum articulare* with reference to the temporo-mandibulary articulation.

Coding: 3 states.

* 0: short rear slope of the *tuberculum articulare* in comparison to the tempero-mandibulary articulation.
* 1: more pronounced rear slope of the *tuberculum articulare* in comparison to the temporo- mandibulary articulation.
* 2: steep rear slope of the *tuberculum articulare* in comparison to the temporo-mandibulary articulation.

- *character 93*: orientation of the mastoidal process.

According to research acarried out by Elyaqtine (1995), it would seem that for the ensemble of *Homo sapiens* and *Homo erectus* specimens the mastoïdal process is in the great majority of cases, oriented medially.

Coding: 2 states.

* 0: the mastoidal process is not oriented towards the interior.
* 1: the mastoidal process is oriented towards the interior.

- ***character 94*: depth of the *incisura mastoidea.***

The depth, width, length and shape of the *incisura mastoidea* section were described by Weidenreich (1943 and 1951). These parameters were questioned by Santa Luca (1980). For MacIntosh and Larnach (1972) the *incisura mastoidea* is large in *Homo erectus*.

Coding: 3 states.

* 0: absence of an *incisura mastoidea*.
* 1: an *incisura mastoidea* exists, but not very deep.
* 2: a deep *incisura mastoidea* exits.

- ***character 95*: antero-posterior length of the l'*incisura mastoidea.***

Coding: 2 states.

* 0: short *incisura mastoidea*.
* 1: long *incisura mastoidea*.
* ?: indetermination linked to state «0» of character 94.

- ***character 96*: medio-lateral width of the *incisura mastoidea.***

Coding: 2 states.

* 0: *incisura mastoidea* is narrow.
* 1: *incisura mastoidea* is wide.
* ?: indetermination linked to state «0» of character 94.

-***character 97*: shape of the transveral section of the *incisura mastoidea.***

It appeared that the shape of the mastoidian incisure systematically described by Hublin in his report Hublin (1978a) could account for the grouping of certain specimens. Three different shapes of the section of the *incisura mastoidea* are here described.

Coding: 3 states.

* 0: *incisura mastoidea* with a round section.
* 1: section of the *incisura mastoidea* in «U» shaped.
* 2: section of the *incisura mastoidea* in «V» shaped.
* ?: indetermination linked to state «0» of character 94.

- ***character 98*: size of the jaxtamastoidian eminence of Rouvière (Figure 38).**

The variations of this structure were studied and described by Weidenreich (1951) and Jacob (1965) under the terminology of paramastoidal crest. For MacIntosh and Larnach (1972) the juxtamastoidan eminence of Rouvière does not exist in *Homo erectus*.

Coding: 3 states.

* 0: absence of Rouvière's juxtamastoidian eminence.
* 1: presence of a small Rouvière's juxtamastoidian eminence.
* 2: presence of a large Rouvière's juxtamastoïdian eminence.

Figure 38: Posterior view of Ngandong XII.

Character 98: state 2: presence of a large Rouvière's juxtamastoïdian eminence.

- ***character 99*: importance of the fissure of Glaser (*fissura glaseri*) (Figure 39).**

When there is no Glaser fissure, the tympanal and the squamosal are joined by a boney weld. An invagination level with this weld is regared as being a large Glaser fissure. As a result of a certain coiling up and by a closeness of the tympanal a sort of juxtaposition results in a narrow space: Glaser's fissure.

Coding: 3 states.

* 0: absence of Glaser's fissure.
* 1: presence of a large Glaser fissure.
* 2: presence of a narrow Glaser fissure.

- ***character 100*: depth of the fissure of Glaser (*fissura glaseri*).**

Coding: 2 states.

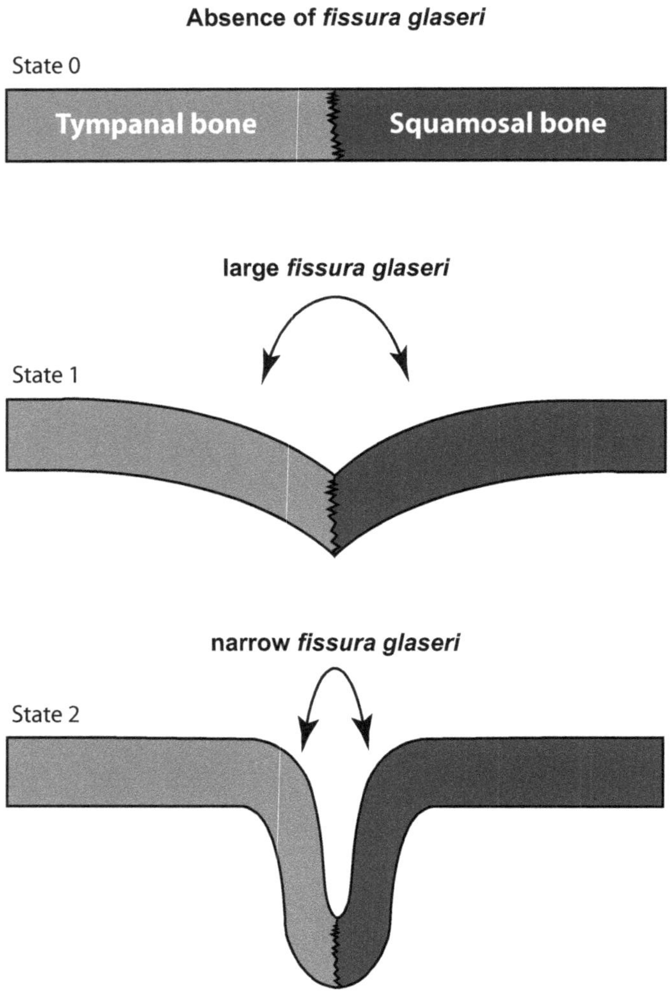

Figure 39

Importance of the invagination between the tympanal plate and the entoglenoidal Formation.

* 0: presence of a fissure of Glaser not deep.
* 1: presence of a deep Glaser fissure.
* ?: indetermination linked to state «0» of character 99.

- *character 101*: division of the fissure of Glaser (*fissura glaseri*).

Picq (1983) highlights that Modern Man distinguishes himself by the existence of a narrow eminence at the roof of the petrosal between the typanal and the squamosal. The *tegmen tympani* in this way divides the median part of the fissure of Glaser.

Codaing: 2 states.

* 0: absence of an elongation of the *tegmen tympani* dividing the median part of the fissure of Glaser.
* 1: presence of an elongation of the *tegmen tympani* dividing the median part of the fissure of Glaser.

* ?: indetermination linked to state «0» of character 99.

- *character 102*: size of the reinforcing between the typmanic plaque and the entoglenoidal formation (*spina entoglenoidalis* or *processus entoglenoidalis*) (Figure 40 and 41).

The endoglenoidal formation is a term employed here to not presume the existence of a *spina entoglenoidalis* rather than a *processus entoglenoidalis.* Elyaqtine (1995) developed a typology in order to describe the variations of this structure. The presence and the expression of the reinforcing between the typanic plaque and the entoglenoïdal has been described in Sinanthropes (Weidenreich, 1943). Rightmire (1984) is adamant of the reinforcement between the tympanic plaque and the entoglenoïdal formation common to Asiatic and African *Homo erectus.* For Andrews (1984), this is an autopomorphy of *Homo erectus*, but Tobias (1965) observed this on specimen OH 24 which he sees as being a *Homo habilis.* Kennedy (1991) affirms that this reinforcing exists in Neanderthalian Man. Elyaqtine's work (1995) and observations carried out here confirm this fact.Bräuer and Mbua (1992, p.103 and 104) demonstrate perfectly the variations of this anatomical structure.

Coding: 3 states.

* 0: tympanic plate and entoglenoidal formation (*spina entoglenoidalis* or *processus entoglenoidalis*) are side by side.
* 1: existence of a groove between the tympanic plate and the entoglenoidal formation

Figure 40: Inferior view of the tympanal of Ngandong VII.

Character 102: state 1: groove between the tympanic plate and the entoglenoidal formation.

(*spina entoglenoidalis* or *processus entoglenoidalis*).

* 2: existence of a space between the tympanic plate and the entoglenoidal formation (*spina entoglenoidalis* or *processus entoglenoidalis*).

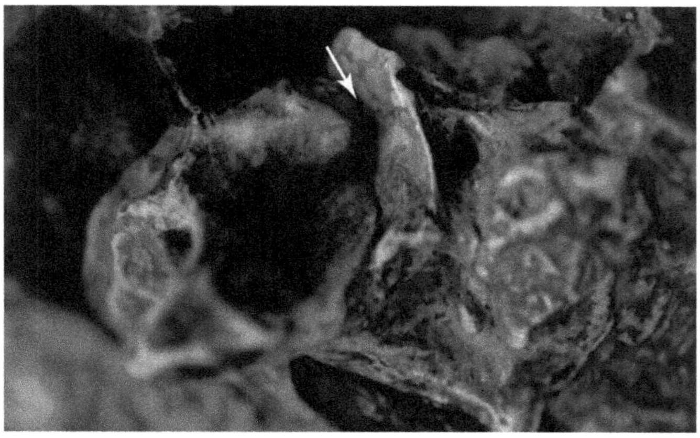

Figure 41: Inferior view of the tympanal of La Chapelle-aux-Saints.

Character 102: state 2: space between the tympanic plate and the entoglenoidal formation.

- *character 103*: obliqueness of the front wall of the *fossa mandibularis*.

Coding: 3 states.

* 0: the front wall of the *fossa mandibularis* is almost horizontal.

* 1: the front wall of the *fossa mandibularis* is oblique.

* 2: the front wall of the *fossa mandibularis* is almost vertical.

- *character 104*: the relative size of the entoglenoidal formation with reference to the *tuberculum zygomaticum anterior*.

The entoglenoidal formation is, in the Ngandong series, (Weidenreich, 1951) even more developed than in Modern Man. The *tuberculum zygomaticum anterior* described by Picq (1983) corresponds to the *processus ectoglenoïdalis* of Weidenreich (1943).

Coding: 4 states.

* 0: the entoglenoidal formation is very large when compared to the *tuberculum zygomaticum anterior*.

* 1: the entoglenoidal formation is large with reference to the *tuberculum zygomaticum anterior*.

* 2: the entoglenoidal formation is is very much the same size as the *tuberculum zygomaticum anterior*.

* 3: the entogelenoidal formation entoglenoïdale is small in comparison to the *tuberculum zygomaticum anterior*.

- *character 105*: sagittal position relative to the entoglenoidal formation in relation to the *tuberculum zygomaticum anterior*.

Picq (1983) compared the size and the position of the entoglenoidal process and of the forward zygomatic tuberculum in Hominides. Thus, in Modern Man the *spina entoglenoidalis* is behind the transveral axis of the *tuberculum zygomaticum anterior*. In the Australopithecines, the *processus entoglenoidalis* is further back with reference to the structure which corresponds to the *tuberculum zygomaticum anterior*. Elyaqtine (1995) mentioned the difficulty of using metrical measures in order to describe the structures of the temporal at the level of the temporo-mandibulary articulation because of not being able to take mesurements between homologous points. For this author, the bearings cannot be relied upon and necessitate the taking into consideration of planes difficult to reproduce when the fragmentary state of the fossils is taken into account. Furthermore, it must be acknowledged that the relative notions employed in order to describe the anatomical structures between each other, limit the use of certain characteristics as a result of too great a subjectivity. The descriptions formulated here have been taken from previous studies and are in conformity with the original definitions.

Coding: 4 states.

* 0: the entoglenoidal formation is at the same level as the *tuberculum zygomaticum anterior*.

* 1: the entoglenoidal formation is behind the *tuberculum zygomaticum anterior*.

* 2: the entoglrnoidal formation is very much to the rear of the *tuberculum zygomaticum anterior*.

- *character 106*: relative size of the entoglenoidal formation and of the *tuberculum zygomaticum anterior* with reference to the ensemble of the *tuberculum articulare*.

Coding: 3 states.

* 0: the entoglenoidal formation and the *tuberculum zygomaticum anterior* are large or very large with reference to the ensemble of the *tuberculum articulare*.
* 1: the size of the entoglenoidal formation and the *tuberculum zygomaticum anterior* is average in comparison to the ensemble of the *tuberculum articulare*.
* 2: the entoglenoidal formation and the *tuberculum zygomaticum anterior* are small in comparison to the ensemble of the *tuberculum articulare*.

- *character 107*: sagittal position relative to the entoglenoidal formation and of the *tuberculum zygomaticum anterior* in comparison to the *fossa mandibularis*.

Coding: 3 states.

* 0: the entoglenoidal formation and the *tuberculum zygomaticum anterior* are median with reference to the ensemble of the *tuberculum articulare*.
* 1: the entoglenoidal formation and the *tuberculum zygomaticum anterior* are forward with reference to the ensemble of the *tuberculum articulare*.
* 2: the entogleoidal formation and the *tuberculum zygomaticum anterior* are very much behind in comparison to the ensemble of the *tuberculum articulare*.

- *character 108*: the antero-posterior convexity of the *tuberculum articulare*.

Following Picq (1983) the antero-posterior convexity is slight in Modern Man whereas is it very marked in the Great Apes.

Coding: 3 states.

* 0: the *tuberculum articulare* is of a flat profile in *norma lateralis*.
* 1: the *tuberculum articulare* has a large rounded profile in *norma lateralis*.
* 2: the *tuberculum articulare* has a small rounded profile in *norma lateralis*.

- *character 109*: the lower transversal concavity of the *tuberculum articulare*.

The lower transversal concavity most often marks the *tuberculum articulare* in Modern Man. A comparative study of African and Asiatic Hominids put forward by Picq (1983) highlights the variation of the concavity. Rightmire (1984) points out that the concavity of the *tuberculum articulare* of the KNMER 3733 specimen is larger than that of the KNMER 3883 specimen.

Coding: 3 states.

* 0: the *tuberculum articulare* has a flat plane in *norma frontalis*.
* 1: the *tuberculum articulare* has a large rounded profile in *norma frontalis*.
* 2: the *tuberculum articulare* has a small rounded profile in *norma frontalis*.

- *character 110*: shape of the rear edge of the *tuberculum articulare*.

Coding: 3 states.

* 0: the rear edge of the *tuberculum articulare* has a sub-rectilinear profile in *norma basilaris*.
* 1: the rear edge of the *tuberculum articulare* has an arched profile in *norma basilaris*.
* 2: the rear edge of the *tuberculum articulare* has a sigmoïdal profile in *norma basilaris*.

- *character 111*: integration of the entoglenoidal formation (*spina entoglenoidalis* or *processus entoglenoidalis*) in the *tuberculum articulare* (Figure 36).

Coding: 2 states.

* 0: the entoglenoidal formation is a distinct relief of the *tuberculum articulare*.
* 1: the entoglenoidal formation is integrated within the *tuberculum articulare*.

- *character 112*: continuity of the rear slope of the *tuberculum articulare* and of the sub-temporal plane.

The transition between the rear slope of the *tuberculum articulare* and of the sub-temporal was described by Picq (1983). Thus, a continuity between the *tuberculum articulare* and the sub-temporal plane exists in the Great Apes, the Australopithecus and *Homo habilis*. A rupture exists in *Homo erectus* and *Homo sapiens*, but this rupture in the slope varies. Thus, the rupture in the slope is more acute in KNMER3883 than in 3733 for Rightmire (1984).

Coding: 2 states.

* 0: a continuity exists between the rear slope of the *tuberculum articulare* and the sub-temporal plane.
* 1: a rupture exists between the rear slope of the *tuberculum articulare* and the sub-temporal plane.

- *character 113*: amplitude of the mandibulary fossa.

Coding: 4 states.

* 0: *fossa mandibularis* very large.
* 1: *fossa mandibularis* large.
* 2: *fossa mandibularis* medium-sized.
* 3: *fossa mandibularis* small.

- *character 114*: antero-posterior concavity of the mandibulary fossa.

Coding: 3 states.

* 0: the antero-posterior cavity of the *fossa mandibularis* has a wide-angled curvature or is flat.
* 1: the antero-posterior cavity of the *fossa mandibularis*is rounded.
* 2: the antero-posterior cavity of the *fossa mandibularis* has a «pinched» rounded shape.

- *character 115*: existence of an ectoglenoïdal crest (*crista ectoglenoïdalis*) on the lateral edge of the *fossa mandibularis*.

Picq (1983) labelled *crista ectoglenoidalis* a crest which, in Modern Man, goes from the *processus postglenoidalis* to join the *tuberculum zygomaticum anterior*. This structure does not exist in the Great Apes.

Coding: 2 states.

* 0: No ectoglenoïdal crest exists on the lateral edge of the *fossa mandibularis*.
* 1: An ectoglenoïdal crest exists on the lateral edge of the *fossa mandibularis*.

- *character 116*: size of the entoglenoidal formation (*spina entoglenoidalis* or *processus entoglenoidalis*) in relation to the sphenoïdal border.

For Weidenreich (1943), the *processus entoglenoidalis* is not a real process but a steep slope entirely constructed by the squamosal. For MacIntosch and Larnach (1972) the *processus entoglenoidalis* in only formed by the squamosal in *Homo erectus*.

Coding: 3 states.

* 0: the entoglenoïdal formation is greater in size than the edge of the sphenoid.
* 1: the entoglenoidal formation entoglénoïdale is the same size as the edge of the sphenoid.
* 2: The entoglenoidal formation is smaller in size than the edge of the sphenoid.

- *character 117*: projection of the entoglenoidal formation.

According to Picq (1983), the *processus entoglenoidalis* is strong in the Great Apes, of medium size in Austalopithecines and projecting considerably in *Homo habilis* whereas it projects only slightly in African *Homo erectus* specimens.

Coding: 2 states.

* 0: The entoglenoidal formation projects considerably.
* 1: The entoglenoidal formationdoers not project.

- *character 118*: sagittal spreading-out of the entoglenoidal formation .

Coding: 5 states.

* 0: the entoglenoidal formation is very spread towards the rear.
* 1: the spread of the entoglenoidal formation towards the rear is average.
* 2: the entoglenoidal formationis hardly extended towards the rear.
* 3: the entoglenoidal formation is presented in the shape of a tuberculum.

* 4: the entoglenoidal formation does not spread towards the rear but presents itself in the shape of a tuberculum.

- *character 119*: size of the *processus postglenoidalis.*

The *processus postglenoidalis* corresponds to the pre-tympanic strip of Vallois described in the Neanderthalians (Picq, 1983). There is no real *processus postglenoidalis* in Sinanthropes, Trinil 2 and the Ngandong specimens; on the other hand it is devoped in Kabwe according to Weidenreich (1943 and 1951). According to Elyaqtine (1995) the *processus postglenoidalis* exists in the Ngandong specimens but is twisted laterally. According to Rightmire (1984), a small *processus postglenoidalis* exists on KNMER 3733.

Coding: 4 states.

* 0: the *processus postglenoidalis* does not exist.
* 1: the *processus postglenoidalis*is small.
* 2: the *processus postglenoidalis* is of average size.
* 3: the *processus postglenoidalis*is large.

- *character 120*: transversal spreading-out of the *processus postglenoidalis* (Figure 36 and 42).

The transversal stretching of the *processus postglenoidalis* is large in Modern Man where it takes up 50 % of the transversal length of the mandibulary fossa (Picq, 1983).

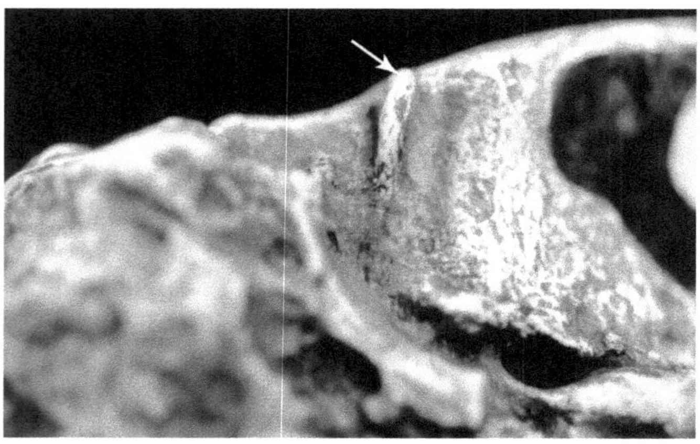

Figure 42: Inferior view of the fossa mandibularis of Sterfontein 5.

Character 120: state 2: large transversal stretching of the processus postglenoidalis.

Coding: 4 states.

* 0: the transversal stretching of the *processus postglenoidalis* is weak or very weak.
* 1: the transversal stretching of the *processus postglenoidalis* is medium-sized.
* 2: the transversal stretching of the *processus postglenoidalis* is large
* 3: the transversal stretching of the *processus postglenoidalis* is very large.

- *character 121*: lateral shifting of the *processus postglenoidalis* with reference to the lateral extremity of the tympanal.

The *processus postglenoidalis* is laterally overtaken by the typanal in the Gorillas whereas this is not the case for *Pan* (Picq, 1983). It is however laterally mishapen in Austropithecus and African *Homo erectus*.

Coding: 3 states.

* 0: the *processus postglenoidalis* does not overtake, laterally, the extremity of the tympanal.
* 1: the *processus postglenoidalis* laterally overtakes the extremity of the typanal.
* ?: indetermination linked to state «0» of character 119.

- *character 122*: shape of the *processus postglenoidalis* in *norma frontalis.*

In Modern Man, the shape of the *processus postglenoidalis* is triangular in *norma frontalis* and *lateralis* (Picq, 1983).

Coding: 3 states.

* 0: the *processus postglenoidalis* is rectangular-shaped.
* 1: the *processus postglenoidalis* has a round shape.
* 2: the *processus postglenoidalis* is flat.
* ?: indetermination linked to state «0» of character 119.

- *character 123*: shape of *processus postglenoidalis* in *norma lateralis.*

Coding: 3 states.

* 0: the *processus postglenoidalis*is triangular shaped.

* 1: the *processus postglenoidalis*has a rounded shape.
* 2: the *processus postglenoidalis*is flat.
* ?: indetermination linked to state «0» of character 119.

II. 4.2. Unprocessed characters.

Three of the characters receiving much discussion concerning *Homo erectus* have not been taken into account. These points were highly controversial in both status and definition of *Homo erectus* by different authors.

II. 4.2.1. The cranial capacity.

Anthropological literature has been much devoted to describing the evolution of cranial capacity (Weideneich,1941; Tobias, 1971; Lestrel and Read, 1973; Pilbeam and Gould, 1974; Holloway, 1975, 1979, 1981, 1983; Lestrel, 1975; Wolpoff, 1980; Henneberg, 1987; Leigh, 1992 and numerous other authors). The general tendency affirms an augmentation in volume of the skull over the time. Thus, Henneberg (1987) studied the cranial capacity of Hominids in their globality with reference to their presumed dating and independently of their supposed phylogenetical belonging. He concluded that the evolution of the encephalic capacity was gradual and followed an exponential growth through geological time. Leigh's (1992) study of the evolution of the cranial capacity in *Homo erectus* and ancient *Homo sapiens* showed that the phylogenetical results depending on the selected chronology, of the constitution of the hypodigms and the analytical methods employed. The craniometrical data is notably subjected to a sizeable variation according to the technique used in order to restitute an incomplete encephalic volume as is the case for the majority of fossils. Miller's study (1991) indicates that the size of the brain and its variation in size do not seem to be operational in distinguishing if specimens usually included in one species belong in fact to several species. This same author also states that for palaeontological samples, geographical, chronological and sexual variations as well as even the choice of samples and the techniques of measurement nihilate the phylogenetical interpretation of the volumetric capacities. As early as 1983, Holloway wrote that studies which concentrated uniquely on brain volume led easily to erroneous findings concerning the process and the method of human evolution (gradual and/or sudden balancing). In addition, Holloway was very skeptical on tests permitting the validation of allometrical relationships between cerebral capacity and bodily volumes. This last point was taken up in Kennedy's (1991) argumentation in clearly stating that, concerning cerebral volume, many arguments accorded to and between specimens attributed to different species or even of different genus exist.

In conclusion, measured or estimated values of cranial capacity seem to give no information of phylogenetical systematics, but the main point which has not permitted the retention the parameters of cranial capacity is the impossibility of carrying out in any standard way which could be reproduced a reliable measure for a large number of specimens taken into account in this study.

II.4.2.2. The thickness of the cranial vault bones.

A particular point observed by Weidenreich (1943) and taken up by numerous authors concerning *Homo erectus* is the thickness of the bones in the cranial vault. Different metrical limits have been put forward (Day and Stringer, 1982; Andrews, 1984; Stringer, 1984; Wood, 1984, Hublin 1978a, 1989; see also Brown, 1994 for a particular point of view). Authors often disagree with each other but for most of the time they are talking of different things. The values compared are not taken from the same point of view. The comparisons deal with means concerning groups *a priori* formed, groups which do not contain the same individuals from one study to another. Unfortunately when Bräuer and Mbua (1992) analyse the variations in thickness of the parietal and the temporal through a series of points of reference which are relatively well defined, their study is applied on groups qualified as being of African *Homo erectus*, Asiatic *Homo erectus* or archaic Chinese or African *Homo sapiens* and the individual data is not readable. Hublin (1978a) had shown the importance of recognising if the greatest thickness affected the external table of the bone or the diploë. In order to use this criterion of anatomical differentiation, it is useful to employ a tool which permits this distinction. Hublin (1989) had instigated an operational mode using tomography which resolved this problem but it was necessary to wait until the development of medical imagery in order to advance in any significant way on this question. Balzeau (2005 and 2006) showed that there was no difference in thickness of the cranium nor in the distribution of dense tables or of diploë between Modern Man and Ngandong, Sambungmacan, Zhoukoudian and Sangiran Man. Preliminary results from studies carried out in this field (Gauld, 1989) had been quoted by Kennedy (1991). However, the results of Kennedy (1991) went much further than those which in reality can be read in the source documents. This extrapolation had been used by Kennedy in order to affirm Hublin's (1986) hypothesis which put forward that the expression of the ensemble of the super-structures of cranium bones is due to general growth in *Homo erectus*. The presumed results of Gauld and one study which do not take into account the differences in thickness between diploë and external table were then used. Research by Balzeau and Badawi-Fayad (2005) highlighted the fact that the frontal sinuses and pneumatisation (Balzeau and Grimaud-Hervé, 2006) had no role in the development of cranial bone superstructures. The ensemble of scanner images of the different fossils studied having not as yet permitted the quantification of this parameter; these have not been intergrated into our study.

Nevertheless mention must be made of the pioneer study by Brown (1994) for the question which, based on the observation of several populations of Modern Man, concluded that it is not possible to distinguish *Homo erectus* from *Homo sapiens* when going uniquely from the thickness of a cranial vault, thereby affirming Weidenreich's initial hypothesis.

II.4.2.3. Separation between the inion and the endinion.

The separation between the *inion* and the *endinion* being dependent on the thickness of the bone is a point which is often discussed with reference to *Homo erectus*. The *inion* is classically situated above the *endinion* in *Homo erectus* whereas these two points coincide in Modern Man (Weidenreich, 1940, 1943 see also Andrews, 1984; Stringer, 1984; Turner and Chamberlain, 1989). But Hublin (1978c) demonstrated that the acceptance of the *inion* varied according to different authors. Furthermore, the *endinion* is not always well defined even if Weidenreich (1951) clearly states what its acceptance is. In subsequent research, definitions of the *inion* and the *endinion* are only rarely clearly explained and measurement techniques between these two points are not indicated. It is impossible to know if the difference was recorded by projection or from another craniometrical point. Bräuer (1990) put forward problems inherent in the different techniques utilised to assess the *inion* and the *endinion*.The employment of measurements between the *inion* and the *endinion* already published are therefore unsure. The method used and described by Kennedy (1991) has the merit of being clear and of relatively easy application, but the specimen fossils of which the *opisthion* is absent were set aside from his study. In order to compensate for this inconvenience, Bräuer and Mbua (1992) proposed a measurement on one of the points of the edge of the rupture of the boney piece. This is where the use of an image scanner would in the future permit the establishing of a very reliable data base in order to assess the pertinence of this anatomical characteristic.

II.5.Metrical data.

As soon as one begins work intrinsically on fossil material, one only has access to a small amount of biological information. It is then necessary to make the best use of this information available and to take into account metrical data. Additionally, it would appear to be useful to integrate as far as possible the variability when trying to establish a classification. To speak of variability or of polymorphism supposes that we have knowledge of the limits of the group studied or the definition of the group itself which is one of the aims of the study. In order to ensure that taxonomic results analysis were not pre-empted, the specimens themselves – tangible qualities – were taken as source material. In the work presented, the only sample which could define population statistics was that which dealt with Modern Man. As the variability of Modern Man is quantifiable, this species was used as the standard within the genus of *Homo*. One study more centred on *Australopithecus* would have necessitated the taking into consideration of different standards (Modern Man, gorillas, chimpanzees…). Mertical data must present certain mathematical properties in order to be processed with adequate statistical tools. The very small number of fossils did not always permit the application of statistical rules which are used in research notably because no checking with conformity to the Law of Normality is possible. In addition, metrical data is so rarely used in cladistics and necessitate appropriate processing and coding methods in order for it to be integrated into computerised analysis. The different stages leading to this prior processing are described here.

II.5.1 Points of reference.

Howells (1969) pointed out that the taking of measurements should not follow a tradition but should serve to find answers to precise questions. First of all it is the ensemble of possible measurements between similar points of reference which must be made. What is essential here is to determine those which can serve to establish an *a posteriori* phylogeny. The precise definition of craniometrical points as given by Martin and Saller (1959) and Howells (1973b) was adopted. The triple points and the points of intersection are the best defined. Those defined by projection on a plane were obtained less rigorously and were not always applicable. This was therefore the case for the application of the Francfort plane on many incomplete fossils. Phylogenetical research requiring, as it does, the taking into account of as many fossils as possible, marking points which permit the taking of measurements on isolated bones were adopted. Thus, markiners on the temporal bone were selected independently of reference to the Frankfurt plane (**Figure 43**).

Three further points were defined on the temporal bone.

S: Considering the temporal bone in *Norma lateralis*, the parallel axis of the antero-posterior elongation of the *processus zygomaticus* of the frontal bone and passing by the *auriculare* determines the basis of a horizontal marker point. Point «**S**» is the **highest point** of the squama situated on a perpendicular right segment at the horizontal marker point.

P: Considering the temporal bone in *Norma lateralis*, the parallel axis is at the antero-posterior elongation of the *procesus zygomaticus oxis frontalis* and passing by the *auriculare* determines the basis of an horizontal marking point. Point «**P**» is the **furthest back** of the squama situated on a perpendicular right segment of the horizontal marker point.

A: Considering the temporal bone in *Norma lateralis*, the parallel axis of the antero-posterior elongation of the *procesus zygomaticus oxis frontalis* and passing by the *auriculare* determines the basis of an horizontal marker point. Point «**A**» is the **furthest to the rear** of the squama situated on a right perpendicular segment to the horizontal marker point.

The other craniometrical points selected for this study were the classical homologous points of the *calvaria* to which were added the Supra-stephanion **(Sst.)** that is the point of the intersection of the coronal suture coronale and the *linea temporalis*

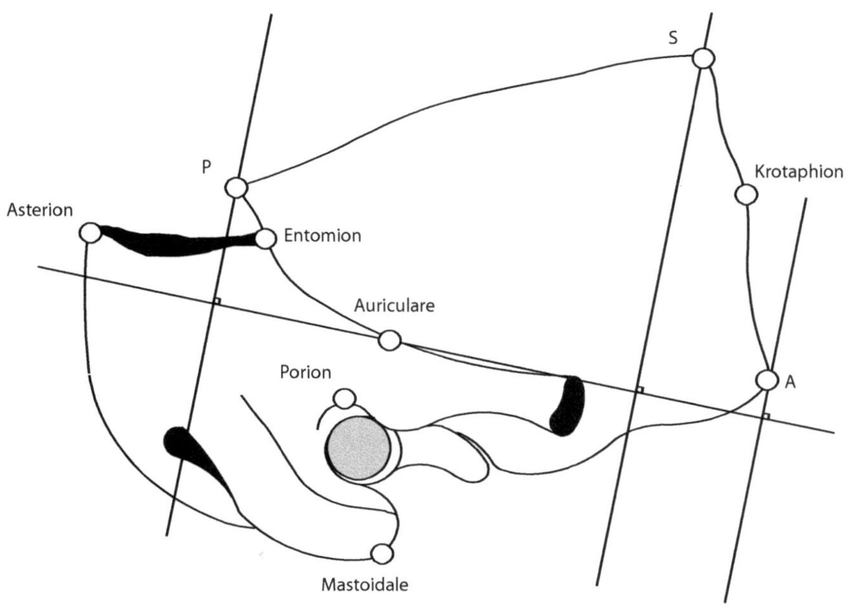

Figure 43: Position of the anatomical reference points S, P and A.

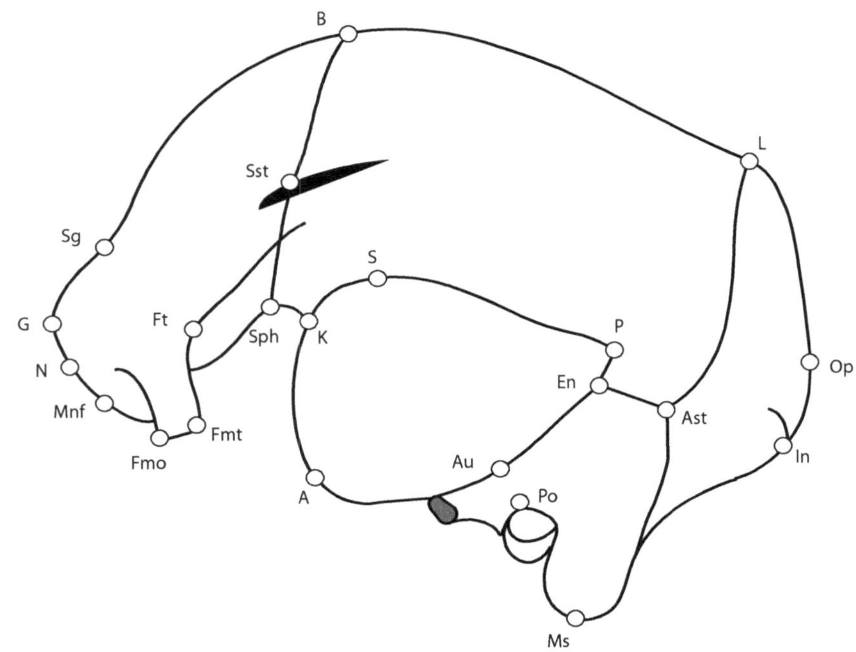

Figure 44: Positon of the landmarks used.

G: *glabella*, **Sg**: *supraglabella*, **B**: *bregma*, **L**: *lambda*, **Op**: *opisthokranion*, **In**: *inion*, **Sst**: *suprastephanion*, **Sph**: *sphenion*, **K**: *krotaphion*, **Ast**: *asterion*, **En**: *entomion*, **Au**: *auriculare*, **Po**: *porion*, **Ft**: *frontotemporale*, **N**: *nasion*, **Mnf**: *maxillonasofrontale*, **Ms**: *mastoideale*, **Fmt**: *frontomalare temporale*, **Fmo**: *frontomalare orbitale*,

A: furthest to the rear point of the temporale squama,
P: posterior point of the temporal squama,
S: upper point of the temporal squama.

superior; this was preferred to the *stephanion*, because, as Grimaud (1982) stated, the *linea temporalis superior* is always more visible than the *linea temporalis inferior*.

The remaining homologous measuring points used as exhaustively as possible are indicated in figure 44. In adapting the measurements to the problems posed, the precepts put forward by Howells (1969, 1973b) were followed.

II.5.2 Prior processing of metrical data.

The ensemble of the chords and the archs existing between all the landmarks were taken into account. In total, 613 measurements were made on each complete skull; a total of 154 solely on the frontal bone, 84 on the parietal alone, 72 for the temporal, 51 for the occipital and 252 concerning the *calvaria* as a whole. An average parameter (a chord or an arch) was calculated from bilateral data for each bone type and for each individual. The ensemble of the data averaged out was then used in order to calculate the indices. Two types of indices were considered. One was supposed to give account of the bellying or the relative relief of a segment:

type I index = (arch AB - chord AB) / chord AB; the other taking into account the relationship beween two segments:

type II index = chord AB / chord CD.

Keeping in mind reduced bilateral data to one single parameter and to bones taken individually, 2264 indices were calculated *per* complete individual. The ensemble of the individuals regroups 30 individuals of Modern Man, 11 chimpanzees, 10 gorillas and 63 fossil specimens. Additional immature individuals (6 individuals of Modern Man of known age) were used for the coding of the metrical in a later stage of analysis.

II.5.3 The metrical indices.

The indices were relatively independent of the format of the ensemble (Olivier, 1974; Mosiman and James, 1979; Darroch and Mosiman, 1985). Thus, even if the refined and unrefined measurements carried out could not be distinguished by the use of simple indices (see Bookstein, 1978), the effect of size is reduced. Measurements between homologous points permitted the reconstitution of tensors by triangulation which used by Bookstein (1978) is useful in the description dof shapes. A simple representation of the mathematical techniques of the representation of these shapes already exists in Read's (1990) study. This present study does not push the technical possibilities of the description of shapes as far. These are taken into account separately in the morphological section. All of the indices were gathered in an exploratory spirit. Correlations could exist between indices or the redondance of information were processed *a posteriori* at the end of the analysis. The question of the independence of information taken into account, whether it is a question of metrical or morphological characters constitute one of the major criticisms directed at cladistics. The redundance of information (Hublin, 1989b) or the over representation of certain parameters or anatomic complexity would seem to generate confusion (Trinkaus, 1990; Strait *et al.,* 1997). Hlusko (2004) asserts that "Cladistics is a powerful tool for reconstruting phylogenetic relationships, but it is a tool whose power is proportional to the number of independent character available for analysis". These objections arise from a quantitative vision which forgets that in cladistics, in all rigour (see Hennig, 1966), a unique autapomorphy suffices to distinguish a taxon. From a practical point of view, it is possible to demonstrate the independence of characters taken into account. If one character is considered several times or if the characters are linked, then their phyletical history will be the same. A checking procedure was put into place for this for the ensemble of the characters at each cladogram node obtained. This precautionary method is here illustrated for a metrical character (**Figure 45**).

The close proximity of the *porion* and the *auriculare* could let it be believed that a diffuseness of information existed on the characters 347 and 362. However, after analysis, if the changes in state of these two characteristcs are identical on two occasions, they do differ at a third node, thereby indicating that the information is not diffuse and that the two characteristics are indeed independent as their histories are different. Regarding the rigorous use of indices, it is necessary to check if their properties regarding the laws of probability allow using the law of Laplace-Gauss. Arguments concerning rules which should be adopted when using indices have arisen between several authors for a long time (Atchley *et al.*, 1976; Atchley and Anderson, 1978; Albrecht, 1978; Dodson, 1978; Hills, 1978). The conclusion from these discussions is that the formulation of indices in logarithmic form generally permits the employing of indices without any problem of respecting the law of Laplace-Gauss. It should here be noted that in their metrical study regarding the phylogeny of the early hominids, Chamberlain and Wood (1987) employed Kazmierczack's formula; which brings us back to the analysis of correlations. The measurement of angles was dismissed for this study as such measurements are linked to particular plane references (frequently unexploitable on isolated anatomical fossils). Additionally, the rigorous processing of angular data necessitates conversions. For the statistical processing of this type of data (indices and angle values) several properties need to be verified before the application of suitable rules. Once these verifications were carried out, a preliminary analysis was launched in order to establish which indices were the most discriminating.

Figure 45: Independence of the characters.

Even if porion and auriculare are relatively close, the transformation of the states of these two characters are not the same all along the phylogenetic history of the taxa taken into account (see the full matrix for detail). At the node between step 67 and Sts 5 the transformation is 0 to 2 and 0 to 1 between the node 62 and KNMER 1813 for both character 362 and 347 but at the node between Sambungmacan and 47 the character 362 changes from the state 0 to 1 whereas there is no change for the character 347. The history of 347 and 362 is different so they are different.

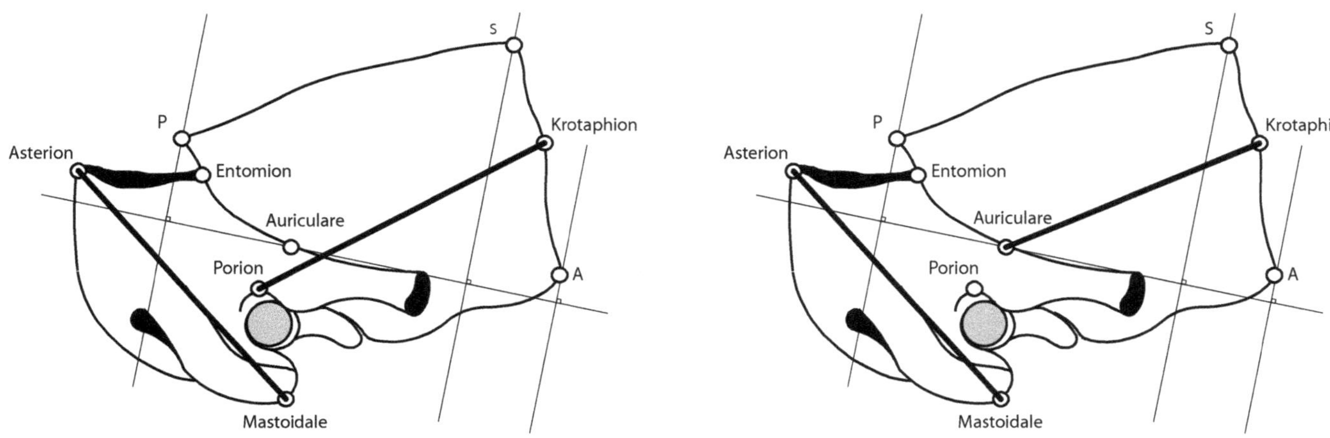

Character 347 : Index = $\frac{\text{[Porion-Krotaphion]}}{\text{[Porion-Mastoidale]}}$

Character 362 : Index = $\frac{\text{[Auriculare-Krotaphion]}}{\text{[Auriculare-Mastoidale]}}$

II.5.4 Method of selection of indices serving to establish the matrix.

Measurements having been carried out between anatomical homologous points, each one can then be put into relation with another (if this were not the case, it would be necessary to pose questions concerning the «inaccurate» precision of certain cranometrical points).Whatever the case, it was most useful to carry out a preliminary processing of indices in order to assess the domain of mathematical applications possible. First of all, it was necessary the check the distribution of these indices of the series of reference actually followed the Law of Normality. In this way it is not necessary to make use of logarithms as their use is in fact much more formal and complex (in fact, the average of logarithms does not correspond with the logarithm of the averages). In order to verify the normality of the distribution of the indices used, an examination of the histogram in the sample of Modern Man which serves as a reference was carried out. Taking into account the large number of indices calculated the graphic method of Henri's straight line which permits a rapid method of verifying the normality of distribution (Grémy and Salmon, 1969) was applied. According to Simons and Pilbeam (1965), the distinction between fossil species must be of the same amplitude than that which exists between living species. This statement supposes, for example, taking the differences between *Pan troglodytes* and *Homo sapiens* of today in order to determine the space in which the fossils should be found. Thus, the evolution is gradual for the ensemble of the characters and no difference would be possible between the fossils. This method was therefore not adopted. We took a population of Modern Man as a reference group. In this population it is possible, for each metrical character (for each index calculated) to determine its amplitude of variation. When there was a superimposition between the population reference group and the specimens taken into account for a given index, this index was not retained for the continuation of the analysis. In fact, this superimposition between all of the specimens is not discriminating and only reflects a symplesiomorphy (the stock common to all primates studied). This superimposition marks an absence of phylogenetical significance within hominids (**Figure 46**). On the other hand, as soon as a fossil specimen was seen to differ from the domain of variation of the population standard it was adopted.

At the end of this selection, only 345 indices remained out of 2264 indices initially calculated. For 1919 metrical indices, it was impossible to distinguish on the one hand the Great Apes and on the other hand the fossil specimens and the complete sample of Modern Man. No descriptive analysis (analysis of correspondances or factorial analysis) was carried out using the gathered

Figure 46: Selection of the metrical data.

In the first case a) there is no difference between the specimens and the human modern sample so this index is rejected for the analysis. In the second case b) there is a significant difference between some of the specimens and the modern human group so the index is kept to undertake the analysis.

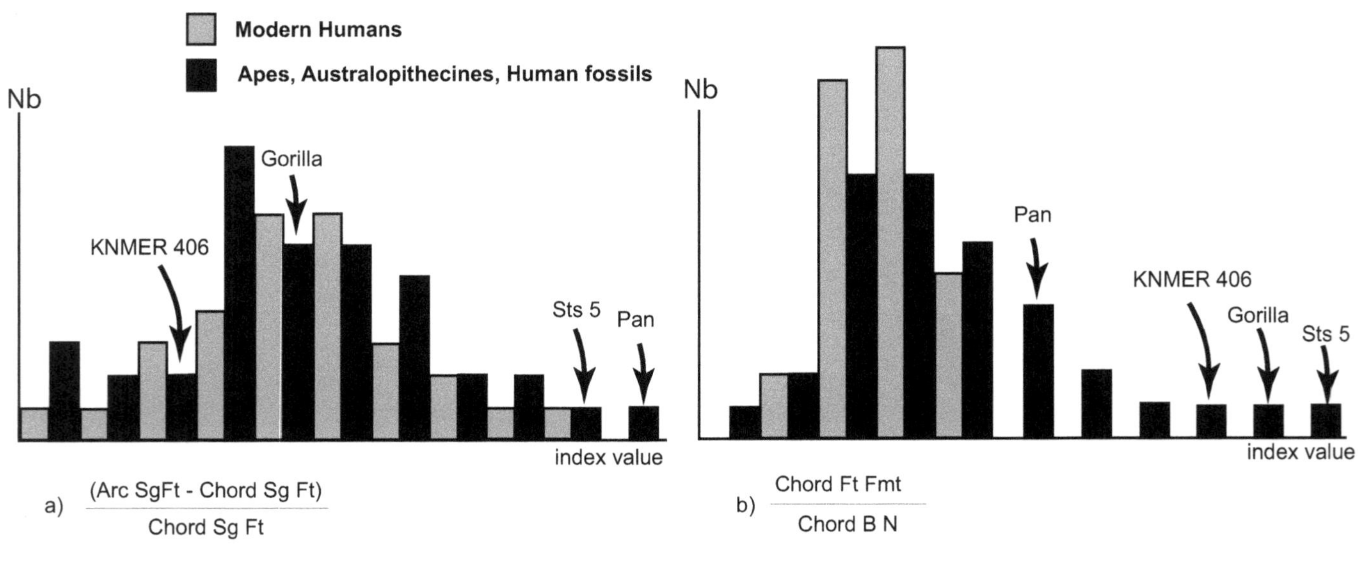

metrical data. In fact, these methods of analysis, with the data being described in relation to the barycentre of the whole of the sample, would only have made up an exercise of « total morphological pattern » of the pheneticians. In addition, the missing data could not have been taken into account and it would have been a question of finding a method of quantification in order to transform the morphological data into continuous quantitative data so that they could be integrated into the analysis. This statistical approach – or more exaclty this « descriptive » approach – is material for further research leading to a phenetic analysis. The purpose of our study is not to classify fossil specimens according to their global similarities but to define their phylogeny. Classification in fact is a subsequent step. The first results allowed us to estimate the symplesiomorphological part as being 84.8 % of the metrical data (1919 out of 2264). The aim was to produce a classification of fossil specimens attributed to *Homo erectus* or to « archaic » *Homo sapiens*; it was therefore a question of working on criteria which led to a differentiation of the specimens taken into account. These criteria were to be sought in the 15.2 % (345 out of 2264) of metrical characters remaining (**Appendix 1**) and on the morphological data.

II.6. Coding of metrical indices.

The coding of metrical indices (**Figure 47**) is inspired from « weight coding » proposed by Thiele (1993). In his study, Thiele proposes a mathematical method in order to integrate metrical data in computerised cladistic analyses. He considered several populations. Each population is characterised, for each index, by the average numerical value calculated. Following this these averages are, centred, reduced, standardised and and transformed into ordinal values. Each average is divided by the shortest gap existing between the two averages of the closest populations. The new numerical value proper to each population is then transcribed into one sequence of the state of characters.

The general problem of « gap coding » techniques resides in the fact that they are based on populations for which the limits are known *a priori* and not on specimens which are to be defined. Chamberlain and Wood (1987) put forward a cladistic analysis which integrated continuous data. Their method was neither « gap coding » nor « weight coding » but « combination coding », *id est*: the standardisation put into practice in order to define the states of the characters for each taxonomic unit rested on the average calculated from the specimens which made up the unit. This method supposes working on taxonomic units made up *a priori*. It is therefore very influenced by the composition of the taxonomic units and by the missing data. This procedure is therefore not ideal in defing a phylogenesis which minimises the taxonomic *a priori*. We followed a course of action where the taxonomic belonging of individuals was not known initially as this was the object of the study. Variability could only be considered for the sample of Modern Man.

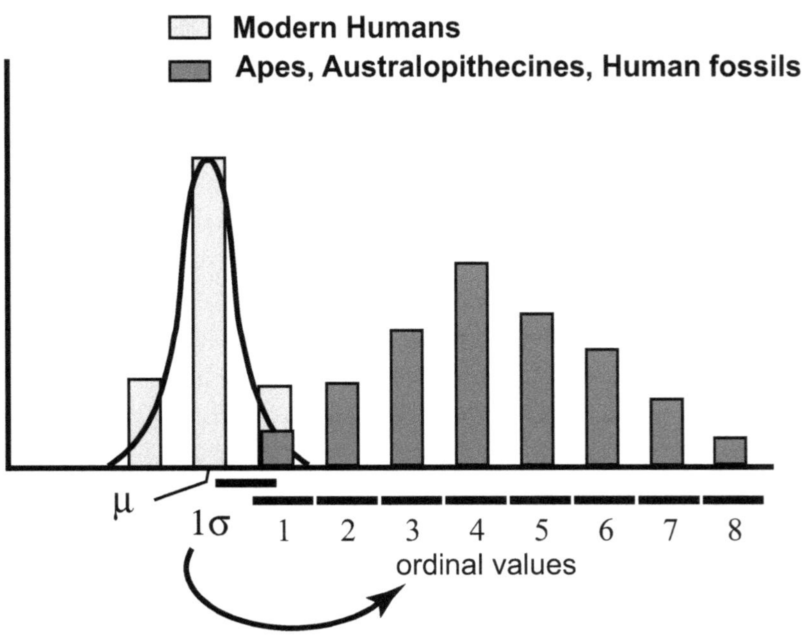

Figure 47: Coding method

For one index, the standard deviation (?) associated to the modern human range is used as a standard and is translated in an ordinal value to make the coding of the information for the « global » sample.

This sample therefore served to standardise the degree of amplitude of the state of the characters for each index. Following this, and going from the extreme value (minimum or maximum) gathered for the sample, the amplitude of variation was divided into segments equivalent to the value of the segments of the standard deviation. Each one of the segments therefore corresponded to a state for the character defined by the index. It is here only a question of strategy for where each parameter distinct from that observed in Modern Man was adopted. A quantitative factor was also integrated into the definition of the characters. The intervals which define the limits between the states of characters are indicated in **Appendix 1**.

II.7. The Processing of Variability.

Studies in variability are extensive. They led MacIntosch and Larnach (1972) to place *Homo sapiens* and *Homo erectus* together demonstrating the retention of characteristic « *erectus* » in Australian Aborigines. If this retention or this continuity, wich could be regional, have been contested by Groves (1989) and Habgood (1989a), it is clear that species defined on the basis of arbitrary limits cannot be utilised in a debate on their chronological development or on their mode of evolution – except in order to accept the legitimacy of a circular reasoning. According to Habgood (1989a) cladistics does not take into consideration individual variability. Habgood maintains that phenetics is the most appropriate method. But from this point of view, it is not possible to know if the phenetic « clusters » obtained are phylogenetical or not. Bräuer and Mbua (1992) showed that cladistics does not touch variability and is inoperative in phenotypical reality. This objection becomes irrelevant with the processing of metrical data proposed here. In an evolutionist or phenetic approach, it is common practice to use coefficients of variation (CV) in order to check if the hypodigmes constituted *a priori* embody more than one species. Cope and Lacy (1992) clearly explained which were the problems and prerequisites inherent in the use of coefficients of variation usually employed in order to affirm the diversity of the fossil samples. The first limitation is linked to the necessity to employ a standard estimated by the whole of the individuals taken into consideration. The second limitation is due to the degree of accuracy of the coefficients of variability which concern samples where the fossils are few in number. In this case the degree of accuracy of a recent monospecific reference is almost always covered. In the application of coefficients of variability when studying the cranial capacity of Hominids in palaeontology, Miller (1991) showed that this « populational » approach was not adapted to fossil taxons presented in small numbers. In fact, coefficients of variation are very dependant on the composition of the samples; they do not permit, for example, the distinction in fossil hominoids *Gorilla gorilla, Pan troglodytes, Pan paniscus* or *Homo sapiens sapiens*. More generally speaking, cladistics is criticised for not taking into account the morphological variability of different individuals. It is suspected of simplifying matters to the benefit of classification (Kennedy, 1991). This accusation also holds for all methods which analyses constituted groups as grouping together tends to level out differences which arise from variability.

II.8. Data Matrix.

The complete matrix of data gathers information concerning **67** taxons (64 specimens + Modern Man +gorillas+chimpanzes) and 468 characters (**Appendix 2**). Immature individuals are marked with the symbol "°". The two anatomical elements of Arago were taken for one single specimen in the final analysis. The initial question concerns the phylogenetical existence of the taxon *Homo erectus* and the robustness of the result obtained. In the matrix of characters only one operational unit named Modern Man figures. This operational unit gathers together 30 identical operational units (30 individuals) for which no polymorphism were noted owing to the definitions of the morphological characters. Concerning character states dealing with numerical data, individual differences were « erased » to the extent in that they are averages of metrical data calculated from the 30 individuals who served in determining a standard (**Figures 46 and 47**). Similarly for the chimpanzee series and the gorillas which appeared finally as two operational taxonomic units.

The metrical and morphological data produced additional information. The redundancy of information was avoided through choosing to highlight metrical data when the homolgous points permitted measurement. Under these conditions, there was no prior reason or criteria for choosing to weight the metrical data in relation to the morphological data. Each characteristic must be considered as being potentially a carrier of information as another.

III. Results of the Analysis.

The cladograms obtained from the ensemble of the specimens studied will be examined first. Finally, an in-depth discussion of the phlogenetical results arising from the analysis of one sample more specifically orientated towards *Homo erectus* will be presented.

III.1. General Characteristics of Different Analyses.

On the whole, analyses which followed the out-group were used in order to root the tree. Four taxons served as an out-group. They were *Gorilla gorilla*, *Pan troglodytes*, and two complete KNMER 406 specimens (generally attributed to *Australopithecus boisei* sometimes to *Paranthropus boisei*) and Sterkfontein 5 (generally attributed to *Australopithecus africanus*). The morphological and metrical data were re-grouped into a single matrix. The complete matrix of data used concerns 66 terminal taxons and 468 characters (The Flores specimen was added to a later analysis). There were 123 morphological characters. Within these, 72 were binary; *id est*: 58.5%. There were 345 items of metrical data. There were 49 binary: 20.8%. In total, 25.8 % of the characters were binary with the others presenting multiple states.

The search for the minimal tree was resolved by the comparison of a large number of trees. According to Darlu and Tassy (1993), on the basis of **n** terminal taxons, $\mathbf{(2n-3)!/(2^{n-2}\,(n-2)!)}$ dichotomic trees could be constructed; thus, for 10 taxons this would make 34,459,425 possible trees. The research algorithms for the shortest tree re-group the terminal taxons together in such a way that with each insertion of a supplementary taxon, the diagram expresses the minimum number of necessary steps in order to account for the distribution of the characters. In this way, the final or global minimum length does not only depend on local relationships between terminal taxa but on all the possible relationships between the terminal taxa. A high number of taxa therefore quickly results in the search for the shortest tree of astronomical proportions. The results presented in this study are all the result of heuristic research. Actually, the number of specimens studied (greater than 20) is too high for exhaustive research work to be carried out in a reasonable time. A heuristic algorithm is one which results in a solution which may or may not be optimal. Algorithms said to be of « step by step addition » gather together two terminal taxa at the root then insert a third terminal taxon. The most parsimonious tree obtained after insertion of this third taxon is conserved and the following taxa inserted in turn. However, this step by step procedure is sensitive to the order in which the taxons are introduced. In order to overcome this problem, several heuristic strategies of « branch rearrangement » exist. The improvement of the initial tree is then carried out by moving the branches. If one rearrangement results in a shorter tree, the latter becomes the subject of a new rearrangement. This practice results in the finding of the shortest tree. Four heuristic research options are proposed by the program PAUP. They are indicated by the following acronym: <General>, <Starting trees>, <Stepwise addition> and <Branch swapping>. Each of these options was applied several times for each of the matrices presented. The results obtained were always identical whatever the method employed.

III.2. Analysis of the Ensemble of the Data.

III.2.1 Non-ordered Processing of Multiple-State Characters.

The analysis of the complete matrix (**Appendix 2**) produced a cladogram (**Figure 48**) where the multiple state characteristics have been processed in a non-ordered method by the option <unord> of the program PAUP. The cladogram presented is the strict consensus of 28 trees. These 28 trees, equally parsimonious are 4,092 steps long. The consensus cladogram has not been resolved. However, two clades appear quite distinctly.

1) The first regroups the specimens of the Ngandong series plus Sambungmacan and Ngawi as well as two African fossils: Ndutu and Omo Kibish 2.

2) A second clade associates the specimens which can globally and without too much argument, be affiliated to *Homo sapiens*. To these individuals

we can add the immature individuals from all taxonomic origins with the exception of specimens Sangiran 10 and KNMWT 15000. These immature specimens are close to Modern Man. The coherence index of each of the 28 trees which make up the consensus is 0.285. In other words, the homoplasy to which it refers is very important. The retention index is higher than the index of coherence but remains weak (0.384).

III.2.2 Ordered Processing of Multiple-State Characters.

The second cladogram (**Figure 49**) was obtained after ordered processing of the multiple state characteristics (option <ord> of the program PAUP). The cladogram presented is the strict consensus of 72 trees. These 72 trees, equally parsimonious, have each a length of 5,119 steps. At the terminal level of the tree of consensus, a « canopy » appears in which it is possible to recognise several clades which are roughly:

1) Modern Man and immature individuals,

2) two groups made up of archaic *Homo sapiens*,

3) a non-hierarchised ensemble made up of Indonesian fossils from Ngandong and the Chinese Sinanthropes and,

4) an ensemble of Neanderthalians.

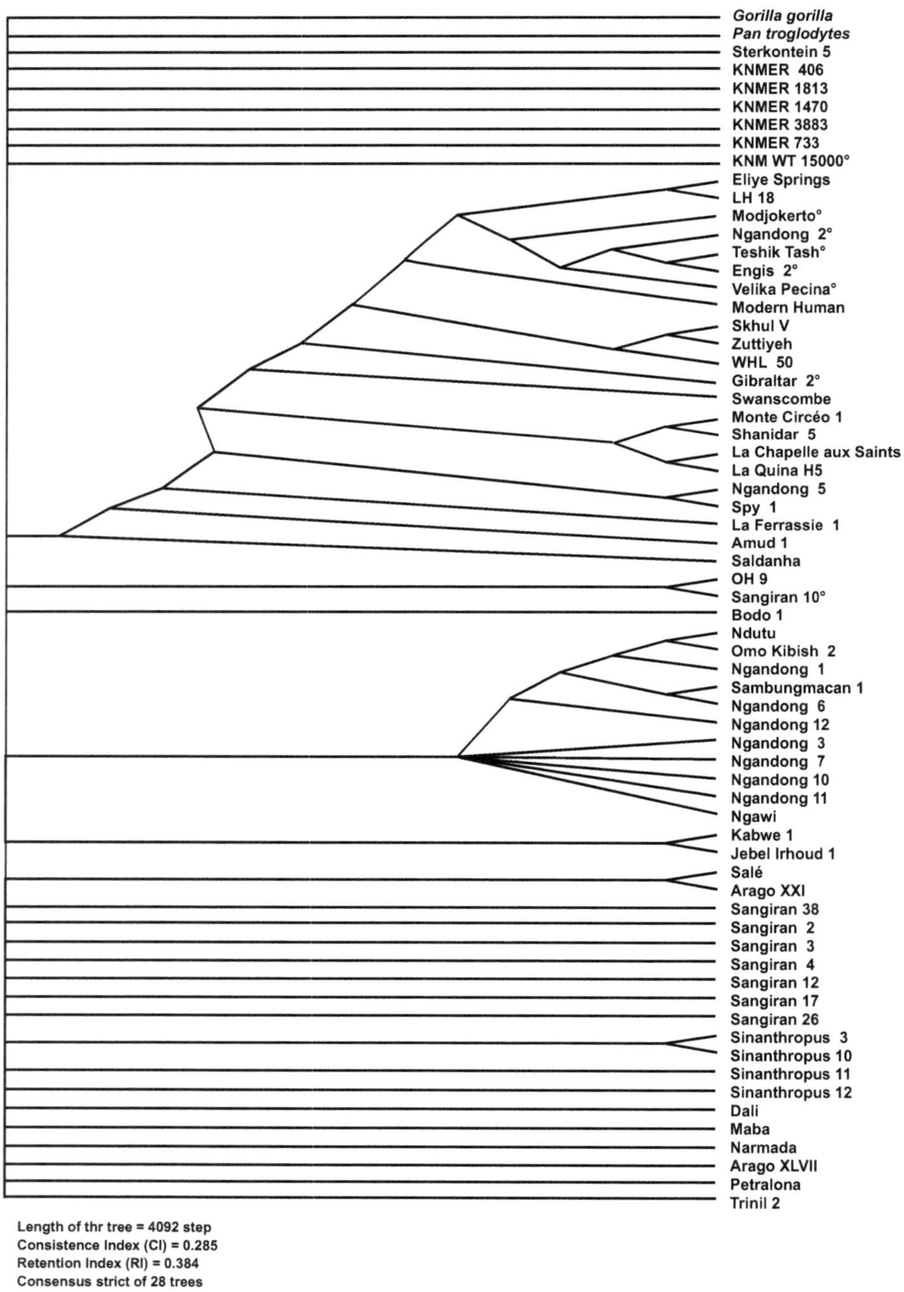

Figure 48

Result of the analysis made using non-ordered treatment for multiple-state characters.

This « canopy » is the sister-group of an ensemble which gathers together Sinanthropes, an Indonesian immature fossil individual and two African fossils. These two groups are in turn the sister-group of a clade which associates the Sangiran fossils at Trinil 2 and the individual from Nariokotome. Sangiran 17 and two very incomplete specimens are associated in one multifurcation with the two clades above. From an internal hierarchical point of view, a paraphyletic sequence exists which only concerns Kenyan specimens: KNMER 1813, KNMER 3883, KNMER 3733 and KNMER 1470, just after the node which marks the clade *Homo*. But the index of coherence (0.235) of each of the 72 trees is worse than in the case of non-ordered processing of characteristic states. On the other hand, the index of retention is better than 0.448.

III.3. Analysis of Cladograms minus Immature Individuals.

When the immature individuals are excluded from the analysis the trees are of a better resolution. The matrix thus constructed then concerns more than 58 operational taxonomic units, but 468 characters remain.

III.3.1 Unordered Processing of Mutiple State Characters.

An non-ordered processing of multiple-state characters shows a consensus of 4 trees (**Figure 50**). Each of these 4 trees, also parsimoniuous, has a length of 3762 steps. Their index of coherence is of 0.308 and the retention index of 0.384. The first bifurcations concerning the genus *Homo*, are

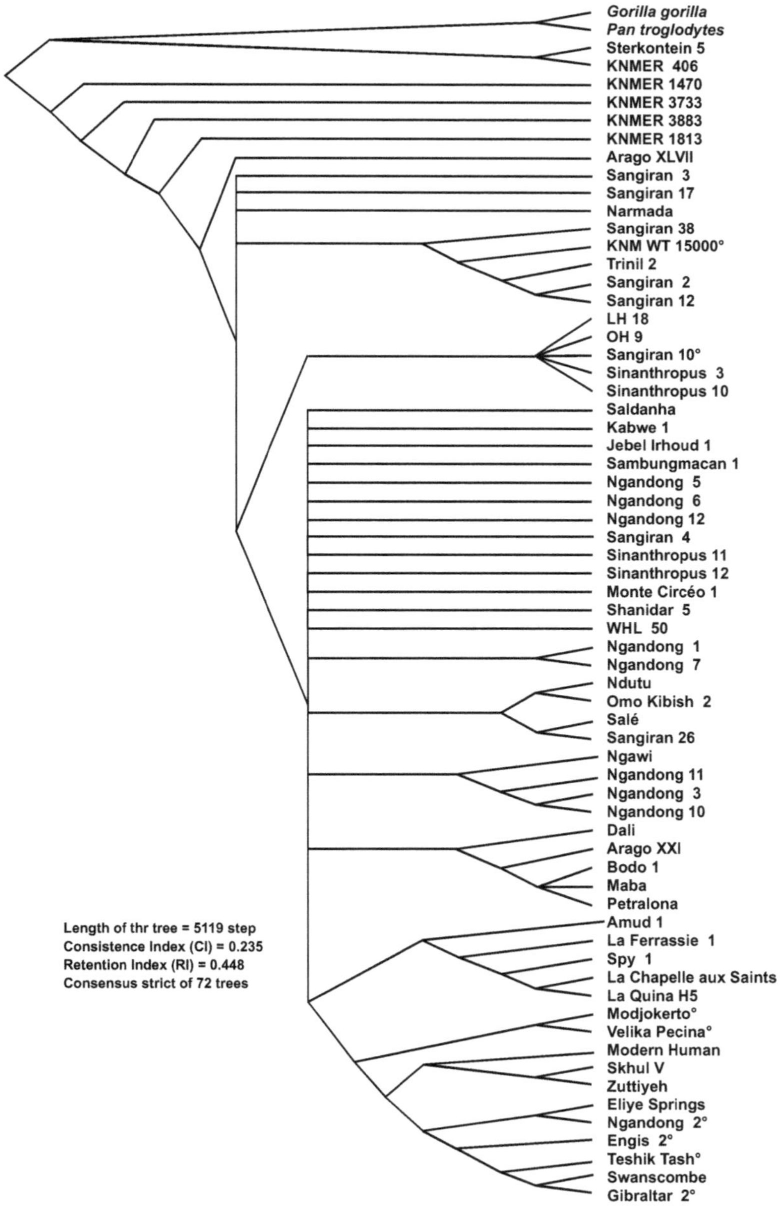

Figure 49

Result of the analysis made using ordered treatment for multiple-state characters.

marked by a paraphyletic sequence isolating the Kenyan specimens KNMER 1470, KNMER 3733, KNMER 3733 and KNMER 1813. Following this, two entities are globally recognisable. One clade reunites the Trinil 2 specimen, the Ngandong series, specimens of Sangiran and some Sinanthropes. A second more extended clad, mostly regroups *Homo sapiens*. Two smaller, numerically-speaking, clades prove to be different before data capture from these two clades in terminal position. These are:

1) a group made up in the majority of African specimens, and

2) a group principally composed of fossils originating from the Sangiran site.

III.3.2 Ordered Processing of Multiple State Characters

The ordered processing of multiple state characteristics of the matrix which excludes immature individuals leads to a consensus of 8 trees (**Figure 51**). Each of these 8 trees, also parsimonious, are 4,697 steps long. The coherence index is slightly superior to that which includes corresponding immature specimens, *id est*: 0.254. The retention index is equal to 0.449. The topology obtained by the consensus of the 8 trees shows the same initial paraphyletic sequence of the genus *Homo* as before. The two principal terminal clades were not very different to those resulting from the non-ordered processing of characters, even if there are numerous individual changes. The strongest represented clade essentially ressembles *Homo sapiens*. The second clade consists especially of specimens from the Ngandong series. To begin with a first clade links Sinanthropes and African fossils together, and then further up it is possible to detect an ensemble essentially made up of Indonesian specimens from the Sangiran site.

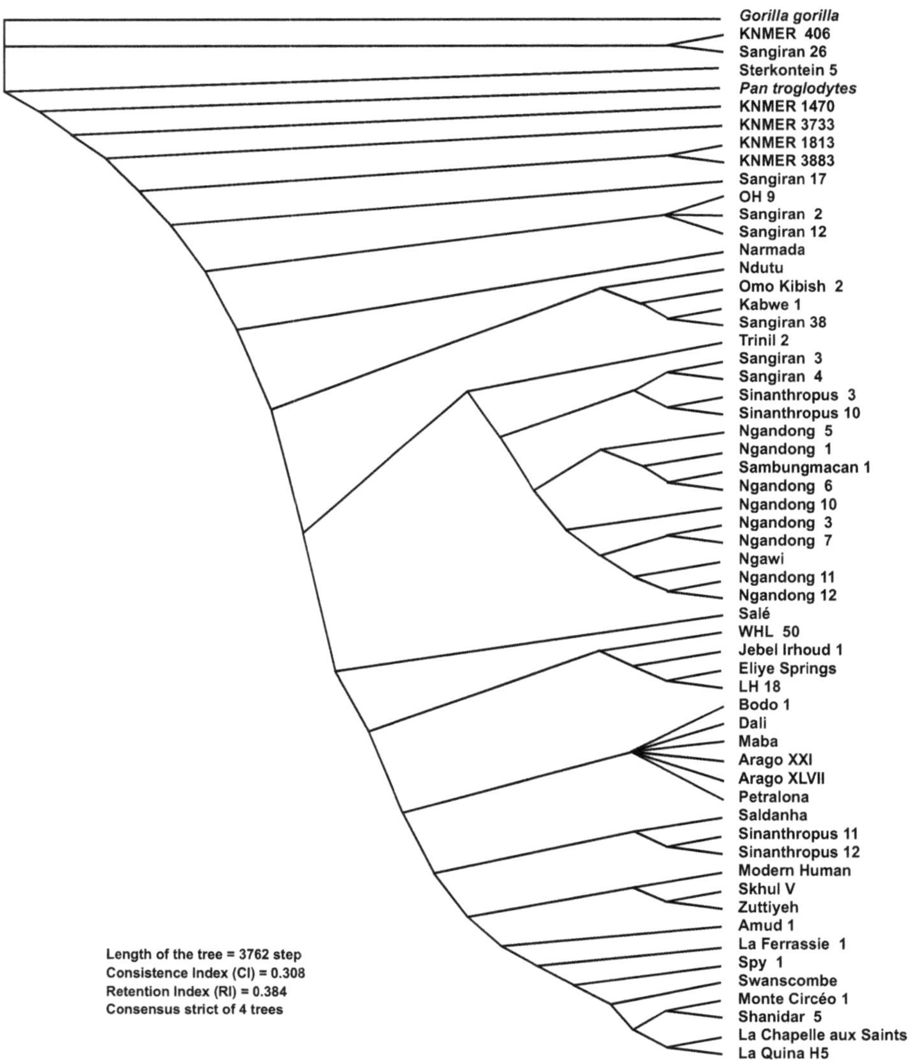

Figure 50

Result of the analysis made using non-ordered treatment for multiple-state characters.

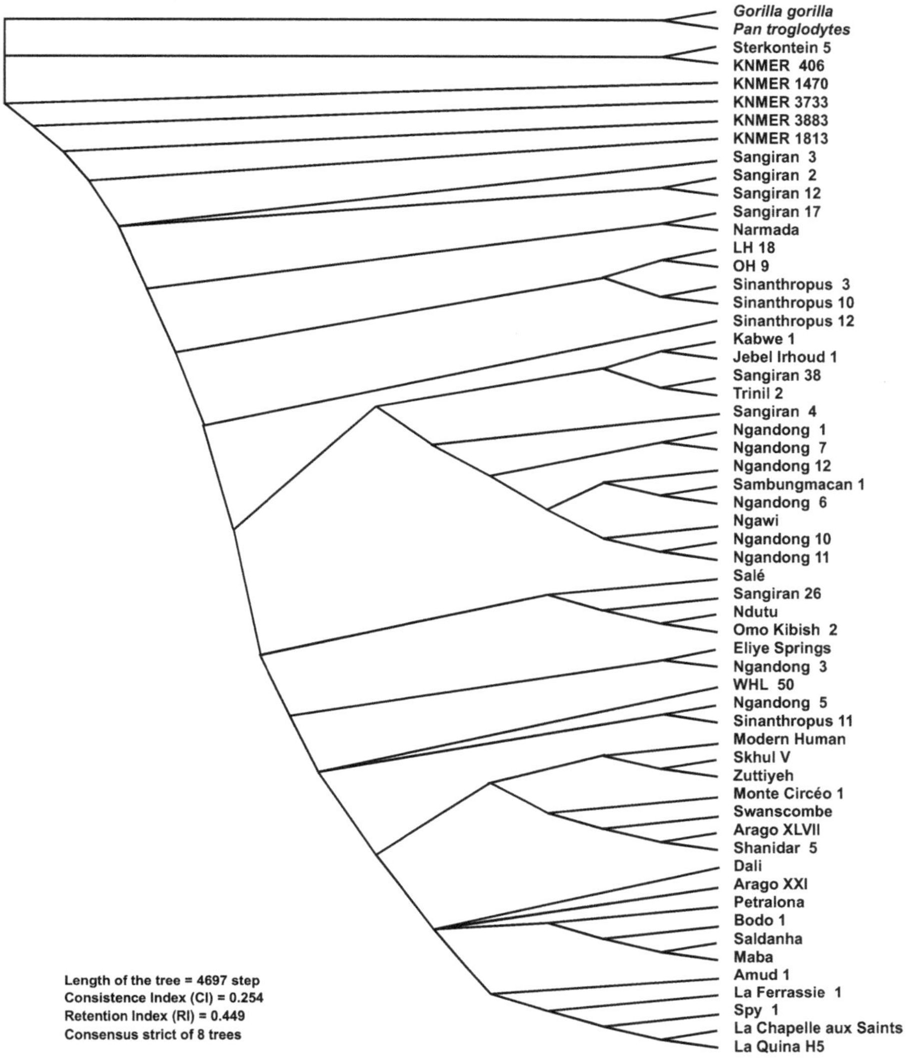

Figure 51

Result of the analysis made using ordered treatment for multiple-state characters.

If ressemblances exist in the principal clades revealed by the two processing methods (ordered and non-ordered of the multiple state charactercs), several groupings of specimens correspond to what is classically put forward by different authors. More closely, numerous differencies exist and some specimens seem to 'escape' from the most current biosystematic classification. Several facts – mostly dealing with homoplasy – are linked to this phenomenon. The absence of information, retranscribed by question marks in the matrix, is due to the incompleteness of the fossils. Following this, in a manner inherent to this fractioning the localisation of criteria on the anatomical items is played out on the remaining information available. If, for example, a portion of the frontal bone is rich in phlogentic information and if this part is defaulting, then the specimen in question could not be situated in an ambiguous or precise way. In the same way, the anatomical analysis of the specimens led us to consider an unequal number of criteria for each bone. (182 criteria for the frontal bone – respectively 25 pieces of morphological data and 157 metrical; 67 for the parietal bone: 16 and 51; 159 for the temporal bone: 62 and 97; 60 for the occipital bone: 20 and 40).

III.4. Rationalisation of the Data through Observation of the Ontogenesis.

The otogenetical information is the most reliable indicator of the polarity of characters according to Nelson (1978). However, as Chamberlain and Wood (1987) indicate, this information is not employed in anthropological analyses which make us of cladistic analysis programs. In anthropology these characters are considered as being primitive uniquely because they are observed on taxons reputed to be the oldest. This is the case of the reinforcement existing between the entoglenoidal formation and the tympanal observed in « early *Homo* » or the Australopithecines (see Kennedy, 1991 and Bräuer and Mbua, 1992). **However, as Weidenreich wrote as early as in 1943 (p: 139), it**

is certainly not allowed to consider one form as primitive uniquely because it is geologically older than another or to consider it as being derived simply because it is more recent.

This chapter proposes an analysis which takes into consideration as much information as possible and in particular to do with ontogenetic information. As a result of different results obtained from ordered and non-ordered processing of multiple state characteristics for a large number of specimens of which several were incomplete, a more restricted sample of specimens is considered here. These were selected with three main criteria in mind. First of all, it is obviously a case of defining *Homo erectus* in order to fulfil the aims of this study. Specimens which offered the most information, *id est*: the most complete were therefore required. Following this, close consideration was given to taxons affiliated to *Homo habilis sensu stricto* on the one hand and to archaic *Homo sapiens* on the other. Finally, a few incomplete specimens, but also situated at the centre of numerous discussions mentioned in literature were also considered. In the case of the Arago site, the two specimens Arago XXI (frontal bone) and Arago XLVII (parietal bone) were placed together into one operational taxonomic unit. (Arago XXI/XLVII). For the 35 specimens so defined, a « mixed » coding of states of the characters was applied. Certain multiple state characters were considered as being ordered and others as non-ordered. The choice supposes prior phylogentic hypotheses.

III.4.1 The Metrical Characters and the Problem of Weighting of Transformations.

The choice of which processing option to adopt (ordered or non-ordered) of the multiple state characters depended on anatomical observations and on analysis carried out on the metrical indices. In the case of morphological data this consisted quite simply of observing immature individuals of *Pan troglodytes*, *Gorilla gorilla* and *Homo sapiens sapiens*. For these taxons there was no great problem in recognising immature individuals unlike for fossils with undefined taxonomic affinities.

III.4.1.1. The Heterochronology of Development.

Some anatomical parts can develop quicker than others during ontogensesis. The growth of different anatomical elements can be expressed through height, length and width, in a greater or smaller way during individual development. When it was possible to follow a gradual ontogenetical development through states of characters, the subsequent polarisation was coded. In this case the states of characters were processed in an ordered manner. If not possible, there was no reason to suppose that a given state of a character be an obligatory intermediary passage in order to go from one state to another. In fact, considering that an index can only follow linear evolution supposes ignoring the heterochronology of the development of anatomical elements. Regarding the skull, this consideration would negate the salutatory development of the skeleton (Lampl, 1993). For the metrical data, in general, this falls back on adopting a gradual vision which infers *a priori* a strong phylogenetical hypothesis. But the indicators, such as they have been defined above, give account of the relative proportions between homologous points. Results of the analysis carried out on individuals still growing up on only 345 metrical indices selected for the analysis, 67could be polarised. The other indices did not follow a linear progression according to individual age. The 67 characters which could be orientated were therefore necessarily ordered. The heterochronology of development was supported by 82.2 % of the indices. The results of this research comforts LØvtrup (1989)'s findings for which it would be an error to consider heretochronology as an important evolutionary agent. For this author, the most significant mechanism in evolution is deviation which leads to changes during early and sub-terminal ontogenesis. In addition, 13 morphological characters were processed in orderly fashion in accordance with observations carried out on immature individuals; *id est*: a total of 80 characters processed in orderly fashion.

III.4.2 Analysis of the Cladogram.

The cladogram (**Figure 52**) is the result of the analysis of a matrix of 35 terminal taxa and 468 characters. A single tree was the result of the analysis. It should first be noted that the reduction in the number of trees (72 and 28 trees respectively for the ordered and non-ordered processing of mutiple state characters of the matrix comprising the whole of the specimens covered) illustrates the rejection of the most incomplete fossils. The number of steps for this tree is 3,092. Its index of consistence is not very high (IC = 0.371), the index of retention is higher (IR = 0.424) and, in any case, the indices are better than those of earlier trees. A high level of homoplasy is still present. Information issuing from analyses which cast aside immature individuals can be seen on this cladogram. In this way, Kenyan specimens classically attributed to African *Homo habilis* or *Homo erectus* remains isolated. They are separated, one from each other, to make up a paraphyletic sequence at the basis of a genus *Homo*. This paraphyletic sequence (nodes 65, 64, 63 and 62) is supported by the large number of characters; they are the more robust nodes. Regarding the other specimens taken into consideration, three principal groups stand out from the rest.

1) A first associated clade: Trinil 2, the fossils of Sangiran and KNMWT 15000.

A second clade, of the same previous sister-group if one ignores two isolated individuals (Narmada and Sinanthropus 3), join together two principal clades.

2) The first clade can quickly be associated to *Homo sapiens*. In fact, it groups together Modern Man, Neandertalians and specimens classically close to archaic *Homo sapiens*.

3) The second clade in its turn can be divided into two clades. One represents African human fossils associated to Kabwe, the other is made up of human fossils linked to the Ngadong series.

A discussion centred on the criteria which the phylogenetical hypothesis presented by this cladogram is put forward. In order not to anticipate the taxonomic conclusions based on the unknown clades, the computerised numbering has been kept within the following discussion in order to be able to verify and follow each stage – notably in the description of diagnoses (**Appendix 4**). The validity of a systematic numbering will be discussed later.

III.4.3 Reminder.

According to processing options put forward by the program PAUP, transformations are optimised differently. Thus, for a same topology, with different homoplasies, the reversions are maximised by the option <Acctran> when it is the convergences which are maximised by the option <Deltran>. Option <Minf> forces the changes at the node of the

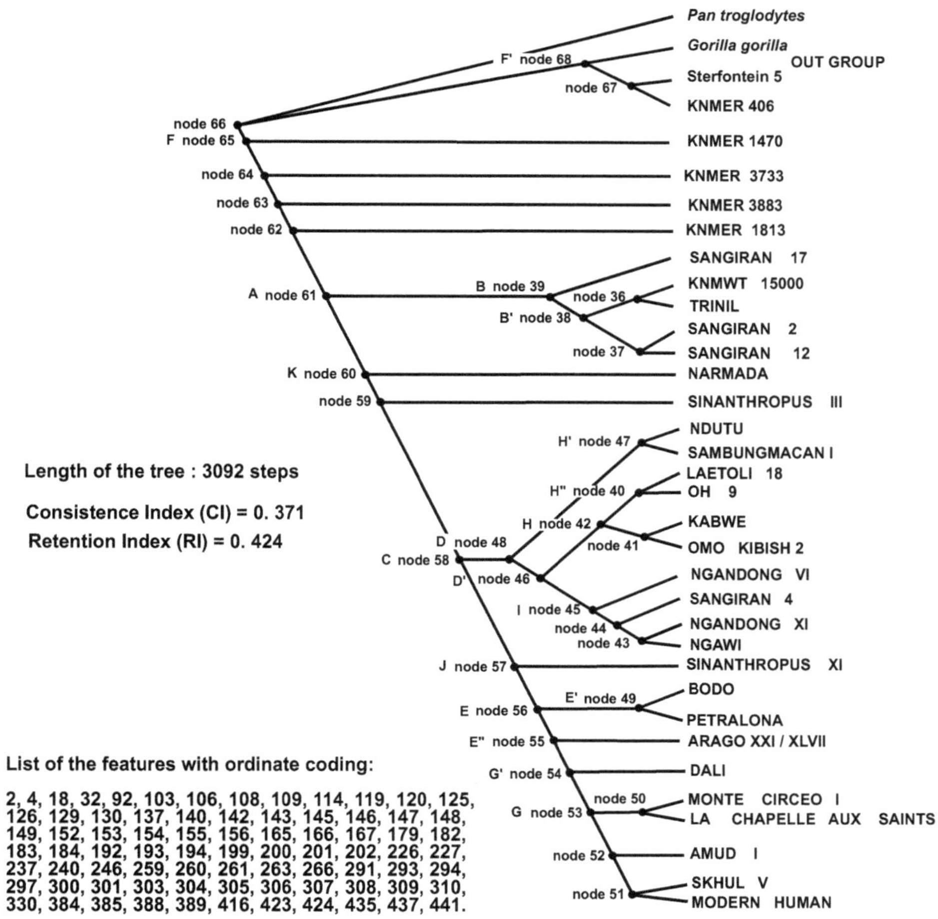

Figure 52

Cladogram resulting from the analysis of the 35 most complete taxa including 468 characters. Letters and numbers indicate the node discussed in the text.

base out-group / in-group and deplaces the transformations towards the terminal branches (Swoffford, 1985). This being the case, this option maximises the autapomorphies in order to avoid augmenting the number of synapomorphies to the internal nodes of the in-group. In this way, a variable number (dependent on the chosen option in the PAUP program: <DELTRAN>, <ACCTRAN> or <MINF>) states of characters support the different nodes. But these differences do not alter the topology of the cladogram.

As a general rule, it is preferable to use the option <Minf> as this is the one which implies the least ambiguity to the internal nodes. But, these nodes are the most interesting for the present study as it is supposed they permit the distinction between the clades « *Homo erectus* » and « *Homo sapiens* ».

III.4.4 The nodes of the Problem.

The characters said to be « fixed », or « non ambiguous » according to Tassy (1990), are those which present for the three options <ACCTRAN>, <DELTRAN> and <MINF> of the program PAUP (**Figure 53**). For a simpler reading of the commentaries, a dual labelling was retained for the different nodes; a letter designates the nodes but these also retain the computerised numbering list. In the nodal diagnosis, the state of the characters is indicated in inverted commas and an asterisk indicates that the character has been processed through an orderly method. The index of coherence of each character is indicated within square brackets. This index of coherence permits the judging of the robustness of the characteristic in the tree (*id est*: the accuracy of the information linked to the distribution of the different states in the tree).

III.4.4.1 Node A (node 61).

For each of the descriptions, the following order of items was established: the list of the states of fixed characteristics which support the node, followed by the remaining states of characteristics proper to the option «Minf». Node A is respectively supported by 58, 36 and 32 apomorphies for the options Acctran, Deltran and Minf. It is supported by 22 « fixed » characters, common to the three options.

Fixed Characters:

- Non – homoplastic Characters:

21(1) presence of a bregmatic eminence [CI=0.250], 112(1) marked rupture of the slope between the sub-temporal plane and the *tuberculum articulare* [CI=0.200], 115(1) existence of an ectoglenoidal crest [CI=0.250], 119(2)* average sized *processus postglenoidalis* [CI=0.182], 142(0)* SstB/SstG [CI=0.571], 143(0)* SstB/SstN [IC=1.000], 146(1)* SstB/SphSg [CI=0.444], 156(1)* SstB/FmtN [CI=0.714], 180(0) FtB/SstSg [CI=0.222], 188(3) FtSg/SphG [CI=0.375], 238(0) FtFmt/BN [CI=0.667], 240(1)* FtFmt/BSg [CI=0.600].

- Homoplastic Characters:

Certain states of characters appear by **convergence**:

27(3) existence of a sagittal keel on the whole of the front half of the *Bregma-Lambda* arc [CI=0.375] and 298(2)LSst/SstAst **with *Pan troglodytes*** [CI=0.400], 72(0) susmastoidian space closed behind [CI=0.143] and 291(1) *BL/SstAst [CI=0.214] **with *Pan troglodytes* and *Gorilla gorilla***, 45(2) existence of a *torus occipitalis* recti-

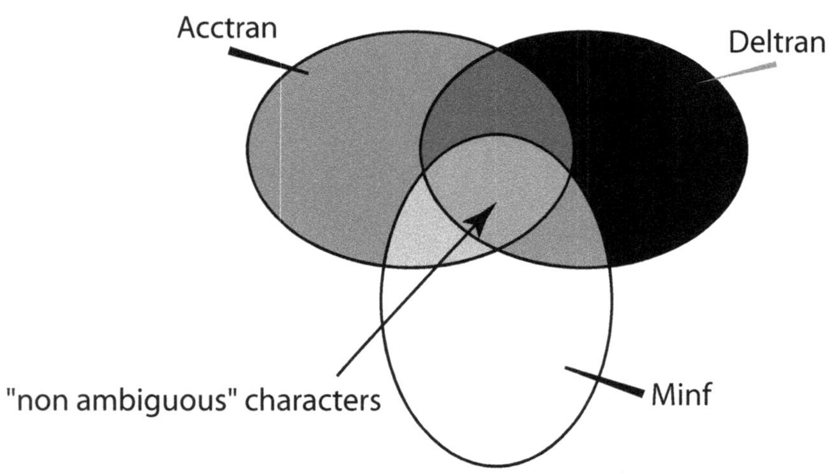

Figure 53

Number of non ambiguous characters for the different options (Acctran, Deltran, Minf)

linear in *norma occipitalis* **with the *Australopithecus*** [CI=0.250], 33(0) large temporal band on the parietal **with KNMER 1470** [CI=0.111], 294(0)* BL/KEn **with Sterkfontein 5 and KNMER 1470** [CI=0.500], 295(3) BL/KAst **with *Gorilla gorilla* and KNMER 1470** [CI=0.375].

Certain states of characters appear by **reversion**:

74(0) postero-anterior divergence of the susmastoidian space [CI=0.286], 191(2) FtB/FmtG [CI=0.500].

Characters which are proper to the Minf option:

- Non homoplastic Characters:

190(4) FtB/FmtSg [CI=0.500], 193(1)* FtB/FmtN [CI=0.571], 297(0)* LSst/SstEn [CI=0.333], 354(1) EnAu/AstMs [CI=0.286].

- Homoplastic Characters:

Certain states of characters appear by **convergence**:

257(4)StFmt/StFt **with *Pan troglodytes*** [CI=0.286], 70(1) forte *crista mastoidea* **with *Gorilla gorilla* and KNMER 406** [CI=0.125], 80(1) great thickness of the tympanal **with KNMER 406 and KNMER 3733** [CI=0.125], 87 (1) a sub temporal plane extends the articular of the gelnoidal cavity **with *Pan troglodytes*, Sterkfontein 5 and KNMER 1470** [CI=0.100].

Certain states of characters appear by **reversion**:

196(2) FmtB/FmtSg [CI=0.286], 235(0) FtFmt/SstFt [CI=0.500].

III.4.4.2 Node B (node 39).

Node B is respectively supported by 39, 18 and 17 apomorphies for the options Acctran, Deltran and Minf. It is supported by 10 « fixed » characters, common to the three options.

Fixed Characters:

- Non homoplastic Characters:

236(2) FtFmt/SphFt [CI=0.600], 342(2) EnPo/AuP [CI=0.400], 399(2) EnS/AuS [CI=0.600].

- Homoplastic Characters:

Certain states of characteristics appear by **convergence**:

120(1)* weak transversal stretching of the *processus postglenoidalis* [CI=0.231] and 397(0) EnS/AstA **with *Pan troglodytes*** [CI=0.200], 463(1) AstOp/MsOp **with *Gorilla gorilla*** [CI=0.333], 384(1)* EnPo/PPo **with *Gorilla gorilla* and KNMER 3883** [CI=0.286], 132(1) BSg **with *Pan troglodytes* and KNMER 3733** [CI=0.167], 130(1)* BFt **with the *Australopithecus*** [CI=0.200], 88(1) existence of a preglenoidal tuberculum **with KNMER 3883** [CI=0.167].

Characters proper to the Minf Option:

- Non-Homoplastic Characters:

13(1) large temporal band on the frontal [CI=0.125], 302(4) LSst/KAst [CI=0.667], 458(0) AstIn/OOp [CI=0.400], 460(1) AstO/OOp [CI=0.667].

- Homoplastic Characters:

Certain states of characteristics appear by **reversion**:

25(0) absence of supratrigonal depression [CI=0.111], 50(1) *torus occipitalis* continuous with the *linea temporalis superior* [CI=0.167], 331(0) Bast/LSph [CI=0.250].

III.4.4.3 Node C (node 58).

Node C is respectively supported by 42, 17 et 28 apomorphies for options Acctran, Deltran and Minf. It is supported by 12 « fixed » characters common to the three options.

Fixed Characters:

- Non- Homoplastic Characters:

116(2) entoglenoidal formation of smaller size than the edge of the sphenoidal edge [CI=0.333], 148(1)* SstB/SphN [CI=0.625], 167(1)* SphB/SstN [CI=0.364], 299(0) LSst/SphEn [CI=0.545].

- Homoplastic Characters:

Certain states of characters appear by **convergence**:

40(1) a depression highlights the *linea temporalis superior* in the infero-posterior of the parietal **with KNMER 406 and Sangiran 17** [CI=0.167], 172(4) SphB/FmtN **with KNMER 1813** [CI=0.462], 187(3) FtB/SphN **with KNMER 3733** [CI=0.625], 223(5) biFt/biFmt **with Sangiran 17** [CI=0.833], 292(3) BL/SphN **with Sterkfontein, KNMER**

1470 and the node B' (node 38) [CI=0.333], 395(0) EnS/AstMs **with Sterkfontein 5, KNMER 3883 and KNMWT 15000** [CI=0.286], 397(0) EnS/AstA **with *Pan troglodytes* and the node B (node 39)** [CI=0.200].

One state of character appeared by **reversion**:

357(2) EnAu/AstMs [CI=0.286].

Characters Proper to the Minf Option:

- Non-Homoplastic Characters:

147(1)* SstB/SphG [CI=0.556], 156(0)* SstB/FmtN [CI=0.714], 217(5) biSst/biFmo [CI=0.800], 239 (0) FtFmt/BG [CI=0.444].

- Homoplastic Characters:

Certain states of characters appear by **convergence**:

107(1) the entoglenoidal formation and the *tuberculum zygomaticum articulare* are in front in relation to the ensemble of the *tuberculum articulare* **with the node F (=node 64)** [CI=0.200], 127(0) biSst **with KNMER 3883** [CI=0.250], 166(1) *SphB/SstG **with 1813** [CI=0.235], 305(4) LSst/KAst **with the node 39** [CI=0.665], 284(3) BAst **with the node B' (=node 38)** [CI=0.300], 326(3) BAst/KEn **with Sterkfontein 5 and the node B' (=node 38)** [CI=0.250], 436(4) LAst/AstIn **with Sterkfontein 5** [CI=0.400].

Certain states of characteristics appear by **reversion**:

72(1) susmastoidian space open behind [CI=0.143], 269(2) SgG/SgN [CI=0.300], 328(0) LAst/LSst [CI=0.250].

III.4.4.4 Node D (node 48).

Node D is supported respectively by 31, 17 and 17 apomorphies for the options Acctran, Deltran and Minf. It is supported by 6 « fixed », characters common to the three options.

The Fixed Characters:

- Non-Homoplastic Character:

154(0)* SstB/FmtSg [CI=0.750].

- Non-Homoplastic Character:

Certain states of characteristics appear by **convergence**:

462(3) AstOp/OOp **with *Pan troglodytes*** [CI=0.500], 130(1) BFt **with *Australopithecus* and node B (=node 39)** [CI=0.200], 369(2) PoK/EnS **with *Australopithecus* and Narmada** [CI=0.429].

Certain states of characters appear by **reversion**:

73(2) large susmastoidian space [CI=0.200], 229(1) SphFt/SstFt [CI=0.400].

Characteris proper to the Minf option:

- Non-Homoplastic Character:

377(4) AstK/AstMs [CI=0.571].

- Non Homoplastic Characters:

Certain states of characteristics appear by **convergence**:

24(1)frontal lump offset medially **with Sinanthropus 3** [CI=0.300], 160(1) SphB/SphSg **with KNMER 1470 and Sinanthropus 3** [CI=0.250], 197(0) FmtB/FmtG **with KNMER 1813** [CI=0.333], 204(0) FmtB/SphSg **with KNMER 1813 and Sinanthropus 3** [CI=0.250], 205(1) FmtB/SphG **with Sterkfontein 5 and Sinanthropus 3** [CI=0.333], 210(1) FmtB/FtN **with Sangiran 17** [CI=0.333], 270(0) AuK/APo **with KNMER 3733 and Sinanthropus 3** [CI=0.429], 460(1) AstO/OOp **with node B (=node 39)** [CI=0.667], 461(2) AstO/MsOp **with *Pan troglodytes*** [CI=0.333].

III.4.4.5 Node E (node 56).

Node E is supported respectively by 37, 25 and 25 apomorphies for the options Acctran, Deltran and Minf. It is supported by 12 « fixed » characters, common to the three options.

Fixed Characters:

- Non-Homoplastic Character:

453(5) LO/AstO [CI=0.385].

- Homoplastic Characters:

Certain states of characteristics appear by **convergence**:

11(1) incomplete *torus supraorbitalis* **with KNMER 1813 and Ngandong 12** [CI=0.333], 65(0) upper edge of the temporal non-rectilinear

with KNMER 1470, Nodes H' and H" (=nodes47 and 41) [CI=0.200], 82(0) the surface of the back wall of the glenoidal cavity is planoconcave towards the front in *norma lateralis* **with the *Australopithecus*** [CI=0.167], 298(0)* LSst/KEn **with the nodes B' and I (=nodes 38 and 45)** [CI=0.267], 317(0) BAst/LEn **with *Gorilla gorilla*** [CI=0.500].

Certain states of characteristics appear by **reversion**:

5(0) absence of a depression at the lateral part of the *sulcus postorbitalis* [CI=0.111], 22(0) absence of coronal reinforcement [CI=0.200], 25(0) absence of supratrigonal depression [CI=0.111], 98(0) absence of a juxtamastoidian eminence [CI=0.167], 137(1)* GFt [CI=0.182], 200(0)* FmtB/SstG [CI=0.222].

Characters proper to the Minf option:

- Non-Homoplastic Characters:

59(0) the lateral walls of the sus-iniac fossa converge above [CI=0.667], 210(2) FmtB/FtN [CI=0.333], 224(4) biFt/biFmo [CI=0.800], 361(3) AuK/AstPo [IC=1.000], 428(0) biAu/AuBAu [CI=0.286].

- Homoplastic Characters:

Certain states of characters appear by **convergence**:

136(1) SgFmt **with *Pan troglodytes*, *Gorilla gorilla* and Sinanthropus 3** [CI=0.267], 194(0)* FtSg/FmtN [CI=0.625], 201(0)* FmtB/StN **with *Pan troglodytes*, Sterkfontein 5 and KNMER 1470** [CI=0.222], 431(2) LAst **with KNMER 1470 and Sinanthropus 3** [CI=0.222], 442(0) LAst/OOp **with *Gorilla gorilla* and node B' (=node 38)** [CI=0.500], 458(0) AstIn/OOp **with node B (=noeud 39)** [CI=0.400].

Certain states of characteristics appear by **reversion**:

102(0) the tympanic plate and the entoglenoidal formation situated close together [CI=0.222], 356(2) EnAu/AuMs [CI=0.250].

III.4.4.6 Node F (=node 64)

Node F is supported respectively by 60, 47 and 41 apomorphies for the Acctran, Deltran and Minf options. It is supported by 25 « fixed » characters, common to the three options.

The Fixed Characters:

- The Non-Homoplastic Characters:

3(1) sizeable relief on the upper edge of the *arcus superciliaris* [CI=0.167], 62(0) high temporal shell [CI=0.111], 143(3)* SstB/SstN [IC=1.000], 146(2)* SstB/SphSg [CI=0.444], 147(3)* SstB/SphG [CI=0.556], 148(3)* SstB/SphN [CI=0.625], 152(1)* SstB/FtG [IC=1.000], 153(2)* SstB/FtN [CI=0.571], 154(2)* SstB/FmtSg [CI=0.750], 156(3)* SstB/FmtN [CI=0.714], 177(3) FtB/FtN [CI=0.300], 183(0)* FtSg/SstG [CI=0.667], 191(1) FtB/FmtG [CI=0.500], 193(2)* FtB/FmtN [CI=0.571], 208(1) FmtB/FtSg [CI=0.429], 212(0) FmtSg/FtN [IC=1.000], 223(3) biFt/biFmt [CI=0.833], 229(0) SphFt/SstFt [CI=0.400], 247(0) SstFt/SstSph [CI=0.667], 260(2)* SstFmt/BG [IC=1.000], 305(0)* SstSph/LAst [CI=0.750], 324(1) BAst/SstAst [CI=0.333].

- Homoplastic Characters:

Certain states of characters appear by **convergence**:

330(0)* LAst/SstEn **with *Gorilla gorilla*** [CI=0.500], 121(1) the *processus postglenoidalis* overhangs laterally the extreme lateral of the tympanal [CI=0.286] and 280(1)FtFmo/FtFmt [CI=0.500] **with the *Australopithecus*,** 308(0)* SstSph/SstAst **with Sterkfontein 5** [CI=0.500].

Characters proper to the Minf option:

- Non Homoplastic Characters:

5(1) existence of a depression on the lateral part of the *sulcus postorbitalis* [CI=0.111], 107(1) the entoglenoidal formation and the *tuberculum zygomaticum anterior* are forward in relation to the ensemble of the *tuberculum articulare* [CI=0.200], 124(1) biMnf [CI=0.429], 151(4) SstB/FtSg [CI=0.429], 211(0) FmtSg/FtG [IC=1.000], 214(3) biSst/biSph [CI=0.800], 216(3) biSst/biFmt [CI=0.667], 220(5) biSph/biFmt [IC=1.000], 230(0) SphFt/BN [CI=0.444], 233(0) SphFt/GN [CI=0.714], 235(2) FtFmt/SstFt [CI=0.500], 323(2) BAst/SstEn [CI=0.667], 433(5) LAst/LIn [CI=0.250].

- The Homoplastic Character:

A single state of character appears by **convergence**:

176(3) FtG/FtB **with *Gorilla gorilla*** [CI=0.400].

III.4.4.7 Node G (=node 53)

Node G is supported respectively by apomorphies 38, 22 et 21 for the options Acctran, Deltran and Minf. It is supported by 13 « fixed » characters, common to the three options.

The Fixed Characters:

- The Non-Homoplastic Characters:

98(0) absence of a juxtamatodinian eminence of Rouvière [CI=0.222].

- Non Homoplastic Characters:

Certain states of characters appear by **convergence**:

51(0) absence of the *sulcus occipitalis* **with *Gorilla gorilla*, KNMER 1470, KNMER 3733, KNMWT 15000 and node D' (=node 46)** [CI=0.182], 55(0) absence of a *tuberculum linearum* **with *Australopithecus*, Sangiran 2, KNMWT 15000 and Sinanthropus 11** [CI=0.182], 89(0) tympanal touching the mastoidal apophyse **with *Australopithecus*, KNMER 3883, Sangiran 17, Narmada and node H (=node 42)** [CI=0.286], 105(1) rear entoglenoidal formation at the *tuberculum zygomaticum anterior* **with Sterkfontein 5, Narmada and node H' (=node 47)** [CI=0.222], 140(2)* NFt **with KNMER 1813 and Ngandong7** [CI=0.200], 218(3) biSt/biMnf **with node 42** [CI=0.400], 455(4) LO/InMs **with KNMER 1813 and Ndutu** [CI=0.429], 456(5) LO/OMs **with KNMER 1813, Ngawi and Ndutu** [CI=0.500].

Certain states of characters appear by **reversion**:

11(2) *torus supraorbitalis* is complete [CI=0.333], 39(0) absence of the *tuber angularis* [CI=0.143], 40(0) absence of depression underlining the *linea temporalis superior* in the lower, infero-posterior part of the parietal [CI=0.167], 60(0) absence of the retromastoidan processus [CI=0.100].

Characters which are proper to the Minf option:

- Homoplastic Characters:

Certain states of characters appear by **convergence**:

166(0)* SphB/StG **with Pan troglodytes, KNMER 3733, Ngandong 7 and Arago** [CI=0.235], 182(0)* FtB/StN **with Arago** [CI=0.444], 434(1) LAst/LO **with Ndutu and node I (=node 45)** [CI=0.300].

Certain states of characteristics appear by **reversion**:

80(0) tympanal hardly thick [CI=0.125], 95(0) *incisura mastoidea* short [CI=0.167], 113(1) *fossa mandibularis* ample [CI=0.231], 136(0) SgFmt [CI=0.267].

III.4.4.8 Node H (=node 42)

Node H is supported respectively by apomorphies 65, 22 and 27 for the options Acctran, Deltran and Minf. It is supported by « fixed » 12 characters common to the three options.

The Fixed Characters:

- The Non-Homoplastic Character:

374(6) AstK/EnMs [CI=0.667].

- Homoplastic Characters:

Certain states of characters appear by **convergence**:

1(1) medial fusion of the two *arcus superciliaris* **with *Pan troglodytes*, *Gorilla gorilla*, KNMER 3883 and node E' (=node 49)** [CI=0.167], 28(2) existence of an obeliac depression **with KNMER 406** [CI=0.167], 89(0) tympanal alongside the mastoidal apophyse **with Australopithecus, KNMER 3883 and Narmada** [CI=0.286], 388(2)* EnMs/AstAu **with *Pan troglodytes*, KNMER 1813 and Sinanthropus 11** [CI=0.429], 391(0) EnMs/AstS **with *Pan troglodytes* and Sinanthropus 3** [CI=0.333].

Certain states of characteristics appear by **reversion**:

52(0) absence of lateral, occipital depression [CI=0.125], 90(1) long *fossa mandibularis* [CI=0.250], 173(2) SphSg/FmtG [CI=0.333], 174(1) SphSg/FmtN [CI=0.500], 194(1)* FtSg/FmtN [CI=0.625].

Character proper to the Minf option:

- Non Homoplastic Characters:

178(2) FtSg/FtG [CI=0.556], 218(3) biSst/biMnf [CI=0.400], 268(0) SgG/BG [CI=0.300], 351(1) EnAu/EnMs [CI=0.400].

- Homoplastic Characters:

Certain states of characters appear by **convergence**:

108(1)* very widely curved radius of the *tuberculum articulare* in *norma lateralis* **with *Australopithecus*** [CI=0.133], 294(0)* BL/SstAst **with KNMWT 15000 and Sangiran 2** [CI=0.214], 431(2) LAst **with Sinanthropus 3, Sambungmachan and node E (=node 56)** [CI=0.222], 432(1) OpIn **with KNMER 3883 and Ndutu** [CI=0.500], 452(6) LO/AstIn **with Ndutu** [CI=0.417].

Certain states of characters appear by **reversion**:

24(0) absence of well individualised frontal lump [CI=0.300], 113(1) wide *fossa mandibularis* [CI=0.231], 172(2) SphB/FmtN [CI=0.462], 185(2) FtB/SstN [CI=0.333], 193(2)* FtB/FmtN [CI=0.571], 196(1) FmtB/FmtSg [CI=0.286].

III.4.4.9 Node I (=node45)

Node I is respectively supported by amorphies 43, 30 and 30 for the options Acctran, Deltran and Minf. It is supported by 12 « fixed » characters, common to the three options.

Fixed Characters:

- Non-Homoplastic Character:

403(2) EnS/APo [CI=0.429].

- Homoplastic Characters:

Certain states of characters appear by **convergence**:

171(4) SphB/FmtG **with *Pan troglodytes*** [CI=0.545], 230(1) SphFt/BN **with KNMER 3733 and Sinanthropus 3** [CI=0.444], 442(2) LAst/OOp **with Sterkfontein 5** [CI=0.500], 443(3) LAst/MsOp **avec KNMER 1813** [CI=0.500], 468(0) LIn/LOp **with node F' (=node 68) and KNMER 3883** [CI=0.250], 284(1) BAst **with KNMER 3733** [CI=0.300].

Certain states of characters appear by **reversion**:

62(1) low temporal squama [CI=0.111], 200(0)* FmtB/StG [CI=0.222], 245(0) SstFt/SphFt [CI=0.200], 435(2)* LAst/LMs [CI=0.222], 459(1) AstIn/MsOp [CI=0.600].

Characters proper to the Minf option:

- Homoplastic Characters:

Certain states of characters appear by **convergence**:

106(2)* small entoglenoidal formation and *tuberculum zygomaticum anterior* in relation to the *tuberculum articulare* **with KNMER 1470, KNMER 1813 and Sinanthropus 3** [CI=0.154], 109(2)* *tuberculum articulare* in *norma frontalis* has a round profile **with KNMER 1470 et Narmada** [CI=0.154], 118(4) the entoglenoidal formation does not stretch out anter-posteriorally **with KNMER 406, KNMER 3883 and node J (node 57)** [CI=0.333], 178(4) FtSg/FtG **with node G' (=node 54)** [CI=0.556], 194(0)* FtSg/FmtN **with node E (=node 56)** [CI=0.625], 197(3) FmtB/FmtG **with KNMER 1470** [CI=0.333], 218(2) biSst/biMnf **with KNMER 3883 and node E" (=node 55)** [CI=0.400], 231(1) SphFt/BG **with KNMER 3733, Sinanthropus 3 and Sinanthropus 11** [CI=0.375], 237(2)* FtFmt/SstSph **with *Australopithecus*, KNMER 3733, KNMER 3883, KNMWT 15000 and Sinanthropus 3** [CI=0.214], 269(0) SgG/SgN **with Laétoli H18** [CI=0.300], 301(0)* LSt/KEn **with nodesB' and E (=nodes 38 and 56)** [CI=0.267], 434(1) LAst/LO **with Sangiran 17 and nodeG (node 53)** [CI=0.300], 458(1) AstIn/OOp **with *Gorilla gorilla* and Sterkfontein 5** [CI=0.400].

Certain states of characters appear by **reversion**:

3(0) absence of relief of the upper edge of *arcus superciliaris* [CI=0.167], 33(1) narrow temporal band on the parietal [CI=0.111], 124(1) biMnf [CI=0.429], 265(2) SstFmt/SgG [CI=0.308], 380(2) AstK/AstMs [CI=0.571].

III.4.4.10 Node J (=node 57)

Node J is supported respectively by amorphies 49, 40 and 38 apomorphies for the Acctran, Deltran and Minf options. It is supported by 31 « fixed » characters, common to the three options.

Fixed Characters:

- Non Homoplastic Characters:

348(0) PoK/AuS [IC=1,000], 349(0) PoK/AuA [CI=0.400], 350(0) PoK/APo [IC=1,000], 358(0) AuK/EnS [CI=0.667], 360(2) AuK/AstAu [IC=1,000], 363(3) AuK/AuMs [CI=0.500], 383(0) AstK/APo [CI=0.800], 366(2) AuK/APo [IC=1,000], 380(0) AstK/AuA [IC=1,000].

- Homoplastic Characters:

Certain states of characters appear by **convergence**:

285(2) LSst **with Sterkfontein 5, KNMER 3883, KNMWT 15000 and KNMER 1813** [CI=0.333], 332(1) EnAu/PoMs **avec Narmada** [CI=0.500], 333(1) EnAu/SPo **with KNMER 1470 and Sangiran 17** [CI=0.429], 337(1) EnPo/AstAu **with KNMER 406 and Sangiran 17** [CI=0.500], 338(0) EnPo/AstPo [CI=0.571] and 342(0) EnPo/AuP [CI=0.400] **with KNMER 1813**, 345(1) PoK/AstAu **with *Australopithecus*** [CI=0.500], 351(2) EnAu/EnMs [CI=0.400] and 365(1) AuK/SPo [CI=0.667] **with *Pan troglodytes***, 357(1) EnAu/AuS **with *Gorilla gorilla* and Ndutu** [CI=0.500], 372(0) AstK/EnAu **with *Pan troglodytes* and *Gorilla gorilla*** [CI=0.333], 373(0) AstK/EnPo [CI=0.500], 374(0) AstK/EnMs [CI=0.667], 378(0) AstK/AuS [CI=0.500], 379(0) AstK/AuP [CI=0.667] and 381(1) AstK/SPo [CI=0.600] **with *Gorilla gorilla***, 375(0) AstK/EnS **with *Pan troglodytes*, *Gorilla gorilla* and OH9** [CI=0.250], 376(0) AstK/EnA **with node F' (=node 68)** [CI=0.600], 382(0) AstK/PPo **with *Gorilla gorilla* and Narmada** [CI=0.500].

Certain states of characters appear by **reversion**:

336(1) EnAu/PMs [CI=0.400], 353(1) EnAu/AstPo [CI=0.444], 355(2) EnAu/AstA [CI=0.400].

Characters proper to the Minf Option:

- Non Homoplastic Characters:

190(5) FtB/FmtSg [CI=0.500], 377(1) AstK/AstMs [CI=0.571].

- Homoplastic Characters:

Certain states of characters appear by **convergence**:

49(0)*torus occipitalis* has no direct link with the *crista mastoidea* **with Sterkfontein 5, KNMER 3883 and Sinanthropus 3** [CI=0.167], 118(4) the entoglenoidal formation does not extend antero-posteriorally **with KNMER 406 and node I (=node 45)** [CI=0.333], 265(2) SgN/BSg **with *Gorilla gorilla* and node I (=node 45)** [CI=0.308], 362(1) AuK/AstMs **with KNMER 1813 and Sambungmachan** [CI=0.500], 461(0) AstO/MsOp **with KNMER 1813 and Sangiran 12** [CI=0.333].

III.4.4.11 Node K (=node 59)

Node K is respectively supported by amorphies 18, 30 and 9 apomorphies for the Acctran, Deltran and Minf options. It is supported by 8 « fixed » characters common to the three options.

Fixed Characters:

- Homoplastic Characters:

53(2) presence of an occipital "corner-shaped" chignon[CI=0.154], 114(2)* antero-posterior cavity of the *fossa mandibularis* which is pinched [CI=0.200].

- Non Homoplastic Characters:

Certain states of characters appear by **convergence**:

14(1) the temporal band projects over the frontal bone [CI=0.111] and 15(1) existence of a tuberculum at the *frontotemporale* [CI=0.111] **with KNMER 3733 and KNMER 3883**, 22(1) existence of a coronal reinforcement on the upper part of the squama **with *Pan troglodytes*, Trinil 2 and Sangiran 2** [CI=0.200], 98(2) large juxtamastoidian eminence of Rouvière **with KNMER 1813** [CI=0.222], 113(2) *fossa mandibularis* is medium-sized **with KNMER 1470** [CI=0.231].

One single state of characters appears by **reversion**:

104(2) entoglenoidal formation practically the same size as the *tuberculum zygomaticum anterior* [CI=0.300].

One character is proper to the Minf option:

- Non Homoplastic Character:

106(1)* the entoglenoidal formation and the *tuberculum zygomaticum anterior* are medium-sized in relation to the ensemble of the *tuberculum articulare* [CI=0.154].

All of the changes of states of characters are to be found in **Appendix 3**. After the description of the characters and of their states at the different nodes of the cladogram, it was discovered that it was especially the metrical data which supported the clades. This renders the reporting of all diagnosis somewhat delicate. In addition, the tree presented is very fragile (CI = 0.371 and RI = 0.424). The computer program PAUP permits a weighting of the characters using the consistence index obtained on the cladogram for each of the characters as a starting point. This procedure permits the reinforcement of the results. In carrying out such weighting on the characters, an identical topological tree is obtained with better indices (CI=0.488 and RI=0.516). The result is not so bad if one bears in mind that it was specimens which were studied.

III.5. Phylogenetical Propositions.

Preliminary Remarks: In the descriptions of nodes, the consistence index attached to each character indicates the reliability of the character for the whole of the tree and not only the node under discussion. In this way, a weak consistence index signifies that the character does not permit us to draw a robust phylogenetical conclusion in the tree. The taking into consideration of non ambiguous synomorphes allows the conception of a phylogram based on groups of the most reliable specimens thus identified (**Figure 54**).

In order to transcribe the cladogram in readable terms and with the aim in mind to express a classification, the position of type-specimens determined the name of the group in which they found themselves following analysis. The relative reliability of the different nodes was also an element taken into account in order to judge the level at which it was possible to assume a specific level for any given clade (**Figure 55**).

Figure 52 at the centre of the genus *Homo* which makes up the «in-group» of our study, it is possible acknowledge a grade «*Homo habilis*» corresponding to the paraphyletic sequence situated between nodes O and A (=nodes 65 and 62). The specimens KNMER 1470 and KNMER 1813 have long been held to be *Homo habilis* by numerous authors (Howell, 1978; White *et al.*, 1981; Tobias, 1983). In this perspective, however, the results of our analysis are in contradiction with the classical affiliation of the specimens KNMER 3733 and KNMER 3883 to African *Homo erectus* as they are situated at the base of the cladogram and clearly outside of the clade which contains the specimen type of *Homo erectus*. One solution consists in taking into account the affiliation of KNMER 1470 to *Homo rudolfensis* Alexeev 1986 and that of KNMER 1813 to *Homo ergaster* (Groves and Mazak), 1975. In addition, Stringer (1986) includes KNMER 1813 in a second group of *Homo habilis* which Wood (1992a) attributes to *Homo ergaster*. When one links these two hypotheses of classification together, we note that in 1989, Groves made the specimen KNMER 1813, the paratype of *Homo*

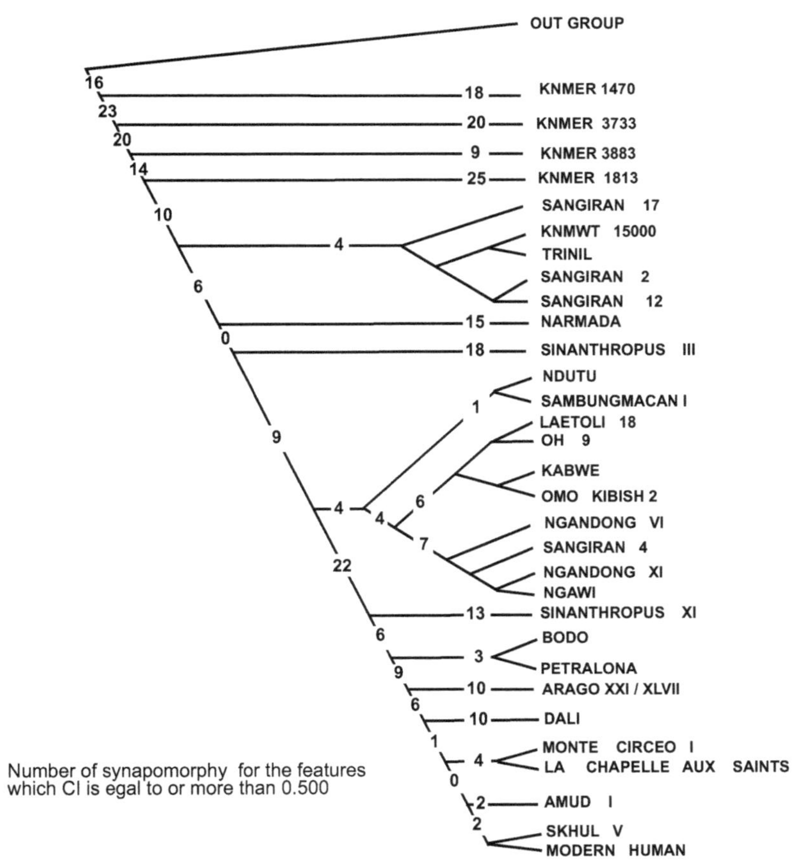

Figure 54

Cladogram resulting from the analysis of the 35 most complete taxa and with 468 characters showing the most robust synapomorphies.

ergaster; This species being defined, originally, through the single mandible of KNMER 992. According to Clarke (1990) this labelling is in conformity with the rules of zoological classification. Wood (1992a) also recommends the useage of this label after having been a supporter of the non-existence of *Homo habilis*as a species. More exactly, he did not make any distinction between specimens affililated to *Homo habilis* and those affiliated to *Homo erectus* in Africa (Wood, 1984). Following Wood's (1984, 1992a and 1994) advice to divide the grade *Homo habilis* into several species by considering on the one hand the affiliation of KNMER 1470 to *Homo rudolfensis* and on the other hand that of KNMER 1813 to *Homo ergaster*, then it is convenient to adopt labelling dedicated to specimens KNMER 3733 and KNMER 3883. In fact, the latter appear to have the same level and the number of robust characteristics which support them are amongst the highest number on the tree. Thus, each taxon from the paraphyletic sequence will be labelled by a term of an identical rank: that of « species ».

The clade supported by node B (=node 39) includes the specimen Trinil 2 which is the type of *Homo erectus* (Dubois, 1893). This clade would therefore be linked to this species. This is not a far-fetched decision as the ensemble of the specimens of which this clade is made up has been attributed to *Homo erectus* by different authors. With reference to the « weighted » cladogram (**Figure 54**), this clade is very close to the rootstock *id est*: it is not very far from clade A =node 61). This suggests the status of plesion and not of the root group (see further pages) for *Homo erectus*; which would confirm the huge difficulty in its definition by so many different authors.

Clade J (=node 57) is associated almost exclusively with *Homo sapiens*: «archaic» *Homo sapiens*, *Homo sapiens daliensis* Wu, 1981, *Homo sapiens sapiens*, *Homo sapiens neanderthalensis* (King, 1864) as well as Sinanthropus 11. This clade could therefore be affiliated to *Homo sapiens*.

Clade H (=node 42) regroups African specimens around the Kabwe 1 specimen which is the holotype of *Homo rhodesiensis* (Woodward, 1921).

Clade I (=node 45) belonging to the fraternal group above, it regroups together Indonesian fossils exclusively and contains the holotype of *Homo (Javanthropus) soloensis* Openoorth, 1932.

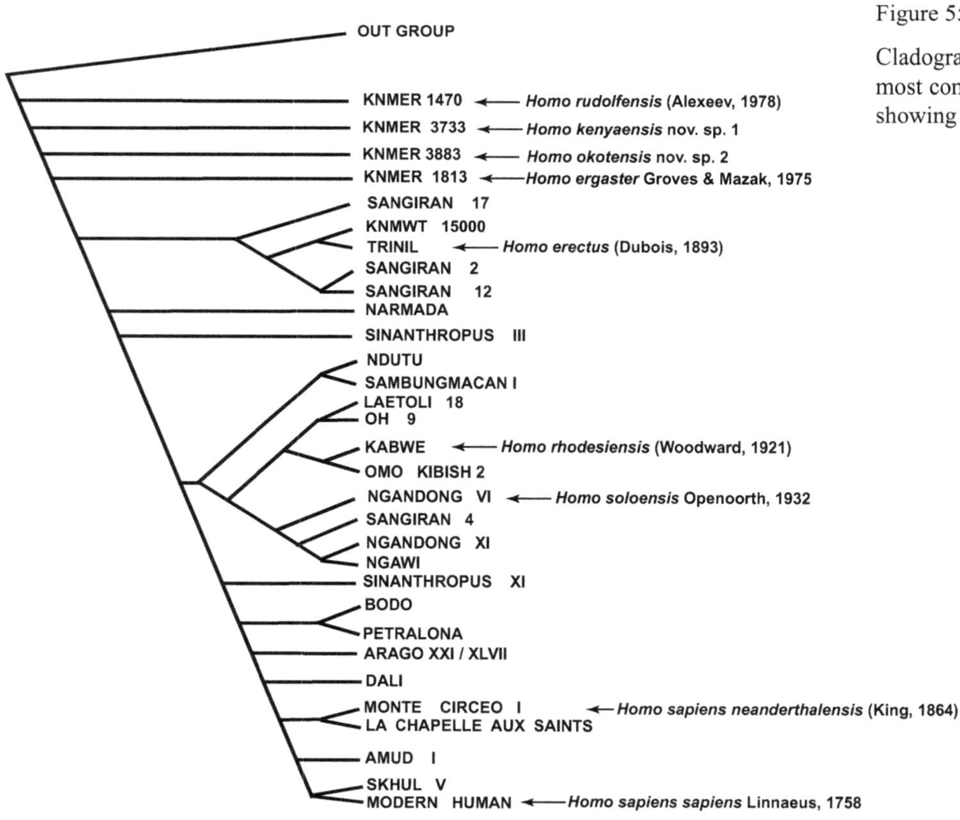

Cladogram resulting from the analysis done with 35 taxa and 468 features.

Figure 55

Cladogram resulting from the analysis of the 35 most complete taxa and with 468 characters showing the type-specimens and holotypes.

These two clades and their related clade (clade H' = node 47) are to be found in clade D (= node 48); But, in the hierarchy of the cladogram, clade D has the same status as clade J (=node 57) which was described as being the one which contained almost only *Homo sapiens*. The problem posed here is that of conformity between taxonomic ranks and clades at the same level. This is an illustration of the problem arising from the harmonisation of the classification and phylogeny. Several arbitrary solutions are possible. One way of resolving the non congruency of the hierarchy of the cladogram and the taxonomic affiliation consists of considering, as suggest most authors, the *Homo sapiens* hypodigm proposed by Howell (1978). In this way, the *calvaria* of Kabwe which was the type of *Homo sapiens rhodesiensis* (Woodward, 1921) suggests altering the position of the clade *Homo sapiens*. Regarding the Ngandong specimens, *id est*: *Homo (Javanthropus) soloensis* Openoorth, 1932 they can also conform with this altering of position. In fact, in 1940, Dubois used the trinôme *Homo sapiens soloensis* in connection with the Ngandong series. This denominating was later adopted at the Wartenstein Symposium in 1962 (Campbell, 1963). In 1980, Howells stressed that the usage of this trinome (which *de facto* excludes the Ngandong specimens from the species *Homo erectus*) had already been criticised by Jacob in 1978. As early as 1976, Jacob brought closer together the Ngandong series due to the association between the Sinathropus and the Pithecanthropus made by Weidenreich (1943). In 1980, Santa Luca confirmed this opinion and, as a result, the trinome *Homo sapiens soloensis* has rarely been used. It should however be noted that Stringer (1987) also excludes the Ngandong series from the group of Asian *Homo erectus* which he used in his cladistic analysis. Regarding this study, the assignation of the *calvaria* of Kabwe and of the Ngandong specimens to the species *Homo sapiens* is the simplest solution and in this case, clade C (=node 58) corresponding, then to *Homo sapiens*.

The case of the Ndutu and Sambungmachan specimens can be explained if one remembers that the quantity of information (the number of characters it was possible to observe) is not very high. It's a little as though, through lack of sufficient information, these specimens found themselves allocated to clade D (=node 48) as a result of their not being able to be allocated to clade H (=node 42) or to clade I (=node 45). One element to be stressed in the choice put forward is that node J (=node 57) is one of the strongest in the cladogram. If one takes note of the robustness of clade D (=node 48) on the phylogramme, one sees that the latter is close to the root (*id est*: node C (=node 58).

In this hypothesis of taxonomic cutting up: one clade corresponding to *Homo erectus* at node A (=node 39) and a clade corresponding to *Homo sapiens* at node C (=node 58), the case of Sinathropus is particular. In fact, if the Sinanthropus 11 specimen can be included in the clade corresponding to *Homo sapiens* (*id est*: node C = node 58), the specimen Sinanthropus 3 is excluded. However, it has never been said, – it would seem – that a taxonomic difference existed within the Zhoukoudian specimens. The few morphological differences are attributed to individual variability. Thus, Sinanthropus 3 has an intermediate phylogenetical position between the clade corresponding to *Homo erectus* and the clade corresponding to *Homo sapiens*. This phylogenetical position is not far-fetched if, according to stratopheneticians, the only chronological dimension of the Zhoukoudian site is taken into consideration. In fact, specimen Sinanthropus 3 is the oldest *calvarium* on the site – as stresses Pope (1992).

III. 5. 1 Assessment and Diagnosis of *Homo erectus*

Clade B corresponds to *Homo erectus sensu stricto* (**Figure 52**) is respectively supported by amorphies 39, 18 and 17 apomorphies for the Acctran, Deltran and Minf options. Amongst these characters 10 are « fixed », *id est*: common to the three options. It is therefore proposed that a much clearer list of characteristics which define *Homo erectus* in the cladogram be put forward. The diagnosis of the species according to the hennigien concept of the species is the list of those synapomorphies which characterise the clade under consideration.

Remark: the metrical characteristics make up the larger part of the characteristics adopted at the node of this clade. In order to express some of these metrical characteristics, the parameters need to be « graduated » with a more global scale created for the ensemble of the hominids studied..

Diagnosis of *Homo erectus*

The frontal bone:

236(2) FtFmt is very slightly superior to SphFt, which corresponds to an average expression of the character within hominids; this corresponds to a post-orbitary constriction such that the triangle (Fmt, Ft, Sph) be isoceles [CI=0.600]; 32(1) BSg, the sagittal profile of the frontal shell is somewhat pro-eminent [CI=0.167], 130(1)* BFt, the frontal shell is somewhat proeminent in an oblique section [CI=0.200], 13(1) large temporal band on the frontal [CI=0.125], 25(0) absence of a supratrigonal depression [CI=0.111].

The parietal bone:

302(4) LSst/KAst is very weak and even LSst<Kast, that is to say that the average upper length of the parietal is less than the lower length of the parietal, *id est*: around 90%[CI=0.667], 331(0) BAst/LSph is very weak, BAst only represents two thirds of LSph, *id est*: the parietal is somewhat squat [CI=0.250].

The temporal bone:

342(2) EnPo represents more than 100% of the AuP [CI=0.400], 399(2) EnS is around 1.6 times the dimension of AuS, that is to say that the temporal shell is rather lengthened and flattened [CI=0.600],120(1)* weak transversal stretching of the *processus postglenoidalis* [CI=0.231], 397(0) EnS/AstA is small, < to 0,77, *id est*: the posterosuperior edge of the temporal shell is short in relation to the ensemble of the temporal [CI=0.200], 84(1)* EnPo/PPo is average and there is a tendency to confuse the *entomion* with the most posterior point of the temporal shell [CI=0.286], 88(1) existence of a preglenoidal tuberculum [CI=0.167].

The occipital bone:

463(1) AstOp/MsOp rather small, around 0,74, the mastoidal apophyse does not project far [CI=0.333], 458(0) AstIn/OOp weak<1,18 [CI=0.400], 460(1) AstO/OOp weak, *id est*: the lower part of the occipital is quite small in comparison with the posterior development of the occipital. [CI=0.667], 50(1) *torus occipitalis* continuous with the *linea temporalis superior* [CI=0.167].

It should be noted that only 4 of the 17 characters present a consistence index above or equal to 0.500. If the phylogenetical links between each operational taxonomic unit (every specimen in the current case) are readable on the cladogram resulting from the analysis, it is a question, after reflection, of deciding which is the taxonomic status of the clades for which the constitution has been discussed in paragraphs above. Specific binomes or trinomes were used in order to name the clades which could be identified on the cladogram resulting from the analysis. This aspect was designed to be illustrative but it is now a question of discussing the taxonomic significance of the clades. In fact, in the present case it is notably a question of knowing if the distinction noted is of the order of species or of sub-species. Several principles can be applied in fact on the cladogram presented (**Figure 55**).

IV. Discussion.

Only one tree linking together the different evolutionary units – past or present – exits: this is the evolutionary tree which explains the history of descendence. Phylogenetical reconstruction has as its aim to be able to distinguish this « real » tree from within the ensemble of trees which we can theoretically construct through different evolutionary units observed (Darlu et Tassy, 1993). If, in conclusion of an analysis of characters, the information contained in the cladogram is sufficient we are reduced to creating a classification for reasons of communication and understanding. Several cladistic classifications using a single cladogram as a starting point are possible (Tassy, 1988). In fact, the taxonomic status of each phylogenitical group can be discussed according to several concepts.

IV.1. Concepts of the species.

In accordance with Kimbel (1991), the pattern of the phylogensis of hominids is considered as the basis without which it would be impossible to find a significant solution to questions concerning *tempo* and the evolutionary mode of hominids. On the other hand, contrary to Kimbel, it is not true that an absence of concept of the species causes a problem when studying extinct hominids. As we have just seen, it is possible to establish phylogenetical links between terminal taxons which are specimens. Certainly the phylogeny put forward reposes on a low index of coherence of the tree. Certainly, the phylogenetical signal is unclear as a result of the strong homoplasy. However, the phylogenetical results presented result in a unique tree, well-supported by anthropological studies. The exact determination of the aims of the study is an essential aspect in all research. The species is both a category and an object of study for which the phylogenetical attributes do not enter into the biological definition (Matile *et al.*, 1987). If the majority of biologists accept the concept of evolution, in reality it is impossible to observe an evolutionary species or to carry out experiments on it. At best, we have at our disposition only a few fragments of populations (this is even more true in the domain of palaeontology). It is not by hypothesis that these fragments become representative of a species. The problem of species relies very much on what is held to be an objective evolutionary unit. In reality, a species is only a mental construction and because of this diverse definitions follow. Regarding this subject, Kimbel (1991) noted that palaeonthrapologists – like palaeontologists in general – have a tendency to consider a species as a fundamental evolutionary unit. This said, this author notes that palaeoanthropologists are more inclined to acknowledge species in Early Pleistocene than in Later of Middle Pleistocene. In these more recent periods, palaeontologists admit more readily that different units acknowledged as being at the heart of the genus *Homo* are situated at an infraspecific rank. The formulation of the different concepts of the species and of the sub-species will not be discussed here. Readers are invited to understand the difference between typological, morphological, biological and evolutionary concepts put forward in the works of Rosen (1984), Mayr (1957 and 1969), Eldredge and Gould (1972), Simpson (1961), Wiley (1978) and Hennig (1966). On the other hand, it is useful to point out a few notions in relation to the concept of speciation.

IV.2. Speciation.

By definition, speciation is linked to the genetic isolation of the new species, (or species in the plural); this is an irresversible process. The study of variation at the heart of a population or between actually or potentially infertile populations is that of a process, by definition, irreversible. Lacking historic dimensions, **this process cannot therefore come from phylogenetical systematics.** This point is the stumbling block of this study as, although going from the principle which refuses interfertility *a priori*, and in the impossibility of proving its existence amongst the fossils, portions of the cladogram reveal relationships between individuals which are potentially fertile. Cladogenesis is a process by which an ancestral species evolve into two sister-species. Anagenitical speciation (= phylogenetical speciation) is generally accepted as being the appearance of a new species through slow evolution from the mother species, without subdivision and without this limiting itself between the two « species » being able to be determined. Phylogenetical systematics does not recognise the status of speciation for such a continuous transformation. In fact, according to Simpson (1961), there is no non-arbirary means to subdivise a continuous line. Gradualism and anagenetism imply an unclear limit to speciation. The species itself is an agregate of populations which, of very differing sizes according to the different species, correspond at best to the sub-species of zoological labelling. In general, populations are not genetically isolated and their relationships are more often reticular-shaped and not only hierarchical. To the extent where the population is the evoluative unit (the object of evolutionary mechanisms), the species is a taxon like another. Infra-specific relationships are called **tokogenic** by Hennig. Relationships between terminal taxons are not then necessarily hierarchical. From a more general point of view, at the level of species, Tassy (1991) stresses the advantages and the limitations of the use of the cladistic method. Thus, if the concept of phylogeny is conceived as being the history of ruptures in genetical « pools » and the apparition of new « pools » isolated between populations, then it would be better to speak of intraspecific **tokogeny**. The aim of a phylogenetical system is to put forward hypotheses between units seen as recognisable ensembles, isolated and closed. These units can be species but not especially identical toi evolutionary units. **The unit on which evolutionary mechanisms express themselves is the population eventually reduced to a few individuals and not necessarily isolated from other populations of the species.** The smallest closed groups do not necessarily correspond to the species in its accepted biological form. The analysis carried out in this study adopts terminal taxa of a specific or infra-specific level. Thus, contrary to what Wolpoff *et al.* (1988) put forward, cladistics can apply even when the comparison deals with infraspecific levels. The first stage consists of determining those specimens which gather together in evolutionary units and for which the tokogenticity can be hierarchical to a certain level. The problem arises in the determining of this level. But it is difficult to point out the level where the phylogenetical relationships are actually cross-linked. In palaeontology, the less arbitrary research strategy is the one which considers each specimen as a unique object. An object which is a reflection of a tangible reality as it is concrete and not theoretical as is a species or any other category. The problem of recognising the taxonomic status is the same as that of the composition of the hypodigm of species. By this, we mean that a level exists where limits are defined arbitrarily. The only recommendation concerning the selection of these arbitrary limits resides in the fact that the criteria for recommending their recognition must be clearly defined, thereby facilitating discussion.

Within a phylogenetical perspective and a historical context, a species is the element situated between the two speciations in the phylogenetical tree. The « time-bio-species » of Bonde (1981) is a genetic « pool » historically bounded by two ruptures. This definition resolves the dilemma between an achronic biological species and a chronological species to the extent where the two definitions satisfy that of the genetically closed « pool ». But in this case the chronological species only answers to the hypothesis of succession of systematic speciations. The particular status of allopatric populations led Nelson and Platnick (1981) to consider a species as being the smallest identifiable ensemble of organisms capable of self reproduction and having an ensemble of unique characters. When the sub-species is included as being a species and an evolutionary unit, populations acquire the formal status of terminal taxa. In the final analysis, what counts is that the distinctive characters are fixed in the terminal taxa. Even if these characters are polymorphic, such a requirement is acceptable as soon as an evolutionary explanation of this polymorphism is possible. According to Darlu and Tassy (1993), a hierarchy can appear from the study of variations of the polymorphism. The existing gap between each o the nodes of the cladogram determines – according to Hennig's concept – a defined evolutionary unit which can be a species. For this study, we employed a nomenclature which takes up the specific or infraspecific terms from anthropological literature. Even if this proposition is not applicable in every case, it must not be forgotten that what counts is the history of the parental links. Theoretically, at the extreme, a unique pair of individuals could determine a new clade.

IV.3. Taxonomic and Phylogenetical Propositions.

An important point resides in the transmission of phylogenetical information through nomenclatural codes. This information is to be found in the nodes of the tree. It is therefore a question of labelling these nodes. For this study, the information is to be found in the genus *Homo* and the specimens are individuals who require grouping into populations, sub-species or species. From a formal point of view, the description of monoplyletic groups particular to cladistics is difficult to transmit through the means of the Linnean nomenclature. Concerning the taxonomic position of clades obtained on the cladogram, we took into account the taxonomic ranks employed by other authors and discussed them. It is essential, however, to bear in mind that this cladogram is fragile and in no measure to support a new classification. In fact, all classification originating from this cladogram and put forward supposes that the phylogenesis of the anatomical parts other than the *calvaria* (the only anatomical part which could be studied) be congruent with the results presented. The cladogram here presented is therefore only a hypothesis which requires verification through the introduction of new taxa and characters (dental, mandibulary, facial and post-cranian). As a general rule in cladistics, the subordination of groups is based on living organisms and terminal taxons are monophylogenetical groups. According to Tassy (1988), two problems present themselves:

- according to hennigian hierarchy, the fraternal groups must have the same taxonomic rank. However, the majority of monophyletic groups are beyond the possibilities of Linnean nomenclature;

- the inclusion of fossils in a stable classification based on living organisms supposes the employment of methods which are not subordinations in *sensu stricto*.

In order to resolve these problems, Tassy (1988) recommends employing the two categories of

plesion and the **group root stock**. According to Patterson and Rosen (1977), **plesion** is the « plesiomophic » sister-group of all living and fossil groups which succeed it. The plesion must not possess any autapomorphies of the following sister-group but must be characterised by at least one automorphy. This latter point differentiates the plesion from the « group root stock » as defined by Hennig (Tassy, 1988). Two hypotheses are put forward using the « weighted » cladogram (**Figure 54**) as a basis (**Figure 56 and 57**). These two hypotheses make use of entries corresponding with that of specific binomes or sub-specific trinomes and constitute a sequence of specific names within the genus *Homo*.

The *Homo habilis* species is paraphyletic grade and Wood's idea (1984, 1992) of discarding *Homo habilis* and of cutting up this taxon into several species is in conformity with the result of this present study. The clades defined by nodes 65, 64, 63, 62 and 57 of figure 46 are the most robust and, each time the KNMER 1470, KNMER 3733, KNMER 3883 and KNMER 1813 specimens were taken into account, the topology of figure 52 is kept to for these taxons whenever the multiple state characters are processed in mixed fashion. It is therefore proposed that the specimens KNMER 3733 and KNMER 3883 be accorded the same taxonomic rank as KNMER 1470 and KNMER 1813. The denomination of species which make up the grade *Homo habilis* is therefore as follows:

Homo rudolfensis Alexeev, 1986 of which the type is specimen KNMER 1470;

Homo kenyaensis nov. sp. of which the type is specimen KNMER 3733 (see appendix 4 for a cladistical diagnosis);

Homo okotensis nov. sp. of which the type is specimen KNMER 3883 (see appendix 4 for a cladistical diagnosis);

Homo ergaster Groves and Mazak, 1975 of which the paratype is KNMER 1813.

The *Homo erectus* species (Dubois, 1893) of which the type is the specimen of Trinil 2 can be seen on the cladogram. This latter species is very close to the rootstock (**Figure 54**) but is characterised by a few automorphies. (It therefore corresponds to the definition of a plesion). This is also ther case for a taxon which we propose calling *Homo pekinensis* (Black, 1927) in line with a new combination but in conformity with the dismissing of the generic term Sinanthropus and for which the type will be the

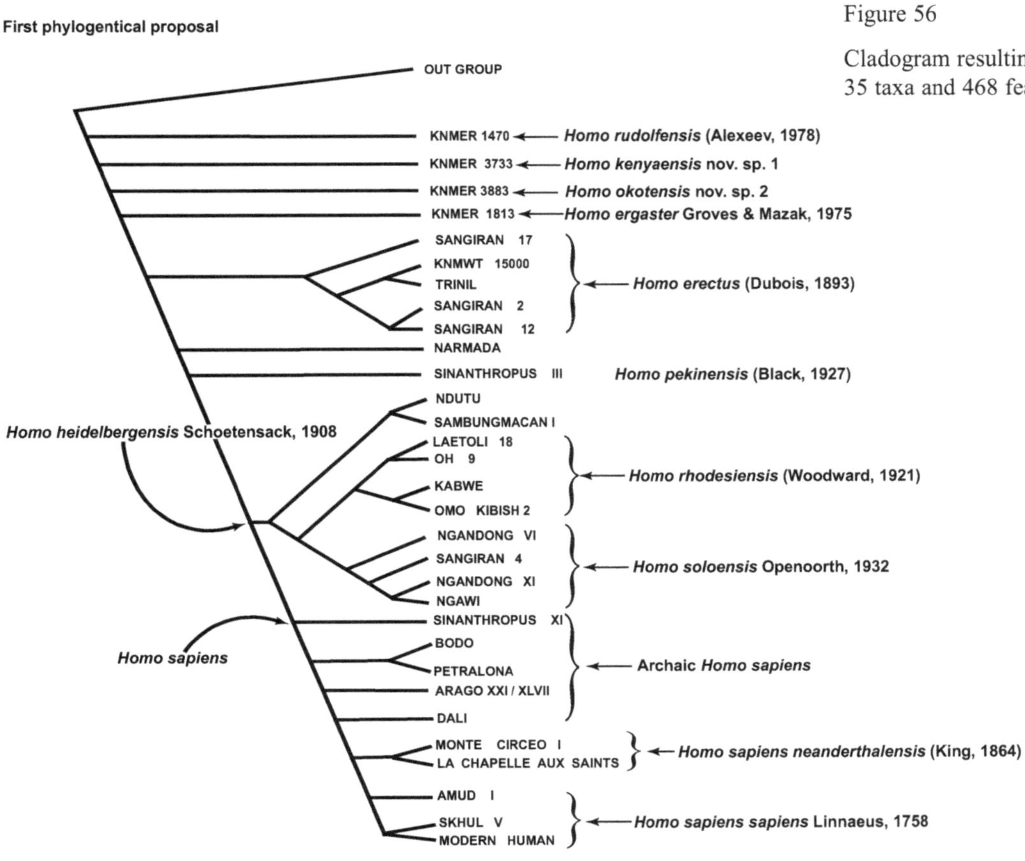

Figure 56

Cladogram resulting from the analysis done with 35 taxa and 468 features.

specimen Sinanthropus 3. We are then faced with an alternative: either the plesion preceding *Homo sapiens* is labelled with the name of the species *Homo rhodesiensis* Woodward, 1921 of which the type is the Kabwe 1 specimen (**Figure 56**); or the plesion is enfeoded to *Homo sapiens* (**Figure 57**) and it then a question of considering it as being sub-specific.

In the second case, the clade conserves the most currently used denomination corresponding to the trinome *Homo sapiens rhodesiensis*. We wish to highlight the fact that Bonde (1989) evoked the trinome *Homo sapiens leakeyi* for specimen OH9, considering that it could represent an ancient form of « time-bio-species » of *Homo sapiens,* in support of the utilisation of trinomes and in accordance with the results of our analysis.

IV.4. Chronological Data.

Several attitudes exist when considering the importance of the chronological data in the phylogenetic reconstruction concerning human fossils. For most palaeoanthropologists chronology is of prime importance and is the determining factor in the acknowledgement of taxons and of their relationships, and therefore their phylogeny. We in no way agree with this practice. In any case, Chamberlain and Wood (1987) highlighted that phylogenetic problems were as much the result of poor labelling of chronological data as of an absence of consensus around the taxonomic affinity of fossil specimens. Contrary to this, according to Jacob (1982) chronometry and biostratigraphy are fundamental as the origins of human fossils in time and space are essential in the construction of hypotheses or theories. Jacob (1983) added that time and space were *sine qua none* factors of evolution. If a great difference exists between evolutionary factors and evolutionary historical inference criteria, it should be noticed that authors often confuse these two aspects. In any case, a relatively tacit indulgent determinism exists around the idea of there being an evolutionary timepiece. Thus, a sufficiently long period of time for a speciation to occur, or, more precisely, that it be acceptable to the different authors. For Sartono (1991a), the problem is to know if a close relationship exists between morphological characters and geological ages of *Homo erectus* which could constitute a basis for differential diagnosis. Sartono supports chronological cutting-up of *Homo erectus*. He believes these subdivisions should obey consensual criteria so that they can be applied coherently to all human fossils.

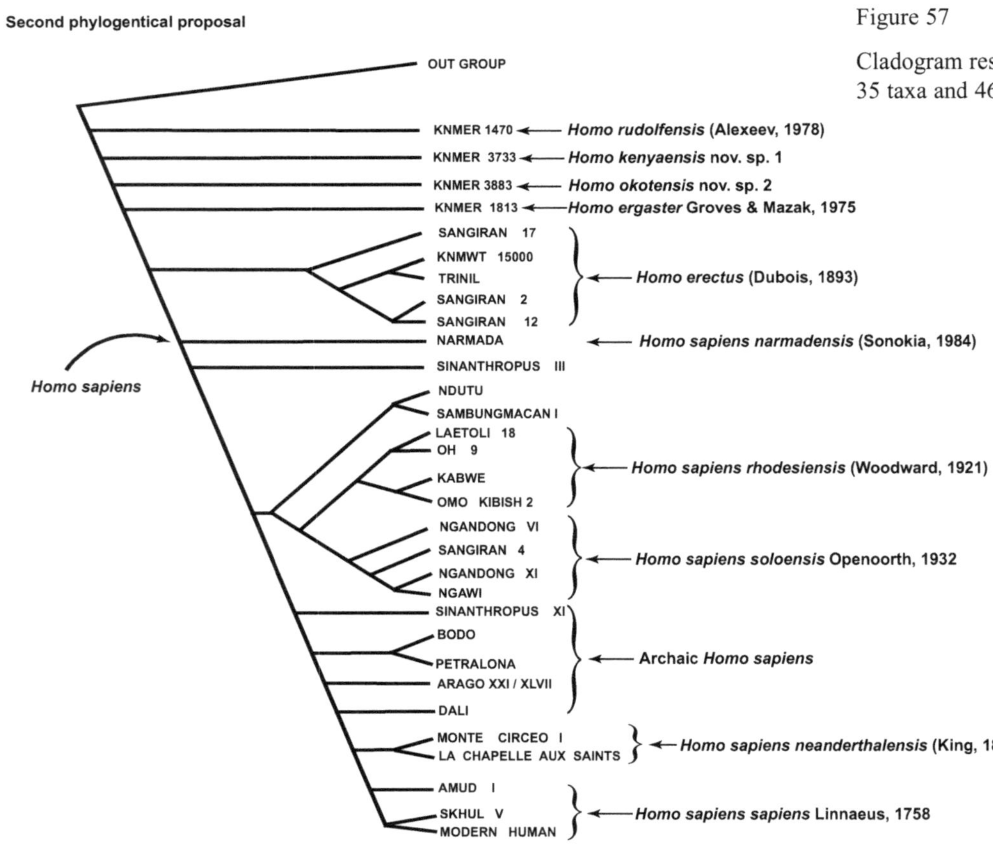

Figure 57

Cladogram resulting from the analysis done with 35 taxa and 468 features.

Stratopheneticians such believe in inserting « total morphological pattern » data into a chronological network. This means that, once established, the stratigraphical or chronological position of fossils those which ressemble each other most (in terms of global similarity) are associated together and regrouped in lines. Cladistics does not consider chronology as being a primordial criterion for phylogenetis inference. This is why Brace (1981) and Habgood (1989a), who believe in using chronology as the principal criterion, affirm that cladists have « problems » with chronology. It is not so much that these authors refute phylogenetical hypotheses put forward by cladistics based on contradictory chronological data, but more a sort of procedure of intent. Stratopheneticians take the cladograms put forward not as hypotheses on relationship structures, but as descriptions of evolutionary processes.

Chronology is certainly a capital factor of organisation and of hierarchisation of phylogenetical events, and, in this way the passage of time plays a fundamental role in the dynamic interpretations in confering a temporal dimension to speciations. If cladistics is in keeping with the fact that evolutionary mechanisms are expressed during time, it is true that it considers the criterion of geological (palaeontological) precedence as an accessory criterion in order to define phylogenetical relationships. It is in opposition with Howells' (1980) opinion of an obvious association with contemporary fossils. More exactly, and notably since Schaeffer *et al.* (1972), cladists state that time does not return, as such, the primitive or derived state of characters. All animals living during the Miocene Period do not possess characters which are primitive compared to those living during the Pliocene Age. To take up once more Nelson and Platnick's (1981) image, the cockroach of today does not only possess derived characters when compared to the fossilised mammoth which is now extinct. Cladistics refutes the stratophenetician point of view which states that evolution can be read in geological strata. At the limit, to counter this, if chronological or stratigraphical data cannot be contested, palaeontological criteria will validate the proposed phylogenetical reconstruction. The phylogenetical hypothesis issuing from the analysis carried out in this study was confronted with the chronological data which most authors accept. In order to construct the figure 58, we made use of a minimalist formula. Thus, the age of the oldest fossil within a clade determined the minimum chronological position of the node which supported the clade. In this way, the clade corresponding to *Homo erectus* exists between 0.6 Ma. and 1.6 Ma.

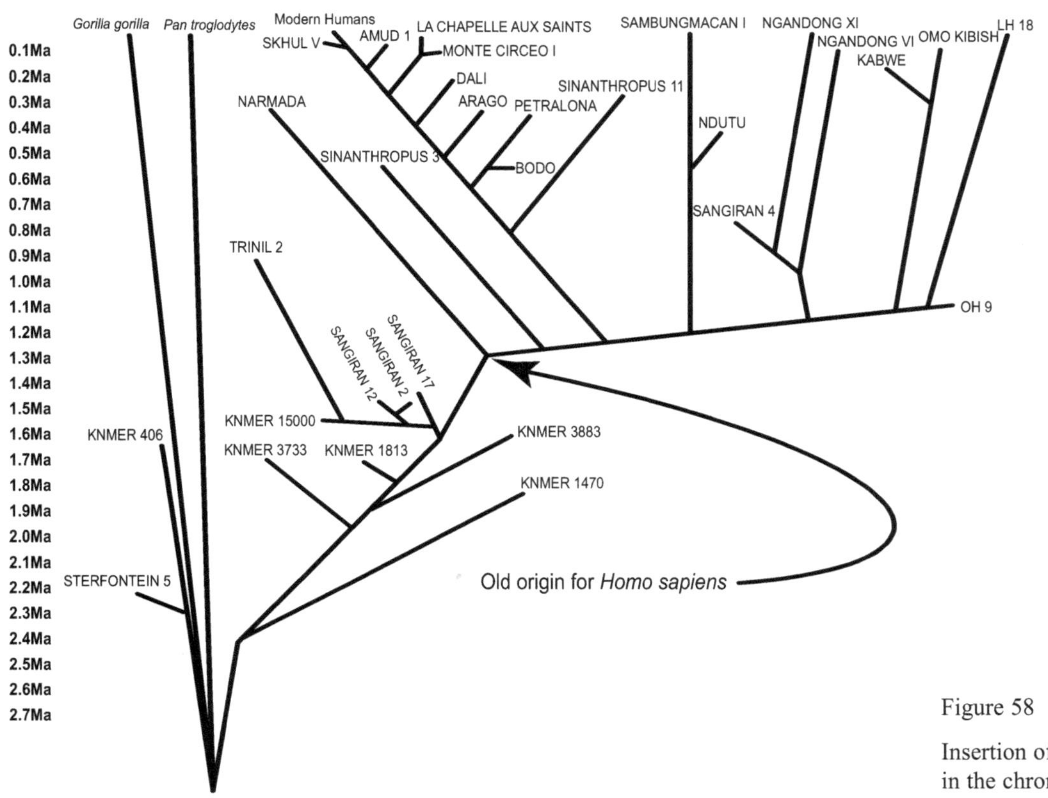

Figure 58

Insertion of the phylogenetic hypothesis in the chronological frame.

In line with dating reported to specimens making it up (Trinil 2 for the most recent and KNMWT 15000 for the oldest). The speciation of *Homo erectus* occured between 1.6 Ma. (date attributed to specimen KNMWT 15000) and, at least, 1.9 Ma. (date corresponding to the age of the specimen belonging to the preceding clade; *id est*: KNMER 1813). Obviously, the date of separation could well be older. This is only an estimate of a limit.

According to this diagram of integration of chronological data, it should be noted that the speciation at the origin of the clade corresponding to *Homo sapiens* is very old as it rests on the presumed age of specimen « OH9 » (1.2 Ma.). Certainly this specimen was discovered out of a stratigraphical context and it is only through hypothesis that it is supposed to belong to the upper part of « Member II » on the « Louis Leakey Korongo » site by Leakey and Hay (1982). If the stratigraphical age of this specimen was cast aside, a more recent origin than that of specimen Sangiran 4 must be considered; *id est*: a date approaching 0.9 Ma. In accordance with this theoretical construction, the origin of *Homo sapiens* would be placed before a million years. In line with the evolutionary model put forward by Eldredge and Gould (1972) phases « exploding with speciation » could have appeared following phases of stasis or of evolutionary grades during moments of isolation of populations following on from environmental changes.

V. Comparative Studies

It was pointed out that the analysis put forward in this book was based on the results of the study of individual specimens and not on groups of specimens constituted *a priori*. This original approach renders the comparison with other studies difficult – including those purporting to be cladist. Since our initial work (Zeitoun, 1996), few other studies have used a similar approach. No doubt this is linked to the fact that the method of using individual specimens as Otus is dogmatically considered to generate results with little meaning below the family level. From the point of view of alpha-taxonomy this dogmatic assertion issues from circular reasoning and supposes that the results are known beforehand. Cladistic analysis carried out by Mallegni *et al.* (2003) is based on specimens but only takes into account characters which are crudely defined. The analysis presented by Asfaw *et al.* 2002 concerning the skull of Bouri remains the one which is closest to our own. Unfortunately, these authors finally decide to choose constituting composite Otus and abandon their initial intention of working on specimens. One should note, however, that there exists a very close relationship between the topology of the tree obtained by these authors and that resulting from our analysis. This convergence is all the more remarkable that several of the specimens taken into account in the two studies differ from each other. Thus, in **Figure 59** one can observe Kenyan specimens (those which we related to *Homo rudolfensis* and *Homo ergaster*) and a

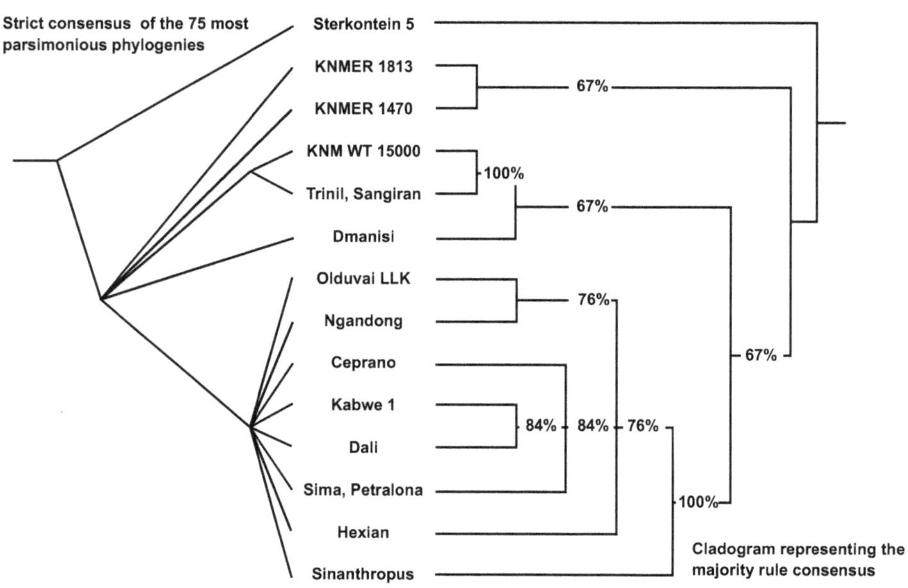

Figure 59

Phylogeny of Bouri according to Asfaw *et al.*, 2002

clade *Homo erectus* with the specimen- type Trinil as well as the Sangiran series and Nariokotome, the Dmanisi series, and finally a series which is paraphyletic with Zoukhoudian, Hexian, Ngandong + Bouri and the association Ceprano, Petralona+Atapuerca, Dali+Kabwe. The Dmanisi series appears as the sister group of *Homo erectus*. The following clade could be named *Homo sapiens*.

This topology, in common, seemingly amounts to a phylogenetic reality. Several hypotheses could be put forward regarding the mechanisms which led to this phylogentic scheme of things once this topology figures as part of a chronological series (**Figure 58**). First of all, we point out the existence of an early spreading within the Homo genus, after which Homo *erectus* appears as being a real species as this taxon is defined by derived characters proper to itself (**Appendix 4**). Finally, a discussion is centred around the status of several species, the conclusions of which are dependant on the agreed interpretation of the taxonomic level of different taxa. In this way, branches situated before the dichotomy of *Homo erectus* can be read according to two ways of thinking:

In the case of the first option – that of situating *Homo erectus* at the species level– questions must be posed concerning the status of *Homo soloensis* and even *Homo pekinensis* in Asia (**Figure 55**).
In the case of the second option – that of situating the species at a sub-specific level (**Figure 56**), – when this resolves the problem of multi-regionalism mentioned by several authors.

V.1. Spreading-Out

V.1.1. Spreading-Out at the Origin of the Genus Homo.

We have seen that several species can be defined at the rootstock of the phylogenetic tree presented. In including this diagram within a chronological frame it is possible to put forward hypothesis based on the evolutionary modalities at the heart of the genus Homo. The spreading-out described in **Figure 60** is in conformity with Stringer's hypothesis (1986, p. 290) which mentions the existence of an initial human line with three species of « early » Homo in the Pleistocene Period. For their part, Schwartz and Tattersall (1999) put forward a hypothesis of at least six human species at this level. One obviously rediscovers here the old debate between « splitter » and « lumper » which in1980 was at the centre of all debates with the coming of cladistics in palaeontology. Situated within the perspective of mammals as a group and in taking into account their specific size, Conroy (2002) posed questions around the actual number of species present within the genus Homo. Even if he conceded that « ecological » limits existed which were particular to the body-weight factor and even of the existence of related contradictory data in biological literature, in comparing the number of species known for

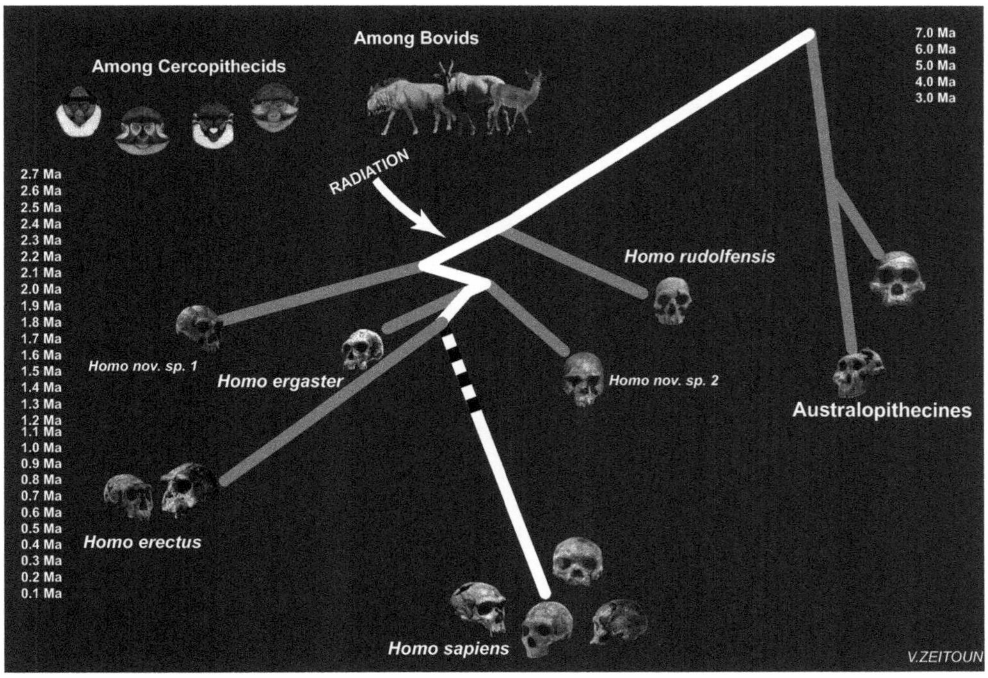

Figure 60

Synchronicity between human spreading and spreading by mammals mammalian taxa of an equivalent weight to that of Man (79 species), he observed that this only concerned 1,8 % of the total species and only 0,63 % of African species. In any case, what these conclusions reveal is that there is a correlation between size and « speciosity ». Thus, Conroy quotes the case of the genuses Gazella and Macropus, of an equivalent weight to Man which is represented by three contemporary species. We could in this way justify this figure (three or more) for mammals including Man and thereby reinforce our hypothesis and those of Stringer, Schwartz and Tatersall. However, this quantitative approach makes no sense from a cladistic point of view. In any case, Conroy's approach is biased. In fact, according to classical zoological classification according to evolutionary Systematics, genuses are defined – amongst others – by their degree of adaptation to an ecological niche but also in accordance with the number of species by genus. There is therefore an artificial number of species imposed in the classical classification which makes no phylogenetic sense. We cannot, therefore, make us of Conroy's conclusions as an argument, even if they do seem to go along with the results of our analysis.

Nevertheless, a high frequency of speciation amongst a group such as *Homo* is not extravagant if the quantity or the rate of change is really high for the initial phase of speciation in isolated and small populations (**Figure 60**). This mechanism is suggested by Vrba (1980 p. 66) and Foley (1993) among non human taxa. The correspondence between environmental shifting and speciation were indeed not a unique phenomenon as demonstrated from faunistic data among Bovids on the one hand (Vrba, 1996), and amongst Cercopithecines (Delson, 1984) on the other. The origin of the genus Homo being situated, as it is, in Eastern Africa, this where it would be most useful to examine the environmental data susceptible to confirm or reinforce hypotheses put forward. Without assessing any conjectural causalities, looking at the environmental data, the described human radiation happened after climatic shifting (Zeitoun, 2000). In fact, an independent assessment of geological, palynological and palaeontological data lead to the conclusion that there was a drought 2.5 million years ago (Azzaroli, 1995; Cooke,1985; Delson,1985; Denton, 1995; Dupont and Leroy, 1995; Deschamps and Maes, 1985; Eck *et al.,* 1987; Geraads, 1987; Lindsay *et al.,*1980; Marshall, 1994; Scott, 1995; Wesselman, 1995). Following on from this, data of similar nature indicates a period of increased humidity – at least in Eastern Africa 1.8 million years ago (Beden, 1987; Bonnefille, 1995; Cooke, 1985; Denys, 1996; Deschamps and Maes, 1985; Dupont and Leroy, 1995; Eisenman, 1987; Geraads, 1994; Geraads and Coppens, 1995; Leakey, 1993; Marshall, 1994) (see **Figure 61**). Basically, the adequacy between evolution and climatic and/or tectonic shifts is not a new idea. As early as 1936, working on the fauna of Ethiopia and Siwalik,

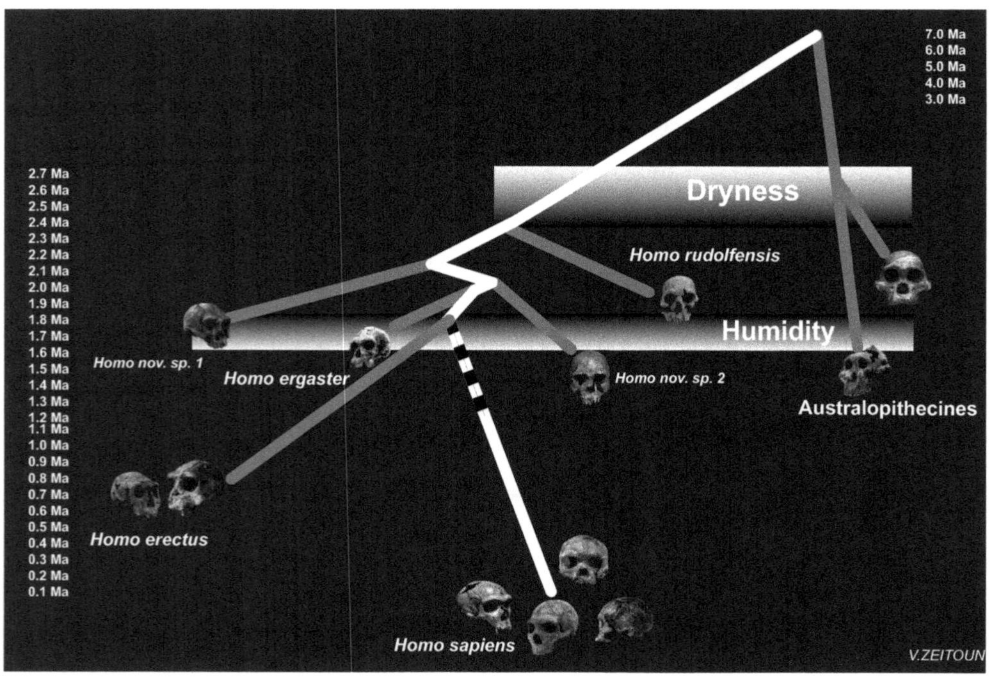

Figure 61

Adequation between human radiation and climatic changes.

Hopwood assessed that the earth-movements which affect the older beds of Olduvai and in the Naivasha area were in some manner connected with the extinction of the older fauna (Hoopwood, 1936 and 1940).

However, Man is a particular animal and, as concerns the genus Homo, in addition to the impact of environmental changes, an important cultural change occurring during this period must be pointed out. As early as 2.34 Ma. in Western Turkana (Roche *et al.,* 1999) an important technical development took place. Such innovations are indicators of cognitive development in certain populations which seemingly contributed to increased mobility. The degree of definition of lithic typo-technological analyses does not, as yet, permit the stating of whether technological advances are the consequences or the causes of biological evolution at this moment and in this geographic area but an important change is certainly indicated. A whole network of elements converges in order to describe a great change in Human development. A partial return to humid conditions around 1.8 million years ago in Eastern Africa, through a modification in environmental constraints and natural resources in several areas possibly led to, or facilitated the first geographical mobility of humans from the African continent (**Figure 62**). These humans could have been *Homo erectus* as is suggested by early dating in Asia (Swisher *et al.*, 1994) or *Homo ergaster* such as those discovered at Dmanisi in Georgia. These humans could quite easily be the conquerors of Eurasia which were at the origin of new migratory models put forward by (Turner, 1999; Bar-Yosef and Belfer-Cohen, 2001). Thus, the earliest human migrations out of Africa would have been triggered – or to be less deterministic – would have been facilitated by environmental changes in line with a series of isolated events which were perhaps also linked to the extinction of certain populations.

In-line with a provisional assessment after considering our phylogenetic hypothesis and recent field discoveries, it would seem that several waves of human migration moving specifically out of Africa took place. In this way, the appearance of the first *Homo* would have taken place around 2.4 million years ago in Eastern Africa, followed by a rapid spreading-out. *Homo erectus* individuals have been discovered as such in Eastern Africa and in Indonesia, *Homo ergaster* in Eastern Africa and in Georgia. The question of the continuity of this population in Eurasia is not dealt with here but it is nevertheless possible to state that at this early stage of human spreading, the specimens of *Homo erectus* which appeared in Eastern Africa migrated towards the rest of Africa and into Asia. These *Homo erectus* evolved and then disappeared from Africa at different periods according to the different territories. Also at this early time, *Homo ergaster* individuals left Eastern Africa and migrated towards Eurasia where they evolved, whereas in Africa they died out in the same way as *Homo rudolfensis*, *Homo okotensis* and *Homo kenyaensis*.

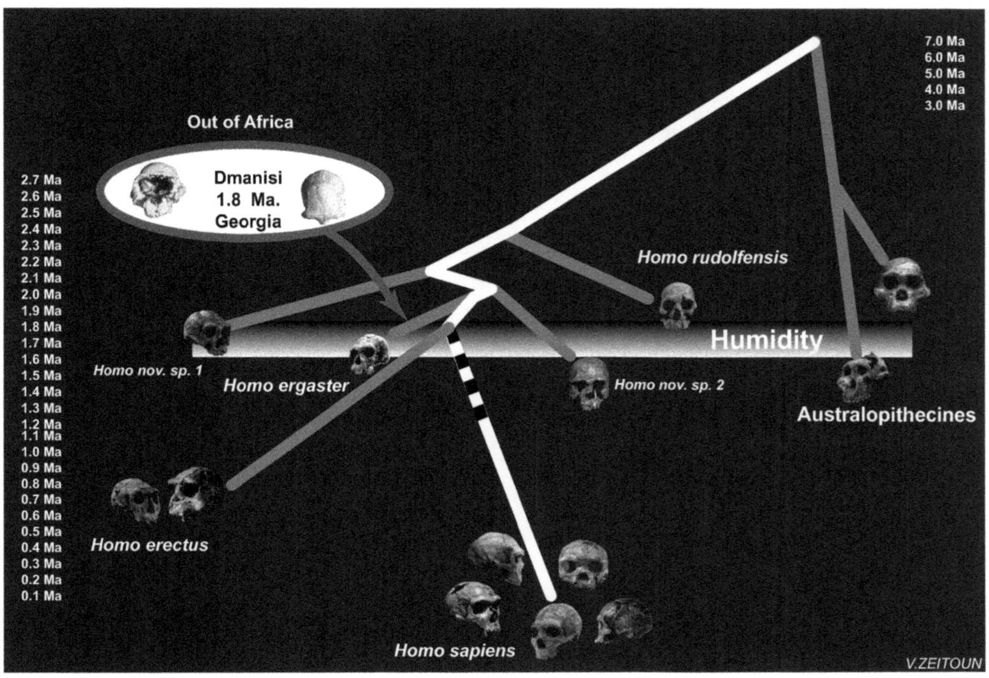

Figure 62

Out of Africa

V.1.2. Antique but authentic *Homo sapiens*

Beyond this spreading-out deduced from the presence of several human species in a restricted period of time, our analysis also supposes a very ancient origin for *Homo sapiens* (**Figure 63**). This hypothesis is based on the fact that specimen OH9 is considered as being an *Homo sapiens leakeyi* by Bonde (1989); From this it can be determined that *Homo sapiens* determines the taxonomic level of the branching which are situated beyond this node. This incomplete Tanzanian *calvaria* is supposed to have belonged to the upper part of « Member II » of which the top is dated between 1.2 and 1.15 Ma. by Leakey and Hay, (1982). This therefore situates the origin of *Homo sapiens* at least in this period. This observation was also made by Asfaw *et al.*, (2002) in their study carried out on the Bouri fossil dated 1.1 Ma. – without the observation being stated in the same way. The Eritrean skull of Danakil dated around 1.0 Ma. also described by its inventors (Abbate *et al.*, 1998) as already having the characters of *Homo sapiens* in its close morphological quality with the Ethiopian skull from Bouri is in any case mentioned by (Asfaw *et al.*, 2002). The logical consequence of discoveries in the field, of anatomical studies and of the phylogenetic analysis put forward here is that the first examples of *Homo sapiens* appeared more than a million years ago.

This formula is only disconcerting through the fact that for some thousands of years, Modern Man: *Homo sapiens sapiens* is the only representative of *Homo sapiens*. According to the data which we have evoked, *Homo sapiens* is the oldest and not *Homo sapiens sapiens*. If the debate around the coexistence of species has fed numerous arguments – in particular those centred around Neanderthal Man – it would appear that from a geological scale, coexistence is the rule. In order to comfort the hypothesis of an ancient origin of *Homo sapiens*, it is sufficient to reconsider the formal point of view the fact that, under the term of *Homo antecessor*, specimens from Atapuerca have been described as being possible ancestors common to Modern Man and to Neanderthal Man (Gibbons, 1997; Bermudez *et al.*, 1997). Is this new taxon legitimate? If one considers here, respectively, Modern Man as *Homo sapiens sapiens* and Neanderthal Man as *Homo sapiens neanderthalensis*, when their common direct ancestors are simply *Homo sapiens= Homo sapiens antecessor?* The oldest specimens attributed to this taxon are 800,000 years old. The Italian specimen Ceprano often associated to Atapuerca is dated at 900,000 years old (Ascenzi *et al.*, 2000).

If we go even further forward, one notes that the cladistic analysis carried out by Mallegni *et al.* (2003) resulted in the most robust closeness between the Ceprano and Bouri specimens. This would indicate an older wave of peopling by *Homo sapiens* coming from Africa to Europe. A previous phenetic analysis (Manzi *et al.*, 2001) demonstrated a strong association between specimens which could have been at the origin of the Neanderthalian line in Europe (*id est* AT-SH-CR4 and AT-SH-CR5,

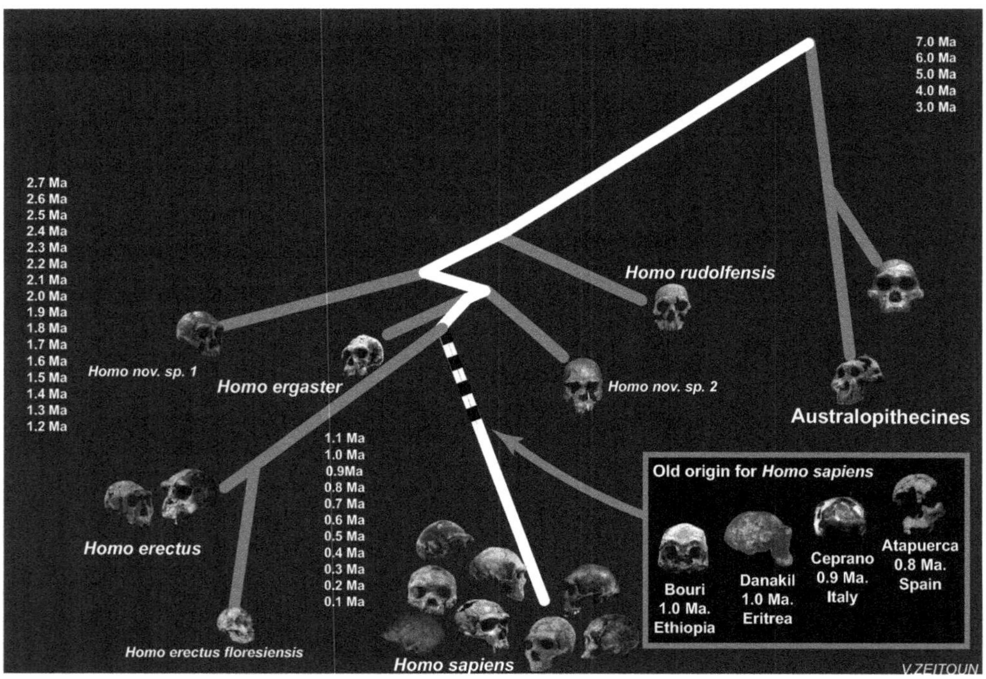

Figure 63

Old origin for *Homo sapiens*.

Arago, Petralona, Steinheim) and African specimens such as Bodo, Kabwe or Saldanha. If it is not a question here of phylogenetic analysis, the cladistic analysis carried out by Gilbert *et al.* (2003) based on data provided by Manzi *et al.* (2001) restitutes a topology which associates Ceprano at the base of a clade which contains both of these series. One rediscovers here in these studies a filiation between the « old » *Homo sapiens* (Bodo etc.) and « more ancient » specimens (Ceprano, Bouri etc.). In the new model which we put forward here **(Figure 64)**, it is possible to insert the nuances introduced by Carbonell *et al.,* (2005). Thus, according to these authors, associated with material from Mode 1, TD6 would have a Eurasian origin and would be at the origin of a non-Neanderthalian line *id est* Modern (*Homo sapiens sapiens*). It would be only a second human wave associated with lithic material from Mode 2 and seemingly coming from Africa that would succeed locally.

This second line would be at the origin of the Neanderthalian: *Homo sapiens neanderthalensis*.
Thus, the first representatives of the *Homo sapiens* species appeared a little more than a million years ago in Eritrea, Eastern Africa. These then spread towards the South and the North and then further into Europe around 900,000 or 800,000 years ago. Populations then acquired regional particularities. This new scheme of things begs a hypothesis of a new multiregionalism and this new multiregionalist model is characterised by its level of reading. In fact, it is no longer a question of seeing regional species as having converged towards a unique speciation in different areas (a model which cannot be possible from a biological point of view as recalled by Vandermeersch in 1989) but of seeing different regional belonging to a ubiquitous species: *Homo sapiens*.

V.2. The human canopy

V.2.1. Coexistence of several human species and a co-existence of several sub-species

A spreading-out can be defined by the occurring of several speciations within a relatively short geological period. On a human scale, this would be equivalent to understanding that several examples of mankind existed at the same time. According to anthropological texts, these would be species dif-

Figure 64

Several Waves of going out of Africa.

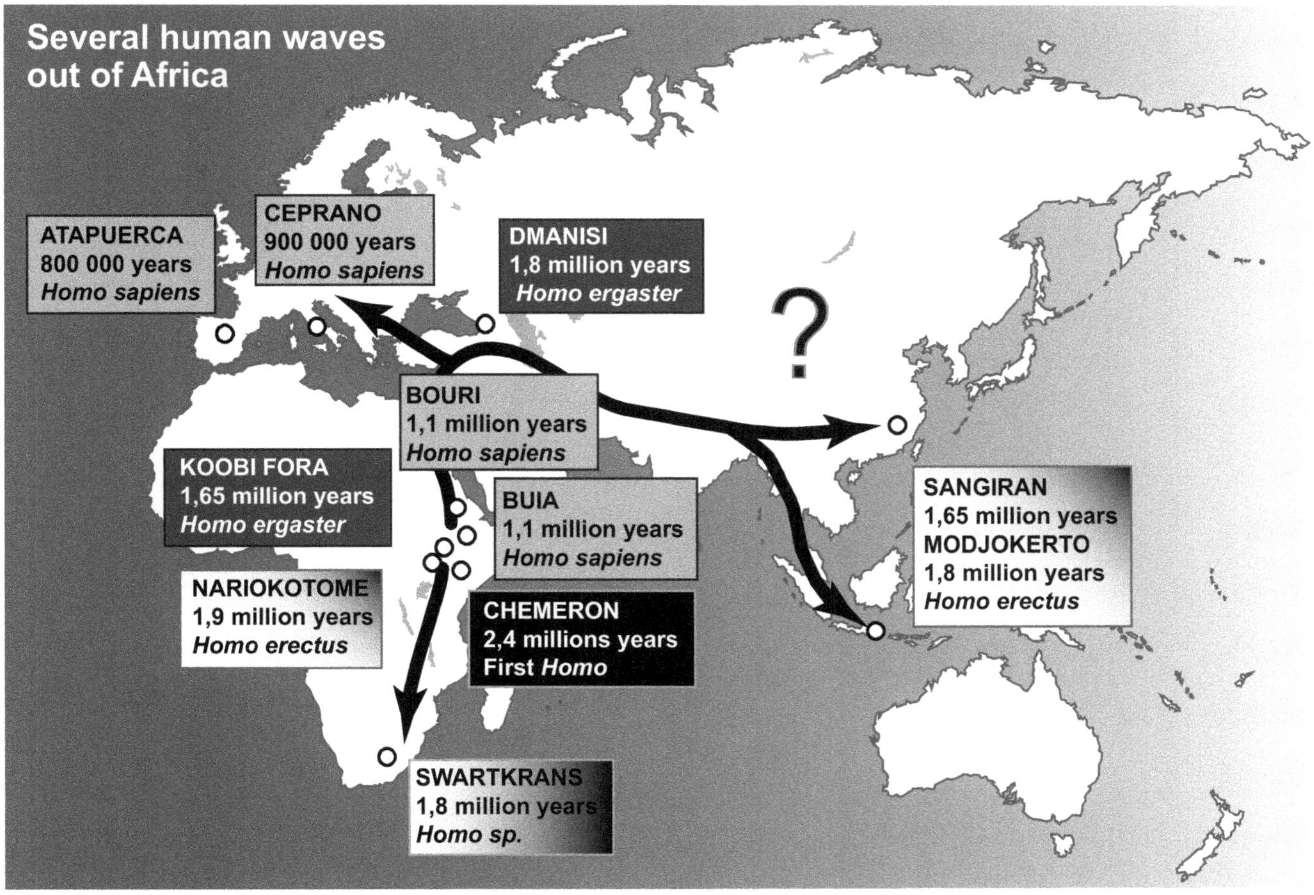

ferent from the other but existing alongside each other during the Lower Pleistocene Period whereas during the Middle or Upper Pleistocene they were sub-species. This contemporaneousness underlies the question of polymorphism at two levels: that of between species and of at the heart of a species. This is a difficult question to answer in palaeontological terms – especially on the basis of the criteria of interfertility utilised in biology.

In comparing the mandibles from the Dmanisi site to taxa reputed for their sexual dimorphism, Skinner *et al.,* (2006) conclude that the whole range of dimorphism at the centre of the genus Homo should be re-considered or even that the possibility of two different species should be considered. Contrary to this, Lee (2005) asserts that the nullified hypothesis which states that the degree of variation in the cranial capacity of the Dmanisi series is within the range variation of a single species in so far as the variation is comparable to that observed in male-female pairs of modern humans, chimpanzees and gorillas.

As soon as one considers size in itself is a criteria in defining a species or not, the questions needs to be posed differently. It is either a question of variation within a taxon or of differences between taxa. The small-brained specimen KNMER 42700 unearthed from Kenya in 2000 is said to show incredible variation in size within *Homo erectus* (Leakey *et al.,* 2003). This small size is comparable to that of specimen D2700 which is the smallest skull from the Dmanisi series and which is considered to be a *Homo ergaster* (Vekua *et al.,* 2002). Our approach within the framework of alpha-taxonomy established coding and definition of characters in order to dismiss the intrinsic parameter of size.

The study of the Dmanisi series is certainly of great importance and we agree with Rightmire *et al.,* (2006) that it is much more reasonable to consider that the preservation and the taphonomy of the site of Dmanisi offer a truly unique opportunity to assess anatomical variation within a palaeo-population.

In-line with the phenomenon of spreading-out which led to the co-existence of several allopatric species, in its initial phase we think that the speciation engendered a polymorphism. This primary polymorphism converged towards disappearance through homogenisation in a given isolated territories through time. It remains to be able to quantify the point of no return beyond which the isolates could no longer mix together: the phenomenon of speciation. Before this stage, sub-specificity could be expressed by different regional facieses.

V.2.2. Towards a new multi-regionalism

Within the scheme of a restricted human presence to a single species – *Homo sapiens*, the two models « Out of Africa » and « Multi-regionalism » have become reconciled or, more exactly, no longer need to exist, as all has become sub-specific. This scheme is identical to that put forward by Bräuer (1984b) under the term of « hybridization and replacement model ». According to this model the fact that a sub-species, unique to *Homo sapiens* is at the origin of Modern Man in one area of the world and then replaces all the other sub-species of *Homo sapiens* and the fact that the genetic proximity of different sub-species of *Homo sapiens* allows a hybridization, is put into the balance. It is the second term of the alternative which differs from Smith's model (1985) as in the latter there would be a convergence of all the sub-species of *Homo sapiens* towards modernity in several parts of the world.

Modernity

The first *Homo sapiens sapiens*es were supposedly present 300,000 years ago in Africa due to the Ethiopian specimen Bodo and, already in the Near East around 250,000 years ago with the Zuttiyeh specimen (Zeitoun, 2001). From the generally accepted point of view, it was only around 35,000 years ago that Modern Man reached Europe and Australia.

The coexistence of Modern Man with other forms of *Homo sapiens* is a condition which seems to be present and reinforces the idea of the species being replaced. The answer to a question concerning chronology will not easily bring a response as it is impossible to quantify or to distinguish the amount of time necessary for the migrations or for the homogenisation of the genetic pool. Highlighting the coexistence between Modern and said archaic forms of *Homo sapiens* can begin to answer some of the questions posed if the origin of the Modern forms is strongly marked in a region and if one is able to follow a morpho-chronological gradient.

We are not convinced that the Pleistocene *Homo sapiens* from Middle Awash (Ethiopia) BOU-VP-16/1 which is said to belong to the subspecies *Homo sapiens idaltu* (White *et al.,* 2003) can be cast aside from anatomical modernity as it does not present any distinctive character with regard to other *Homo sapiens sapiens* fossils. We are more inclined to believe that it is a question of the most complete and best dated specimens of Modern Man between Bodo, on the one hand, and Qafzeh on the other. In fact, specimen BOU-VP-16/1 is dated back to being between 160,000 and 154,000 years by Clark *et al.,* (2003) whereas Braüer (1989) assigned Bodo to the isotopic stage 11 and that Qafzeh is dated back to 100,000 years ago by Mercier (1992).

South Asian prehistory which is characterised by gradual rather than abrupt transitions in technological developments in stone tool manufacture is also characterised by the presence of a more antique modern anatomy. As far as Eastern Asia is concerned, Narmada Man, acknowledged as being a *Homo sapiens* (Kennedy *et al.,* 1991), is now dated as being between 125,000 and 250,000 years old and Liujang Man, related to Modern Man would be older than 153,000 years old (Shen and Michel, 2007).

Thus, palaeontology does bring some tangible elements over the time-frame of 100,000 years, the moment that some molecular biologists would have us believe that modern humans from Africa began their displacement from other hominids in Europe and the Near-East (Kennedy, 1998). Palaeo-geneticians spread their biological clock over the Age of human fossils. It would therefore be inappropriate for palaeo-anthropologists to create a chronology based on geneticians data, excepting in adopting a determinist and gradualist approach.

V.2.3. Regional Parallelism

V.2.3.1 The case of **Homo soloensis**

If there is here a mass of arguments supporting an ancient origin of *Homo sapiens* and even for a greater ancient origin of Modern Man, this is also true for the most recent chronological data: between 25,000 and 52,000 years of age (Swisher *et al.*, 1996; Falgueres *et al.*, 2001) which concerns Asian specimens considered to be of *Homo erectus* such as Ngandong, Sambungmacan or Ngawi. We would then be in a situation of ancient Modern Man and of recent *Homo erectus*. If the co-existence of these species is not in itself a problem, we have carried out an analysis which excluded incomplete fossils or of uncertain stratigraphic positioning such as Sangiran 4, Sangiran 12 and Narmada for Asia and Ndutu and OH 9 in order to test the topology of the unique tree which came to light in our earlier analysis (**Figure 52**). A unique tree also resulted from the analysis thereby supporting the earlier analysis (**Figure 65**). The number of synapomorphies for which the CI is higher or equivalent to 0.5 allows for an appreciation as to where the most reliable nodes are situated. One re-discovers a paraphyletic sequence at the base of the tree, followed by the species *Homo erectus*. The following clade regroups two coherent branches – African and Asiatic. Finally we find a paraphyletic sequence from Pekin Man to Neanderthal Man via Modern Man.

The question which remains is that of what status to accord to the different branches. In dealing with Solo men, it should be remembered that during their anthropological career, these fossils changed from one state to another very quickly. Thus, Oppenoorth described the species *Homo (Javanthropus) soloensis* from the observations of the human skulls discovered in the deposits of the banks of the Solo River (Oppenoorth 1932). He stated their phylogenetic place as a geographic variant similar to African *Homo rhodesiensis* (Oppenoorth, 1937). But Dubois called them *Homo sapiens soloensis* (Dubois, 1940) and Von Koenigswald considered them as tropical Neanderthalians (Von Koenigswald, 1958). Tobias is the only researcher who regularly used the official trinomen *Homo sapiens soloensis* (Tobias, 1985) but several authors (Jelinek, 1981; Stringer, 1987; Bonde, 1989; Braüer and Mbua, 1992) still considered them as archaic *Homo sapiens.*

The rehabilitation of this trinomial nomenclature is in conformity with the hypothesis of a polymorphous *Homo sapiens* line with several sub-species. This has not, however, been the option accepted for many years. Jacob, the curator of the fossils, disagreed with the grouping of Solo men in *Homo sapiens* (Jacob,1978) and was, indeed, followed by the main authors who preferred to call them « advanced » or « evolved » *Homo erectus*. This trend was certainly influenced by the work of Weidenreich (1943, 1951) who considered them as a geographic morph and relatives of Pekin man, and also by the study of Santa Luca (1980) considering their rank as *Homo erectus* although the work of McIntosch and Larnach (1972) brought strong reservations because of the thickness of the cranial vault – a so-called typical trait of *Homo erectus*. This work on such a trait of « *erectus* » was confirmed by Brown (1994) and recently by a CT-scan study (Balzeau, 2005). The Ngandong series = Solo men are now extended to the specimens from Ngawi and Sambungmacan (Widianto and Zeitoun, 2003; Baba *et al.*, 2003). Solo men are also considered as demonstrably persistent representatives of *Homo erectus* well into Upper Pleistocene (Clark Howell, 1999) where they are the descendants of « classic » *Homo erectus*: Trinil-Sangiran series and nothing else is proposed for them in terms of evolutionary process.

Nevertheless, all works on the Indonesian specimens propose classifying the human fossils of this geographical area into at least two different categories (Sartono and Grimaud-Hervé, 1983; Bräuer and Mbua, 1992; Widianto, 1993; Grimaud-Hervé, 1997 and many others). Is it possible to invoke independent data in order to settle the question of the taxonomic status of these specimens? In pursuing this method, what can be said of *Homo soloensis* ?

V.2.3.2 Chronological considerations

With a long chronology (from 1.8 Ma to 30,000 years) instead of a short (traditional) period (from 1.0 Ma to 250,000 years), there is a wider chronological gap between the « classic » and the « evolved » populations of Asian *Homo erectus*. As said Langbroek and Roebroeks (2000), the establishment of the chronology for the hominid fossils from Java is a long story, full of confusion and controversies due to a complex geological setting and the absence or loss of information on key fossils. Thus, problems still remain because of uncertainty about the stratigraphical position of the fossils (Laitman and Tattersall, 2001). The uncertainty is also due to different interpretations or readings of the same data (Barstra, 1994; Hyodo *et al.*, 1993; Sémah *et al.*, 1992; Sémah *et al.,* 2000). Another kind of objection could be raised because of the use of indirect radiometric dating but, for instance, Falguères *et al.* (2001) applied direct methods on the skull of Sambungmacan 1, which provided dating that is consistent with the former work of Swisher *et al.* (1996). For older Javanese «classic» *Homo erectus*, arguments from field observations and literature put forward by Huffman (2001) are convincing enough to support the data of Swisher *et al.* (1994). Thus, following the hypothesis of a more « elastic » chronological framework, the evidence of a phylogenetic splitting between two kinds of

Figure 65

Phylogenetical Hypothesis considering the score of CI over 0.5

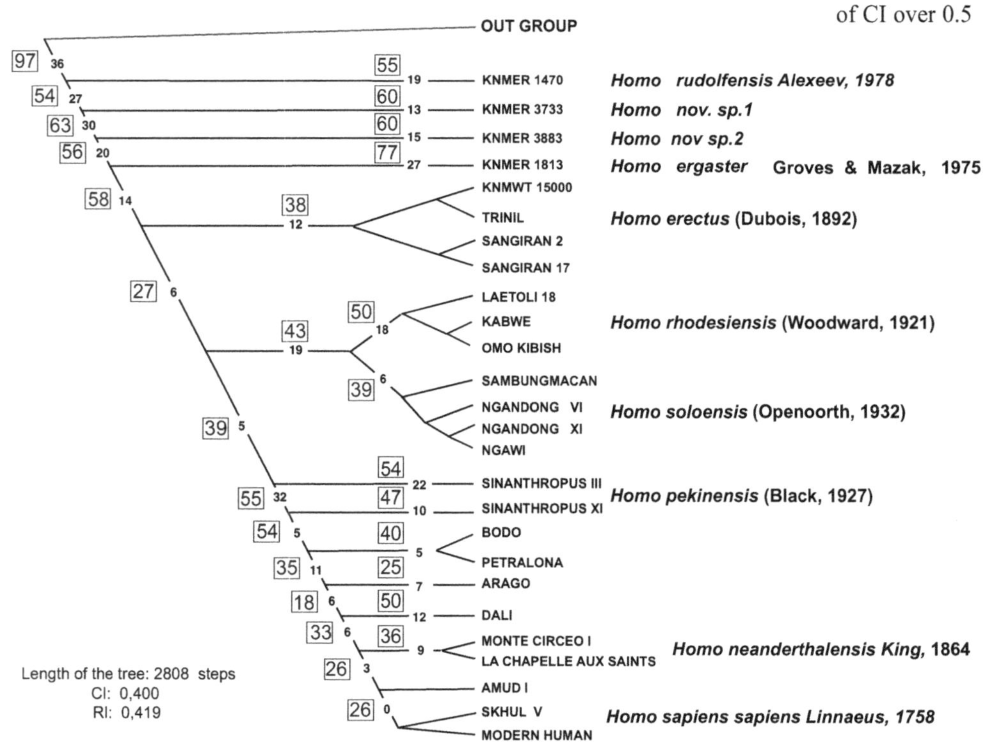

Single Tree with the number of synapomorphies and the number of synapomorphies which CI > 0.500

« *Homo erectus* » prompts us to consider the significance of biological continuity between both « classic » and « evolved » *Homo erectus*.

V.2.3.3 Environmental considerations

Catastrophic events occurred around the Brunhes/Matuyama geomagnetic reversal (Fontaine, 1976; Schneider *et al.,* 1992; Bunopas *et al.,* 1999). They are known as the Australasian tektite strewn field that is dated at about 770, 000 or 803,000 years ago by Hou *et al.* (2000). Such an event must have triggered shifting among the Asian populations and their environment.

Another environmental event that happened 71,000 years ago, the explosion of the Toba volcano in Sumatra (Chesner *et al.,* 1991; Zielinski *et al.*, 1996), spewed ash which fell as far away as India (Acharyya and Basu, 1993). This explosion, recorded up to the northern polar ice sheet (Lang *et al.*, 1999) is synchronous with the fall of 16° recorded in the ice core of Greenland. This volcanic event is considered to have had drastic global consequences on human ecology and the evolutionary process (Ambrose, 1998). If this scenario is true at a global level, it is even more crucially important for South-East Asia. **Consequently, both Plio-Pleistocene events (comet and giant volcano) could have triggered an intermittent disappearance of humans in Java or at least a strong selective pressure.**

The Solo populations, which form a unique, homogeneous series were initially referred to either *Homo soloensis*, or *Homo sapiens soloensis*. It is difficult – if not impossible – to choose appropriate criteria to define fossil species, except for the criterion of monophyly. But if we are following traditional anthropological procedure, the way to split the clusters of fossils is based on a legal definition of the taxa, as in Classical Evolutionary Systematics. The validity, or at any rate, the consistency of such a result by comparison with chronological and environmental data can be tested. The question of bio-geographical vicariance among humans is thus the only relevant level of discussion. But while the data show a chronological gap with drastic environmental events, the morphological differences are the strongest. Yet the boundary between palaeo-species or palaeo-subspecies is randomly defined. It is nevertheless noteworthy that relatively recent works (Clark Howell, 1999; Durband, 2002; Detroit, 2002; Kidder and Durband, 2004) consider Solo men as possibly belonging to another taxon. *Homo soloensis* could be the most appropriate name.

V.3 The case of *Homo pekinensis*

Focusing on Asia one strange but nowadays more acceptable result must be briefly commented on. Indeed, the issue of the phylogenentic analysis suggests that Pekin men are isolated as a different clade from « classic » *Homo erectus* (Trinil-Sangiran: series) but also from « evolved » *Homo erectus* (Ngandong, Sambungmacan, Ngawi: series). This result is far removed from the classical studies that considered these fossils as a geographic morph of *Homo erectus.*

However, differences – as much morphological as metrical – between Solo Man and Pekin Man have been commented on in the works of Weidenreich (1943, 1951), Santa Luca (1980) and more recently, Détroit (2002), Kidder and Durband (2004) and Terhune *et al.,* (2007). Pekin Man would seem to have a unique pattern for the temporal bone pneumatization, whereas contemporaneous and younger Asian and European fossils are closer to the modern human pattern (Balzeau *et al.,* 2008). Thus it could be appropriate to consider the formal recognition of *Homo pekinensis* as a continental taxon by comparison to the peninsular series (see also Zeitoun, 1996a for a former similar suggestion). This point of view is consistent with the suggestion made by Clark Howell (1999) and more demonstratively by the work of Kidder and Durband (2004) that is, for instance, followed by Carbonell *et al.* (2005) in order to explain a much greater closeness between European and Asian fossils than between African fossils. In this way, Pekin Man would be a completely distinct category. As soon as characters become proper to him, this makes him a true species. Suggested in a paper for a very scientific journal twelve years ago, this proposal was rejected because « if it was true, it will be already known for a long time…».

V.4. The case of Flores Man.

Small in size but huge in its effect on Palaeo-anthropologists, is *Homo floresiensis* (Brown *et al.,* 2004; Morwood *et al.,* 2004) just a new name or is it really a new taxon?

Following (Morwood *et al.,* 2004), this new fossil illustrates that:

1- several kinds of humans existed together

much more recently than thought before *id est* -18 000 years instead of -30 000 years ago when the Western Neanderthalians and Eastern Solo Man vanished and left our direct ancestors *Homo sapiens sapiens* alone in the Ancient World.

2- Humankind remained malleable to environmental constraints despite the development of technological skills and culture.

Regardless of all the controversy concerning the true nature of the Flores Man, the first rule of Systematics is to fully examine the pathological state of the specimen prior to any discussion of its convenient classification.

If one considers the anatomical data arising from pathologies affecting the height of individuals, we can first of all invoke hypophyseal nanism.

Hypophyseal nanism (Murray *et al.,* 2000) is a very rare syndrome that leads to an allometrically conservative reduction of the whole body (skull and body). This pathology is known in Asia and is suspected to occur in the Pacific. This rare disease occurs at higher frequencies in endogamous populations (Maheshwari *et al.,* 1988). Endogamy is favoured by geographical isolation such as on an island.

Published data and pictures from Liang Bua (Brown *et al.*, 2004) show a flexion of the tibia: a feature known in relation to nanism. What happens if the hypothesis of nanism is confronted with palaeontological data ? Let us begin with a relatively simple calculation. Considering measurements of pathological cases among present populations (Maheshwari *et al.,* 1988) and considering a population of prehistoric man (Trinil Man or Solo Man) who's height has been estimated at 1.50 metre, by a simple calculation (1.20 for pathology/1.63 for mean population x 1.50 for fossils) we get 1.10 metre.

With a mean cranial capacity value of 904 cm^3 for *Homo erectus* (Rightmire, 2004) and a mean value of 1350 cm^3 for Modern Humans (Beals *et al.,* 1984) the same calculation gives (1000/1350 x 904) = 670 cm^3.

Concerning the stature, these values are not so far from those estimated from the Flores bones:

- 1 metre and 380 cm_3 (Brown *et al.,* 2004) upgraded to 417 cm_3 (Falk *et al.,* 2005);

- 1.20 metres and a little more than 430 cm_3 (Jacob *et al.,* 2006).

Today, people of short stature, mean values of 1.49 for males and 1.44 for females of Rampasasa (Rahmawati *et al.,* 2007) still live around the Luang Bua cave. Even if these authors shed light that in terms of variation (5,5 SD) LB1 below the combined sex mean of the people of Rampasasa, there is no doubt that the most striking trait of Flores is his minute cranial volume. This data seem to fit well with the Laron syndrome (Hershkovitz *et al.,* 2007). Obendorf *et al.,* (2007) mentioned by them as being possibly indicative of human endemic cretinism. But some authors argue that LB1 can not be considered as a microcephal (Falk *et al.,* 2005) because it is different from their studied specimens, whilst others (Weber *et al.*, 2005; Martin *et al.*, 2006) consider that the studies carried out were not extensive enough to fully reject this hypothesis. Indeed, many kinds of microcephaly exist. At least, according to Conroy and Smith (2007), LB1 brain could not be functionally equivalent to a modern human or a chimpanzee brain.

Morwood *et al.* (2005), followed by Lieberman (2005), proposed that the discovery of new individuals provides evidence that Flores is a species. It must be stressed that several individuals had already been discovered during the first excavation (Brown *et al.*, 2004) and, as a result, this argument cannot stand on its own merit.

Finally, there are two major proponents in the debate – researchers who believe it is a microcephalic modern human and those who consider it as being a new species. But this is a question of opinions and little analysis has been carried out. Only Argue *et al.* 2006 have undertaken a phylogenetic analysis which supports the hypothesis of a new species. Tocheri *et al.* (2007) recently brought evidence of postcranial anatomical data that will also support the hypothesis of a new species but, these authors acknowledge the tenuousness of such data as far as there is no comparable material for *Homo erectus*.

We undertook a cladistic analysis of the characters of the Luang Bua 1 calvaria by including them into our previous analysis. If LB1 was a modern it will be better because in this case the debate was closed. The latter analysis that we undertook shows **(Figure 66)** that LB1 is close to Trinil and different from the Solo series and modern humans. Our phylogenetic positioning, therefore, demonstrates that Flores is *Homo erectus*. A third point of view.

According to the fact that there is no comparable material for *Homo erectus* in their study, the work of Tocheri *et al.* (2007) does not exclude that LB1

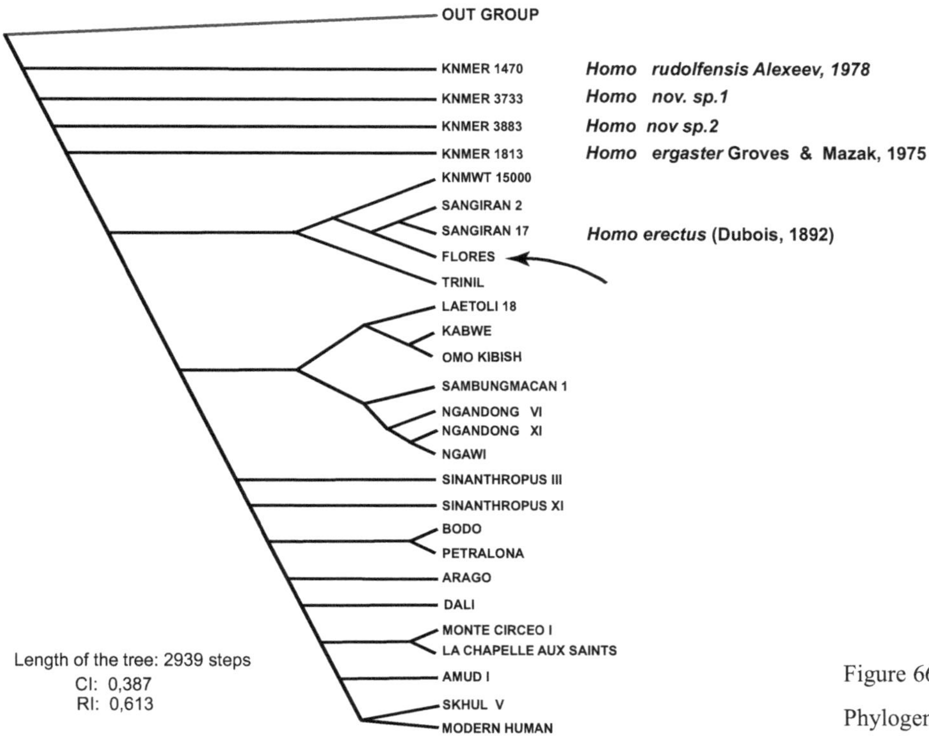

Figure 66

Phylogenetic tree including the case of Flores

could also be *Homo erectus*. Some authors could consider that a small *Homo erectus* is a new species. But this concept should therefore become more generalised. Thus, specimen KNMER 1813 considered as the paratype of *Homo ergaster* by Groves (1989) has a brain size of 510 cm^3 and Fergusson (1995) believes that this is another species: *Homo microcranous*. This latest interpretation has not, generally speaking, been accepted by other authors. In any case, amongst human fossil series, others have a small cranial volume that does not systematically lead authors to propose new species. The third skull from Dmanisi (D2280) has a cranial capacity of about 600 cm^3 (Vekua *et al.*, 2002) and KNMER 42700 a cranial capacity of 691 cm^3 (Spoor *et al.*, 2007).

Due to skull size, which was not taken into account by the coding of the characters, whatever the causality of the process is (hypophysary nanism or insular nanism), we propose that the Flores specimens were small, possibly dwarf or at least microcephalic *Homo erectus*.

VI. The Human canopy

After the conclusion of an analysis and of an assessment of the published morphological data, 123 morphological characters were retained in order to describe the human *calvaria*. A study of 345 metrical indicators, calculated from the ensemble of craniometrical points of each element of the *calvaria,* was suggested. An analysis applied to specimens permitted the putting forward of phylogenetic hypotheses of which the impact was great with reference to our reading of the evolution of the genus Homo.

A first spreading of Kenyan specimens KNMER 3733 and KNMER 3883, respectively *Homo kenyaensis*, *Homo okotensis* and the species *Homo ergaster* Groves and Mazak, 1975 and *Homo rudolfensis* (Alexeev, 1978) took place in Eastern Africa. This paraphyletic sequence was a prelude to the coming of the species *Homo erectus* which exists in terms of a species. The definition of its hypodigm reduces to quite a restricted number of fossils: the specimen type (Trinil) to those from the Indonesian site of Sangiran and to the Kenyan specimen from Nariokotome. In spite of the difficulty evoked for the introduction of a taxon in a linnean classification, it was confirmed that the species *Homo erectus* previously defined by different authors was but a grade. *Homo erectus* is in fact a fossil species which is very close to the root which led to *Homo Sapiens*. This confers the definition of plesion to this terminal taxon. Additional plesions corresponding to *Homo pekinensis* (Black, 1927), *Homo rhodesiensis* Woodward, 1921 or *Homo soloensis* (Oppenoorth, 1932) or to sub-species of *Homo sapiens* give a new perspective to multiregional models and « Out of Africa » usually put forward by palaeonthropologists. One origin of the *Homo sapiens* species previous to a million years ago appears as a theoretical phylogenetic hypothesis which data from the field in the future will certainly contribute in confirming this. Our interest is more particularly concerned with Asia which, for having been the veritable cradle of Palaeonthopology with the pioneering work of Eugène Dubois, remains the leading light in the understanding of the phenomenon of « buissonnement ».

Even when the existence of two robust Asian humanoids: *Gigantopithecus* and *Meganthropus*, is disregarded, at least two categories of fossil Human remain agreed upon by Palaeontologists: Trinil Man = Classic *Homo erectus* and Solo Man evolved *Homo erectus*, Archaic *Homo sapiens* or *Homo soloensis*.

Between 40,000 and 27,000 years ago (Swisher *et al.*, 1996), it is plausible that Modern Man, Solo Man and Flores Man lived in neighbouring areas. Based on our analysis Modern Man appears to have coexisted with another species until much more recently than was previously thought: *Homo erectus* of Flores. Our findings suggest that *Homo erectus* is not the direct ascendant of Modern Man. Indeed, in theory it is not possible to recognize a new species and its mother-species simultaneously in the fossil record, except at the beginning of the speciation process. In the case of Flores we are chronologically far from this case with dating between 90,000 and 17,000 years. In China, Liujang Man would be more than 153,000 years old (Shen and Michel, 2007). Modern Man would even have gone as far as Australia beyond 116,000 years in the Northern part of this area (Fullagar *et al.*, 1996). For this relatively recent period of history in the evolution of the human species, the « buissonnement » which results from East African spreading leads to a canopy of species and sub-species. This phenomenon is not restricted to the debate between Modern and Neanderthal Man but also concerns in a much more tangible way in South-East Asia several human entities. That is why since the beginning of the Palaeoanthropology created by Dubois, the research of human origin must go on under the Asian tropical canopy.

Appendix 1

Character	type I	type II	0	1	2	3	4	5	6	7	8
124	**biMnf**		<1.45	1.46 to 1.95	1.96 to 2.38	>2.38					
125	**biFmt**		>0.30	0.30 to 0.25	0.24 to 0.23	0.22 to 0.14	0.15 to 0.10	<0.10			
126	**biFt**		>0.05	< 0.06							
127	**biSst**		>0.10	< 0.11							
128	**biSph**		<0.40	0.40 to 0.45	> 0.45						
129	**BSst**		<0.06	>0.05							
130	**BFt**		>0.05	<0.06							
131	**BSph**		>0.12	0.12 to 0.06	<0.05						
132	**BSg**		<0.05	>0.06							
133	**BG**		>0.06	<0.05							
134	**FtFmo**		<0.12	0.12 to 0.18	0.19 to 0.22	> 0.23					
135	**SgSst**		<0.06	>0.05							
136	**SgFmt**		<0.10	0.10 to 0.14	0.15 to 0.18	0.19 to 0.20	0.21 to 0.25	>0.25			
137	**GFt**		>0.15	0.15 to 0.11	<0.10						
138	**GSph**		<0.21	0.21 to 0.25	>0.25						
139	**GFmt**		<0.13	0.13 to 0.19	0.20 to 0.25	>0.25					
140	**NFt**		>0.20	0.20 to 0.15	0.14 to 0.12	<0.11					
141	**NSst**		<0.15	>0.14							
142		**SstB/SstG**	>0.50	0.50 to 0.45	0.44 to 0.30	0.29 to 0.25	0.24 to 0.20	<0.19			
143		**SstB/SstN**	>0.50	0.50 to 0.45	0.44 to 0.35	0.34 to 0.30	0.29 to 0.25	0.24 to 0.20	<0.19		
144		**SstSg/SstG**	<0.75	0.75 to 0.85	>0.85						
145		**SstSg/SstN**	>0.83	0.83 to 0.75	0.74 to 0.65	<0.65					
146		**SstB/SphSg**	>0.72	0.72 to 0.65	0.64 to 0.55	0.54 to 0.35	< 0.35				
147		**SstB/SphG**	>0.71	0.71 to 0.65	0.64 to 0.55	0.54 to 0.40	0.39 to 0.25	<0.25			
148		**SstB/SphN**	>0.70	0.70 to 0.65	0.64 to 0.55	0.54 to 0.40	0.39 to 0.25	<0.25			
149		**SstSg/SphG**	<0.85	0.85 to 0.94	0.95 to 1.04	>1.04					
150		**SstSg/SphN**	<0.85	0.85 to 0.94	0.95 to 1.04	1.05 to 1.08	>1.08				
151		**SstB/FtSg**	<0.50	>0.49							
152		**SstB/FtG**	>0.78	0.78 to 0.60	0.59 to 0.50	0.49 to 0.45	0.44 to 0.30	<0.30			
153		**SstB/FtN**	>0.89	0.89 to 0.80	0.79 to 0.60	0.59 to 0.50	0.49 to 0.30	< 0.30			
154		**SstB/FmtSg**	>0.81	0.81 to 0.70	0.69 to 0.50	0.49 to 0.29	<0.29				
155		**SstB/FmtG**	>0.84	0.84 to 0.75	0.74 to 0.65	0.64 to 0.55	0.54 to 0.30	<0.29			
156		**SstB/FmtN**	>0.85	0.85 to 0.75	0.74 to 0.65	0.64 to 0.55	0.54 to 0.35	0.35 to 0.25	<0.25		
157		**SstSg/FmtG**	<1.00	1.00 to 1.09	1.10 to 1.19	1.20 to 1.33	>1.33				
158		**SstSg/FmtN**	<1.05	1.05 to 1.20	1.21 to 1.30	1.31 to 1.35	>1.35				

Character	type I	type II	0	1	2	3	4	5	6	7	8
159		**SstG/FmtN**	<1.30	1.30 to 1.39	1.40 to 1.44	>1.44					
160		**SphB/SphSg**	<1.05	1.05 to 1.14	1.15 to 1.21	<1.21					
161		**SphB/SphG**	<0.95	0.95 to 1.04	1.05 to1.19	>1.19					
162		**SphB/SphN**	<0.90	0.90 to 0.99	1.00 to 1.09	1.10 to 1.17	>1.17				
163		**SphSg/SphG**	>0.93	0.93 to 0.85	<0.85						
164		**SphSg/SphN**	>0.97	0.97 to 0.95	0.94 to 0.85	<0.85					
165		**SphB/SstSg**	>1.24	1.24 to1.02	1.01 to 0.95	<0.95					
166		**SphB/SstG**	>0.94	0.94 to 0.90	0.89 to 0.85	0.84 to 0.80	<0.80				
167		**SphB/SstN**	>0.89	0.89 to0.85	0.84 to 0.80	0.79 to 0.75	<0.75				
168		**SphB/FtSg**	<2.15	2.15 to2.60	2.61 to 3.60	>3.60					
169		**SphG/FtN**	<1.51	1.51 to 1.60	>1.60						
170		**SphB/FmtSg**	<1.05	1.05 to 1.15	1.16 to 1.25	1.26 to 1.30	1.31 to 1.47	>1.47			
171		**SphB/FmtG**	<1.10	1.10 to 1.19	1.20 to 1.29	1.30 to 1.39	1.40 to 1.49	1.50 to 1.55	>1.55		
172		**SphB/FmtN**	<1.10	1.10 to 1.19	1.20 to 1.29	1.30 to 1.39	1.40 to 1.49	1.50 to 1.58	>1.58		
173		**SphSg/FmtG**	<0.80	0.80 to 0.94	0.95 to 1.06	>1.06					
174		**SphSg/FmtN**	<0.90	0.90 to 1.05	>1.05						
175		**FtB/FtSg**	<1.40	1.40 to 1.55	1.56 to 1.61	1.62 to 2.16	>2.16				
176		**FtB/FtG**	<1.00	1.00 to 1.15	1.16 to 1.25	1.26 to 1.38	>1.38				
177		**FtB/FtN**	<1.00	1.00 to 1.15	1.16 to 1.25	1.26 to 1.38	>1.38				
178		**FtSg/FtG**	<0.55	0.55 to 0.64	0.65 to0.74	0.75 to 0.84	0.85 to 0.90	>0.90			
179		**FtSg/FtN**	>0.85	0.85 to 0.75	0.74 to 0.65	0.64 to 0.55	0.54 to 0.45	<0.45			
180		**FtB/SstSg**	>1.03	1.03 to 0.95	0.94 to 0.85	<0.85					
181		**FtB/SstG**	>0.96	0.96 to 0.85	0.84 to 0.75	0.74 to 0.65	<0.65				
182		**FtB/SstN**	>0.93	0.93 to 0.85	0.84 to 0.75	0.74 to 0.65	<0.65				
183		**FtSg/SstG**	>0.47	0.47 to 0.40	<0.40						
184		**FtSg/SstN**	>0.47	0.47 to 0.40	<0.40						
185		**FtB/SphSg**	>1.16	1.16 to 1.05	1.04 to 0.95	<0.95					
186		**FtB/SphG**	>1.14	1.14 to 1.05	1.04 to 0.95	0.94 to 0.75	<0.75				
187		**FtB/SphN**	>1.16	1.16 to 1.05	1.04 to 0.95	0.94 to 0.85	0.84 to 0.75	<0.75			
188		**FtSg/SphG**	>0.62	0.62 to 0.55	0.54 to 0.45	<0.45					
189		**FtSg/SphN**	>0.62	0.62 to 0.55	0.54 to 0.45	<0.45					
190		**FtB/FmtSg**	<0.90	0.90 to 1.04	1.05 to 1.14	1.15 to 1.24	1.25 to 1.34	1.35 to 1.41	>1.41		
191		**FtB/FmtG**	<0.95	0.95 to 1.15	1.16 to 1.48	>1.48					
192		**FtB/FmtN**	>1.53	1.53 to 1.35	1.34 to 1.20	1.21 to 0.95	<0.95				
193		**FtSg/FmtG**	>0.79	0.79 to 0.75	0.74 to 0.65	0.64 to 0.55	0.54 to 0.45	<0.45			

Character	type I	type II	0	1	2	3	4	5	6	7	8
194		FtSg/FmtN	>0.82	0.82 to 0.75	0.74 to 0.65	0.64 to 0.55	0.54 to 0.45	<0.45			
195		FtG/FmtN	<0.64	0.64 to 0.74	0.75 to 0.85	>0.85					
196		FmtB/FmtSg	<1.46	1.46 to 1.55	1.56 to 1.65	1.66 to 1.75	>1.75				
197		FmtB/FmtG	<1.46	1.46 to 1.55	1.56 to 1.65	1.66 to 1.75	1.76 to 1.84	>1.84			
198		FmtB/FmtN	<1.45	1.45 to 1.56	1.55 to 1.66	1.65 to 1.76	1.75 to 1.86	>1.86			
199		FmtB/SstSg	>1.52	1.52 to 1.24	<1.24						
200		FmtB/SstG	>1.10	1.10 to 1.05	<1.05						
201		FmtB/SstN	>1.09	1.09 to 1.05	<1.05						
202		FmtSg/SstG	>0.80	0.80 to 0.74	<0.74						
203		FmtG/SstN	>0.69	<0.70							
204		FmtB/SphSg	<1.44	1.44 to 1.80	>1.80						
205		FmtB/SphG	<1.25	1.25 to 1.34	1.35 to 1.41	>1.42					
206		FmtB/SphN	<1.20	1.20 to 1.34	1.35 to 1.44	>1.44					
207		FmtG/SphN	<0.98	>0.97							
208		FmtB/FtSg	<1.97	1.97 to 2.49	2.50 to 2.80	2.81 to 3.90	>3.90				
209		FmtB/FtG	<1.65	1.65 to 1.79	>1.79						
210		FmtB/FtN	<1.60	1.60 to 1.70	1.71 to 1.80	>1.80					
211		FmtSg/FtG	<1.22	1.22 to 1.30	1.31 to 1.40	>1.40					
212		FmtSg/FtN	<1.22	1.22 to 1.35	>1.35						
213		FmtG/FtN	<1.17	1.17 to 1.25	1.26 to 1.35	1.36 to 1.45	>1.45				
214		biSst/biSph	<-0.85	.-0.85 to -0.50	.-0.49 to -0.30	.-0.29 to -0.16	>-0.15				
215		biSst/biFt	<-0.50	.-0.50 to -0.15	.-0.14 to -0.05	>-0.05					
216		biSst/biFmt	<-2.10	.-2.10 to -1.51	.-1.50 to -1.00	.-0.99 to -0.35	.-0.34 to -0.25	>-0.25			
217		biSst/biFmo	<-1.75	.-1.75 to -1.35	.-1.34 to -0.50	.-0.49 to -0.35	.-0.34 to -0.25	>-0.25			
218		biSst/biMnf	<0.50	0.50 to 0.65	0.66 to 0.73	>0.73					
219		biSph/biFt	<0.23	0.23 to 0.30	0.31 to 0.60	>0.60					
220		biSph/biFmt	<-0.50	.-0.50 to -0.41	.-0.40 to-0.31	.-0.30 to -0.21	>-0.21				
221		biSph/biFmo	<-0.40	.-0.40 to -0.25	.-0.24 to -0.15	.-0.14 to -0.10	>-0.10				
222		biSph/biMnf	<0.65	0.65 to 0.73	>0.73						
223		biFt/biFmt	<-2.60	.-2.60 to -1.00	.-0.99 to -0.60	.-0.59 to 0.35	.-0.35 to -0.20	>-0.20			
224		biFt/biFmo	<-2.10	.-2.10 to -1.00	.-0.99 to -0.20	>-0.20					
225		biFt/biMnf	<0.60	0.60 to 0.69	>0.69						
226		SstSph/BN	<0.57	0.57 to 0.64	0.65 to 0.70	0.71 to 0.80	0.81 to 0.90	>0.90			
227		SstSph/BG	<0.59	0.59 to 0.64	0.65 to 0.70	0.71 to 0.80	0.81 to 0.90	>0.90			
228		SstSph/BSg	<0.63	0.63 to0.90	0.91 to 1.40	>1.40					

Character	type I	type II	0	1	2	3	4	5	6	7	8
229		**SphFt/SstFt**	<0.68	0.68 to 0.80	0.81 to 0.90	0.91 to 1.00	>1.00				
230		**SphFt/BN**	<0.30	0.30 to 0.35	0.36 to 0.50	0.51 to 0.60	>0.60				
231		**SphFt/BG**	<0.30	0.30 to 0.35	0.36 to 0.45	>0.45					
232		**SphFt/BSg**	<0.35	0.35 to 0.39	0.40 to 0.45	0.46 to 0.50	0.51 to 0.60	0.61 to 0.70	0.71 to 0.8(	0.81 to 0.9(	>0.90
233		**SphFt/GN**	<3.71	3.71 to 5.00	5.01 to 6.00	6.01 to 7.00	>7.00				
234		**FtFmt/SstFmt**	<0.42	0.42 to 0.50	0.51 to 0.60	>0.60					
235		**FtFmt/SstFt**	<0.58	0.58 to 0.69	0.70 to 0.79	0.80 to 0.89	0.90 to 0.99	1.00 to 1.09	>1.09		
236		**FtFmt/SphFt**	<0.95	0.95 to 1.04	1.05 to 1.09	1.10 to 1.14	1.15 to 1.19	>1.19			
237		**FtFmt/SstSph**	>0.89	0.89 to0.80	0.79 to 0.67	<0.67					
238		**FtFmt/BN**	<0.26	0.26 to 0.35	0.36 to 0.40	0.41 to 0.50	>0.50				
239		**FtFmt/BG**	<0.25	0.25 to 0.35	0.36 to 0.45	0.46 to 0.55	>0.55				
240		**FtFmt/BSg**	<030	0.30 to 0.40	0.41 to 0.50	0.51 to 0.60	0.61 to 0.70	0.71 to 0.80	>0.80		
241		**FtFmt/SgN**	<0.76	0.76 to 1.53	1.54 to 1.90	>1.90					
242		**FtFmt/GN**	<4.20	4.20 to 5.00	5.01 to 5.50	5.51 to 6.50	6.51 to7.50	>7.50			
243		**SstFt/SstFmt**	<0.55	0.55 to 0.67	0.68 to 0.83	0.84 to 0.95	>0.95				
244		**SstFt/FtFmt**	<1.00	1.00 to 1.33	>1.33						
245		**SstFt/SphFt**	<1.24	>1.23							
246		**SstFt/SphFmt**	<1.00	1.00 to 1.09	1.10 to 1.19	1.20 to 1.29	>1.29				
247		**SstFt/SstSph**	>1.02	0.91 to 0.86	0.85 to 0.76	0.75 to 0.70	<0.70				
248		**SstFt/BN**	<0.60	>0.59							
249		**SstFt/BG**	<0.53	0.53 to 0.60	0.61 to 0.65	>0.65					
250		**SstFt/BSg**	<0.59	0.59 to 0.70	0.71 to 0.75	0.76 to 0.80	0.81 to 0.90	0.91 to1.00	>1.00		
251		**SstFt/SgG**	<1.30	1.30 to 1.39	1.40 to 1.49	>1.49					
252		**SphFmt/SstFt**	>0.90	0.90 to 0.81	<0.81						
253		**SphFmt/BN**	<0.30	0.30 to 0.35	0.36 to 0.40	0.41 to 0.45	>0.45				
254		**SphFmt/BG**	<0.32	0.32 to 0.35	0.36 to 0.40	0.41 to 0.45	>0.45				
255		**SphFmt/BSg**	<0.35	0.35 to 0.40	0.41 to 0.45	0.46 to 0.50	0.51 to0.60	0.61 to 0.70	>0.70		
256		**SphFmt/GN**	<3.25	3.25 to 3.50	3.51 to 4.00	4.01 to 4.50	4.51 to 5.50	>5.50			
257		**SstFmt/SstFt**	>1.80	1.80 to 1.75	1.74 to 1.65	1.64 to 1.50	<1.50				
258		**SstFmt/FtFmt**	>1.97	1.97 to 1.85	1.84 to 1.70	1.69 to 1.50	<1.50				
259		**SstFmt/BN**	<0.70	0.70 to 0.85	0.86 to 0.94	0.95 to 1.04	>1.04				
260		**SstFmt/BG**	<0.67	0.67 to 0.79	0.80 to 0.89	0.90 to 0.99	>0.99				
261		**SstFmt/BSg**	<0.79	0.79 to 0.85	0.86 to 1.05	1.06 to 1.35	1.36 to 1.50	>1.50			
262		**SstFmt/SgG**	<1.60	1.60 to 2.50	>2.50						
263		**SgG/BSg**	<0.18	0.18 to 0.29	0.30 to 0.34	0.35 to 0.39	0.40 to 0.44	0.45 to 0.49	>0.49		

Character	type I	type II	0	1	2	3	4	5	6	7	8
264		**BSg/BG**	<0.40	0.40 to 0.50	0.51 to 0.70	0.71 to 0.81	>0.81				
265		**SgN/BSg**	<0.27	0.27 to 0.30	0.31 to 0.40	0.41 to 0.50	>0.50				
266		**BSg/BN**	<0.60	0.60 to 0.69	0.66 to 0.70	0.70 to 0.77	>0.77				
267		**GN/SgG**	<0.20	0.20 to 0.30	0.31 to 0.47	>0.47					
268		**SgG/BG**	<0.22	0.22 to 0.29	0.30 to 0.39	0.40 to 0.49	>0.49				
269		**SgG/SgN**	<0.71	0.71 to 0.80	0.81 to 0.90	>0.90					
270		**SgG/BN**	<0.21	0.21 to 0.30	>0.30						
271		**GN/SgN**	<0.20	0.20 to 0.33	>0.33						
272		**SstSph/SstFt**	<1.10	1.10 to 1.20	1.21 to 1.30	1.31 to 1.50	>1.50				
273		**SstSph/SstFmo**	<0.76	0.76 to 0.85	>0.85						
274		**SstSph/SstFmt**	<0.81	0.81 to 0.85	>0.85						
275		**SstFt/FtFmo**	<0.90	0.90 to 1.03	>1.03						
276		**SstFmo/FtFmo**	<1.05	>1.04							
277		**SstFmo/FtFmt**	<1.60	1.60 to1.74	>1.74						
278		**SphFt/FtFmt**	<1.90	1.91 to 2.19	>2.19						
279		**FtFmo/FtFmt**	<0.90	0.90 to 1.01	>1.01						
280	**BL**		<0.95	0.95 to1.06	>1.06						
281	**BK**		<0.06	>0.05							
282	**BEn**		<0.06	0.06 to 0.10	0.11 to 0.13	>0.13					
283	**BAst**		<0.11	>0.10							
284	**LSst**		<0.10	0.10 to 0.14	0.15 to 0.19	>0.19					
285	**SstSphn**		<0.06	0.06 to 0.09	0.10 to 0.13	>0.13					
286	**Kast**		<0.09	0.09 to 0.13	>0.13						
287		**BL/LSph**	<0.11	0.11 to 0.13	>0.13						
288		**BL/LK**	<0.70	>0.69							
289		**BL/SstK**	<0.74	>0.73							
290		**BL/SstAst**	<1.50	1.50 to 1.90	>1.90						
291		**BL/SphEn**	<0.80	0.80 to 0.85	0.86 to 0.90	0.91 to 0.94	>0.94				
292		**BL/SphAst**	<0.85	0.85 to 1.02	1.03 to 1.18	>1.18					
293		**BL/KEn**	>0.98	0.98 to 0.93	0.92 to 0.80	<0.80					
294		**BL/KAst**	>1.22	1.22 to 1.05	<1.05						
295		**LSst/LSph**	>1.05	1.05 to 0.96	0.95 to 0.86	<0.86					
296		**LSst/SstEn**	>0.80	0.80 to 0.70	<0.70						
297		**LSst/SstAst**	>1.22	1.22 to 1.10	1.09 to 0.97	<0.97					
298		**LSst/ShEn**	<0.90	0.90 to 0.95	>0.95						

Character	type I	type II	0	1	2	3	4	5	6	7	8
299		LSst/SphAst	>1.45	1.45 to 1.35	1.34 to 1.25	1.24 to 1.15	1.14 to0.95	<0.95			
300		LSst/KEn	<1.03	1.03 to 0.95	0.94 to 0.85	<0.85					
301		LSst/KAst	<1.56	1.56 to 1.40	1.39 to 1.25	1.24 to 1.10	<1.10				
302		SstSph/LK	>1.19	1.19 to 1.10	1.09 to 1.00	0.99 to 0.90	<0.90				
303		SstSph/LEn	<0.48	0.48 to 0.54	>0.55						
304		SstSph/LAst	<0.66	0.66 to 0.77	>0.77						
305		SstSph/StEn	<0.79	0.79 to 0.92	0.93 to 1.04	>1.04					
306		SstSph/StAst	<0.63	0.63 to 0.70	>0.70						
307		SstSph/SphEn	<0.71	0.71 to 0.81	>0.81						
308		SstSph/KAst	<0.81	>0.80							
309		SphK/StSph	<0.70	0.70 to 0.80	<0.80						
310		LAst/KAst	>0.15	<0.16							
311		SphK/EnAst	>0.78	0.78 to 0.69	<0.69						
312		EnK/LEn	<0.99	0.99 to 1.41	>1.41						
313		EnK/LAst	<0.92	0.92 to 1.00	>1.00						
314		EnK/SstEn	<0.92	0.92 to 1.12	1.13 to 1.23	>1.23					
315		EnK/SstAst	<0.88	0.88 to 0.94	>0.94						
316		EnK/KAst	<0.78	0.78 to 0.85	>0.85						
317		LAst/KEn	<0.85	0.85 to 0.90	>0.90						
318		BAst/LSph	<0.80	0.80 to 0.85	0.85 to 0.90	>0.90					
319		BAst/LK	<0.84	0.84 to 0.89	0.90 to 0.95	>0.95					
320		BAst/LEn	<1.24	1.24 to 1.48	1.49 to 1.55	>1.55					
321		BAst/SstSph	<1.80	1.80 to 2.12	>2.12						
322		BAst/SstK	<1.46	1.46 to 1.90	1.91 to 2.33	>2.33					
323		BAst/SstEn	<1.20	1.21 to 1.31	>1.31						
324		BAst/SstAst	<1.15	>1.14							
325		BAst/SphAst	<0.90	0.90 to 1.09	1.10 to 1.28	>1.28					
326		BAst/KEn	<1.20	1.20 to 1.45	1.45 to 1.74	>1.74					
327		BAst/KAst	<1.15	1.15 to 1.24	1.25 to 1.34	>1.34					
328		LAst/LSst	<0.72	0.72 to 0.78	0.79 to 0.85	>0.85					
329		LAst/LEn	<0.95	>0.94							
330		LAst/SstEn	>0.80	0.80 to 0.70	<0.70						
331		LAst/SphEn	<0.66	0.66 to 0.78	0.79 to 0.90	>0.90					
332		EnAu/PoMs	<1.26	1.26 to 1.39	1.40 to 1.53	>1.53					
333		EnAu/SPo	<0.71	0.71 to 0.79	0.80 to 0.88	>0.88					

Character	type I	type II	0	1	2	3	4	5	6	7	8
334		**EnAu/APo**	<0.90	0.90 to 0.96	>0.96						
335		**EnAu/SMs**	<0.44	0.44 to 0.49	0.50 to 0.61	>0.61					
336		**EnAu/PMs**	<0.75	0.75 to 0.84	0.85 to 0.94	>0.94					
337		**EnPo/AstAu**	<0.70	0.70 to 0.76	0.77 to 0.83	<0.83					
338		**EnPo/AstPo**	<0.49	0.49 to 0.71	0.72 to0.77	0.78 to 0.88	>0.88				
339		**EnPo/AstMs**	<0.75	>0.74							
340		**EnPo/AuMs**	<1.19	1.19 to 1.35	>1.35						
341		**EnPo/AuS**	<0.87	0.87 to 0.96	>0.96						
342		**EnPo/AuP**	<0.78	0.78 to 1.03	>1.03						
343		**EnPo/PoMs**	<1.19	1.19 to 1.34	1.35 to 1.50	1.51 to 1.66	>1.66				
344		**EnPo/SPo**	<0.77	0.77 to 0.84	0.85 to 0.92	>0.92					
345		**PoK/AstAu**	<0.78	0.78 to 1.19	>1.19						
346		**PoK/AstPo**	<0.83	0.83 to 1.25	>1.25						
347		**PoK/AstMs**	<1.19	1.19 to 1.28	>1.28						
348		**PoK/AuS**	<0.99	>0.98							
349		**PoK/AuA**	<0.88	0.89 to 0.99	>0.99						
350		**PoK/APo**	<0.87	>0.86							
351		**EnAu/EnMs**	<0.84	0.84 to 0.92	>0.92						
352		**EnAu/AstAu**	<0.67	0.67 to 0.75	>0.75						
353		**EnAu/AstPo**	<0.68	0.68 to 0.73	0.74 to 0.79	0.80 to 0.85	>0.85				
354		**EnAu/AstMs**	<0.69	0.69 to 0.74	>0.74						
355		**EnAu/AstA**	<0.41	0.41 to 0.45	>0.45						
356		**EnAu/AuMs**	<1.02	1.02 to 1.25	>1.25						
357		**EnAu/Aus**	<0.88	>0.87							
358		**AuK/EnS**	<0.55	0.55 to 0.62	>0.62						
359		**AuK/EnA**	<0.56	>0.57							
360		**AuK/AstAu**	<1.15	>1.14							
361		**AuK/AstPo**	<1.08	>1.07							
362		**AuK/AstMs**	<1.16	1.16 to 1.36	>1.36						
363		**AuK/AuMs**	<1.66	>1.65							
364		**AuK/AuA**	<0.84	0.84 to 0.94	>0.94						
365		**AuK/SPo**	<1.08	1.08 to 1.18	>1.18						
366		**AuK/APo**	<0.82	>0.81							
367		**PoK/EnAu**	<1.26	1.26 to 1.50	>1.50						
368		**PoK/EnPo**	<1.24	1.24 to 1.45	>1.45						

Character	type I	type II	0	1	2	3	4	5	6	7	8
369		**PoK/EnS**	<0.64	0.64 to 0.74	>0.74						
370		**PoK/EnA**	<0.54	0.54 to 0.59	>0.59						
371		**EnK/APo**	<1.89	>1.88							
372		**AstK/EnAu**	<2.21	2.22 to 2.57	>2.57						
373		**AstK/EnPo**	<2.01	2.02 to 2.34	>2.34						
374		**AstK/EnMs**	<1.53	1.54 to 2.05	2.06 to 2.27	2.28 to 2.49	2.50 to 2.72	>2.72			
375		**AstK/EnS**	<1.23	>1.24							
376		**AstK/EnA**	<0.95	0.96 to 1.06	1.07 to 1.17	>1.17					
377		**AstK/AstMs**	<1.41	1.42 to 1.53	1.54 to 2.05	>2.05					
378		**AstK/AuS**	<1.75	>1.74							
379		**AstK/AuP**	<1.70	1.70 to 2.04	>2.04						
380		**AstK/AuA**	<1.45	1.46 to 1.62	>1.62						
381		**AstK/SPo**	<1.43	1.43 to 1.55	1.56 to 2.01	2.02 to 2.13	>2.13				
382		**AstK/PPo**	<1.94	1.94 to 2.40	2.41 to 2.74	>2.74					
383		**AstK/APo**	<1.42	1.42 to 1.60	1.61 to 2.40	2.41 to 2.59	>2.59				
384		**EnPo/PPo**	>1.08	1.08 to 1.00	<1.00						
385		**EnPo/PMs**	<0.78	0.78 to 0.87	0.88 to 0.97	>0.97					
386		**EnMs/EnS**	<0.51	>0.50							
387		**EnMs/EnA**	<0.51	>0.50							
388		**EnMs/AstAu**	<0.50	0.50 to 0.59	0.60 to 0.68	>0.68					
389		**EnMs/AstPo**	<0.59	0.58 to 0.68	>0.68						
390		**EnMs/AstMs**	<1.04	1.04 to 1.11	1.12 to 1.24	>1.24					
391		**EnMs/AstS**	<0.44	0.44 to 0.75	>0.75						
392		**EnMs/AstP**	<2.17	>2.16							
393		**EnS/AstAu**	<1.60	>1.59							
394		**EnS/AstPo**	<1.69	>1.68							
395		**EnS/AstMs**	<1.38	1.39 to 1.47	1.48 to 1.62	>1.62					
396		**EnS/AstP**	<3.01	3.01 to 3.58	3.59 to 4.09	>4.09					
397		**EnS/AstA**	<0.77	>0.76							
398		**EnS/AuMs**	<2.09	2.09 to 2.99	3.00 to 3.42	3.42 to 3.84	3.84 to 4.06	>4.06			
399		**EnS/AuS**	<0.95	0.95 to 1.52	1.53 to 1.68	>1.68					
400		**EnS/AuP**	<2.08	2.09 to2.31	2.32 to 2.54	>2.54					
401		**EnS/PoMs**	<2.29	2.29 to 2.49	2.50 to 2.70	2.71 to 2.91	>2.91				
402		**EnS/SPo**	<1.55	1.55 to 1.69	>1.69						
403		**EnS/PPo**	<2.11	2.11 to 2.35	2.36 to 2.60	>2.60					

Character	type I	type II	0	1	2	3	4	5	6	7	8
404		EnS/APo	<1.62	1.62 to 1.80	>1.80						
405		EnS/SMs	<0.87	0.87 to 0.93	0.94 to 1.02	>1.02					
406		EnS/PMs	<1.54	1.54 to 1.73	1.74 to 1.90	>1.90					
407		EnA/AstAu	<1.60	1.60 to 1.86	>1.86						
408		EnA/AstPo	<1.75	1.75 to 1.90	>1.90						
409		EnA/AstMs	<1.75	1.75 to 1.90	>1.90						
410		EnA/AstP	<3.57	3.57 to 4.10	4.11 to 5.17	>5.17					
411		EnA/AstA	<0.84	0.84 to 0.91	0.92 to 0.99	>0.99					
412		EnA/AuMs	<2.46	2.46 to 2.70	2.71 to 2.95	>2.95					
413		EnA/PoMs	<2.97	>2.96							
414		EnA/SPo	<1.61	1.61 to 1.73	1.74 to 1.98	>1.98					
415		EnA/PPo	<2.34	2.34 to 2.57	2.58 to 2.81	>2.81					
416		AstAu/AstMs	<0.90	0.90 to 1.10	1.11 to 1.31	>1.31					
417		AstAu/AstS	<0.60	>0.59							
418		AstAu/AstP	<2.96	2.96 to 3.36	>3.36						
419		AstAu/AuS	<0.85	0.85 to 0.95	>0.95						
420		AstAu/AuP	<1.13	>1.12							
421		AstAu/PPo	<1.10	>1.09							
422		AstAu/APo	<0.79	0.80 to 0.93	0.94 to 1.07	>1.07					
423		AstAu/PMs	>1.41	1.41 to 1.28	1.28 to 1.15	<1.15					
424		AstPo/AstMs	>1.21	1.21 to 1.11	1.10 to 1.01	1.00 to 0.91	<0.91				
425		AstPo/AuP	<0.92	0.92 to 1.15	>1.15						
426		AstPo/PPo	<1.07	>1.06							
427		AstPo/APo	<.78	0.78 to 1.49	1.49 to 1.63	1.64 to 1.78	>1.78				
428		biAu/AuBAu	<0.46	0.46 to 0.55	>0.55						
429	LO		<0.24	0.24 to 0.30	0.31 to 0.35	0.36 to 0.40	>0.40				
430	LIn		<0.16	0.16 to0.20	0.21 to 0.25	0.26 to 0.30	>0.30				
431	LAst		<0.11	0.11 to 0.15	0.16 to 0.20	>0.20					
432	OPIn		<0.15	0.15 to 0.20	0.21 to 0.30	>0.30					
433		LAst/LIn	>2.21	2.21 to 1.96	1.95 to 1.70	1.69 to 1.44	1.43 to 1.31	<1.31			
434		LAst/LO	<0.93	0.93 to 1.01	1.02 to 1.10	>1.10					
435		LAst/LMs	>0.75	0.75 to 0.71	<0.71						
436		LAst/AstIn	<0.90	0.90 to 1.07	1.08 to 1.16	1.16 to 1.24	>1.24				
437		LAst/AstO	>1.09	1.09 to 1.00	0.99 to 0.90	<0.90					
438		LAst/AstOp	<0.93	0.93 to 1.00	1.01 to 1.09	>1.09					

Character	type I	type II	0	1	2	3	4	5	6	7	8
439		**LAst/InO**	<1.00	1.00 to 1.42	1.43 to 1.91	>1.91					
440		**LAst/InMs**	<0.65	0.65 to 0.74	0.75 to 0.84	>0.84					
441		**LAst/OMs**	<0.98	0.98 to 1.11	1.12 to 1.26	>1.26					
442		**LAst/OOp**	<1.31	1.31 to 1.47	1.48 to 1.65	1.66 to 1.82	>1.82				
443		**LAst/MsOp**	<0.62	0.62 to 0.70	0.71 to 0.83	0.84 to 0.90	>0.90				
444		**LIn/LMs**	<0.28	0.28 to 0.41	>0.41						
445		**LIn/AstIn**	<0.62	0.62 to 0.77	>0.77						
446		**LIn/AstO**	<0.42	0.42 to 0.60	0.61 to 0.78	>0.78					
447		**LIn/AstOp**	<0.71	>0.70							
448		**LIn/InO**	<0.51	0.52 to 0.84	>0.84						
449		**LIn/InMs**	<0.54	>0.53							
450		**LIn/OMs**	<0.43	0.43 to 0.56	0.57 to 0.73	0.74 to 0.88	>0.88				
451		**LIn/MsOp**	<0.32	0.32 to 0.48	>0.48						
452		**LO/AstIn**	<0.85	0.85 to 0.96	0.97 to 1.07	1.08 to 1.18	1.19 to 1.29	1.30 to1.44	>1.44		
453		**LO/AstO**	<1.00	1.00 to 1.10	1.11 to 1.20	1.21 to 1.30	1.31 to 1.40	>1.40			
454		**LO/InO**	<1.20	1.20 to 1.40	1.41 to 1.50	1.51 to 1.63	>1.63				
455		**LO/InMs**	<0.60	0.60 to 0.80	0.81 to 0.90	0.91 to 1.03	>1.03				
456		**LO/OMs**	<0.90	0.90 to 1.00	1.01 to 1.20	1.21 to 1.30	1.31 to 1.44	>1.44			
457		**AstIn/AstOp**	<0.97	0.97 to 1.02	1.03 to 1.08	>1.08					
458		**AstIn/OOp**	<1.18	1.18 to 1.30	>1.30						
459		**AstIn/MsOp**	<0.65	0.65 to 0.75	0.76 to 0.85	0.86 to 0.95	>0.95				
460		**AstO/OOp**	<1.18	1.18 to 1.33	1.34 to 1.48	1.49 to 1.63	>1.63				
461		**AstO/MsOp**	<0.68	0.68 to 0.79	>0.79						
462		**AstOp/OOp**	<1.21	1.21 to 1.25	1.26 to 1.31	1.32 to 1.37	1.38 to 1.48	1.48 to 1.58	>1.58		
463		**AstOp/MsOp**	<0.71	0.71 to 0.74	0.75 to 0.81	>0.81					
464		**InO/OOp**	<0.86	0.86 to 0.98	0.99 to 1.10	1.11 to1.22	>1.22				
465		**InO/MsOp**	<0.45	0.45 to 0.55	0.56 to 0.77	>0.78					
466		**InMs/OOp**	<1.12	1.12 to 1.50	1.51 to 1.70	1.71 to 1.90	1.91 to 2.10	>2.10			
467		**InMs/MsOp**	<0.94	>0.93							
468		**LIn/LOp**	<1.16	>1.15							

Appendix 2

Taxon	Character 111111111122222222223333333333444444444455555555556 123456789012345678901234567890123456789012345678901234567890
Gorilla gorilla	10000000202000010???000011????021000000000011111011000011 10?0
Pan troglodytes	100201112021000002000100003000021000000000101111112000111 0?1
Sterkfontein 5	021001102022000002110020001110021001100001012110002110 0?10?1
Knmer 406	0?000001202100010???000001?20102100000001001121101110000?10?0
Knmer 1813	0111011120110000011100001001 1?11100010100101111010110020 10?0
Knmer 1470	00000111200300000001?00000011?10000?121??10?0?????0?001010??
Knmer 3883	1?111111202301100111000010202021110012000 10?111000211010 10?0
Knmer 3733	011211100023011004 01?22000011?321100020001001010100000101 0?1
Knmwt 15000°	????1????????????????000200?0?020110002000100200???00?10?1120
Eliye Springs	???11???????0110?211110 2?032002210011200011020 0???00102010??
Laetoli H18	1211001100230000020111220032101000000101010 00?0???0010?010?1
OH 9	11111011212300 0????????0???????10?00101101?021100000?0?01??0
Bodo 1	11110111 0?1301100201101000 3??1100?0????????????????????????
Saldanha	?0121?112?2?1??0020000221032101000 0????0?11?211000200020????
Ndutu	?????????????????0?????????11???????1?01111?2110102?111??0?0
Kabwe	12100111212201100211101010322012110102100101311011?011???0?1
Omo Kibish 2	????????????0??0010111??0030001?????1211?0003?1?????202010??
Jbebel Irhoud 1	021210111121011 0??01000200111010000 0120001010???????100?10?0
Salé	????????????0??0?2111120?0311012100 0?21001001010002 1?02000?1
Sambungmacan 1	?0001111????11100201110110312011000001111110101010112 01010?1
Modjokerto°	?20011?1??011000?201?01000312010?????2???100200??000100?00??
Ngandong 1	1001110110230110040111110031201100000211110111 0??021102110?1
Ngandong 2°	01000101200110000211001 20??????????????????????????????????
Ngandong 3	??????????????0????1????0320011 1?10??1??10020??????10??????
Ngandong 5	10001101112100000 0??01011030201010 00?111??0????????????????
Ngandong 6	10111111112110000401021110312012000001110101111000201020 10?0
Ngandong 7	0002111100200110040011111031200110100111000020 0???0120211200
Ngandong 10°	0101111111211000010110011032201100 10?1110101211010112 0111201
Ngandong 11	11011000211000000200111100302011100112110100110??12110111111
Ngandong 12	01001110211110000400121110312021110010110110111000212 0211121
Ngawi	000111100?211000020112111031212200101211010011101000101112 2?
Sangiran 2	????1???????1000??00110000112021000000200010121101121000?00?0
Sangiran 3	?????????????????????????020?1211?0??2???10?21????0?00???0??
Sangiran 4	?????????????????????????0?021111?00121001111110100100 2010?1
Sangiran 10°	????1???????????0????1???0030212100 00?110?0002110?1000020?20?
Sangiran 12	???????????????????122??01120210?01121001?0211111210 01010?0
Sangiran 17	10100111202310000200102000300012000101110100201011200010 10?0
Sangiran 26	??
Sangiran 38	????????????0?????011????030001211011010101 0210???10000?0220
Sinanthropus 3	011211102123111002111121103121110000111001012110012120 2010?1
Sinanthropus 10	01121010212110000211101?103110201?0011100100211001 0?102110?1
Sinanthropus 11	01021010202111100211112210302110000001110 10?21100111100??0?0
Sinanthropus 12	0102111021221000020 11?1100301110101011110110211001 20?02010?0
Dali	0011011121110000020100020 03??011101012111111111000212 02010?1
Maba	??1201112??301100????0?30??1???0100??????1??????????0???????
Narmada	?01?0???????000???0?101?003??11??1000?1001??????????0???????
Arago XXI et XLVII	011101112113000002 11??0000011?011?1??1110??????????????????
Petralona	12100111211311100201000100322011001 0?111001030000020202010?1
Monte Circeo	011001112023011003010202002120111000?2000100301000 0?200?0310
Swanscombe	?????????????????????????0022?120?00?100010 0??????00000?0200
La Chapelle aux Saints	00020101202310001301000200002?2000001000010121 0???0110100310
La Quina H5	?10111012023100012010012001020200010020001002 00???00200?0310
Spy 1	000211012023100000??0013103020100?10?20101??2010000?100?120?
La Ferrassie 1	000011012023000000??002300321010101002000100301000011 0100200
Amud 1	011111012?2311100201001 31???1?0010111200010011100000100?0200
Shanidar 5	000210012113011000??00030????0101?0?010??????????????????????
Skhull V	?11211010?23011000??00030032101010111200011010100 00?101010?1

```
              Character           111111111122222222223333333333444444444455555555556
Taxon                    123456789012345678901234567890123456789012345678901234567890
-------------------------------------------------------------------------------------
WHL 50                   ?21011110?23?1100?0?1002003110001?00?111?10?110??000200?00?0
Gibraltar 2°             00001101000 3?00000??0012000110 10????1000??????????????????
Teshik Tash°             001101010023?000110110120020201011010001102 00??0000?0?0320
Engis 2°                 021201110003?000110100030021 2?20?????1???1000?????0??00?0200
Velika Pecina°           021111110001?0000?00??121???????????????????????????????????
Trinil                   ??1?10?1????100002?111?01031112211 0??2???1??210?????1??????
Zuttiyeh                 010201110013?1101?1?00031???????????????????????????????????
modern humans            021201112103000002010002003210100000010001000????00011110?0
Flores                   ??1001?????1010?211??1000122?1101000111010011000011 21?010?0

              Character                                  11111111111111111111
                         666666666777777777788888888889999999999000000000011111111112
Taxon                    123456789012345678901234567890123456789012345678901234567890
-------------------------------------------------------------------------------------
Gorilla gorilla          010110??010020001020010110001102111 1000??000000000000000032
Pan troglodytes          010110111000100011200100101011021 0???00??0010000000000000031
Sterkfontein 5           0101101100??21101010000110100102010110 0??2021001010011000032
Knmer   406              0101101001111100002100111100011100???00??2120022121011001432
Knmer  1813              011011101001111010001101000011001101120??1112220110011 0?0033
Knmer  1470              ?10001101?????0?1???????001???00??????????0?12002?0020?00133
Knmer  3883              00111010000121000000111110010?1211002021011?2010121010?01433
Knmer  3733              0001110010112110000111000000?1102120100110002201001001000001 2
Knmwt 15000°             010111011100 10?0001111000 01?210011010021001020100100101100 20
Eliye Springs            00000???1?????1?01101010111000 00?1111221001222122111 21?11410
Laetoli H18              000010??100000000111101100 1?010012111220??121?121201111?0232
OH 9                     010110101110201010011101101?0110121110 10?1120?12110132?10032
Bodo 1                   ????????????????????????????????????????????????????????????
Saldanha                 ????????????????????????????????????????????????????????????
Ndtu                     000000??111121?0100111100000101??2111221?1121110220032 1?1222
Kabwe                    ?00000??111110100020111100000112121111200013021112111102103 2
Omo Kibish 2             ?00000??11?121??10211111100101 12?????????1?2001??10?????0032
Jebel Irhoud 1           110000111011101111?????110010?11212 1111??0?0?121222?11101?232
Salé                     00???01111?120?010201011000010001111122112120012111 02202120?
Sambungmacan 1           ?0100011101120001111110100111010110012200 01011?22?00311?0310
Modjokerto°              ????????????????10111?10000?11???????21?0021202201011 1?1121
Ngandong  1              100??1111101200011011111 0?1?201?120111200 11????02??132??0330
Ngandong  2              ????????????????????????????????????????????????????????????
Ngandong  3              ?????01111?121??1?????11?????????1002???????????????????????
Ngandong  5              ?0??1???????????????????????????????????????????????????14??
Ngandong  6              110011101011101110011010000110 10?1?110200221002 02?1?32120310
Ngandong  7              110010101001210010011111001110121210122011120200221132?21432
Ngandong 10              1????110110120001001111100111011111102201 12?200022?1221?140?
Ngandong 11              100000101100200111111101000?1011????1021?10?001022?132??1410
Ngandong 12              0110111011011001102111110011101 2?21012200002011021012 20?1410
Ngawi                    ?000101011012000?02011110010101012?111???020210?1112?0?1410
Sangiran  2              0100101011000000010111000011210211011 0???11?200012?13010?010
Sangiran  3              ????????????????????????????????????????????????????????????
Sangiran  4              11???1101101201000011011111?11121211122 1??1202122111221 21432
Sangiran 10°             110??01?10?0?0?1?????????????1???????????????????????????????21
Sangiran 12              11?????01100?010??????01???????????????????????????????????
Sangiran 17              1011111010001000?0211101000101011201102 10?11002211010 1?21021
Sangiran 26              ?????00111?11010111100110000100 2?1110221101202002200310?0213
Sangiran 38              010??1??11?1101?????????????????????????????????????????????
Sinanthropus  3          ?101111111001010110111000 01?101212111221111202021111221110 22
Sinanthropus 10          100??1111000100?11011101????1???12111 2????1?????????????????
Sinanthropus 11          1111111011001000010011010010101211111 2?0??2????1211 1221?1412
Sinanthropus 12          110111111111100010211 1?1??1?1112?2111220?12?????22?1?22??10??
```

```
                 Character                                        111111111111111111111
                        666666666777777777788888888889999999999000000000011111111112
Taxon                   123456789012345678901234567890123456789012345678901234567890
----------------------------------------------------------------------------------------
Dali                    10000111110110001021101110101112121110 21?0130020021001021433
Maba                    ????????????????????????????????????????????????????????????
Narmada                 ?0?0111011?02000101011000010001 2?0???0210111100022111 0111223
Arago XXI et XLVII      ????????????????????????????????????????????????????????????
Petralona               000000101?????1?1???00??0000?012?10100???0120000111 1221?0121
Monte Circeo            000110001101101001100001000001020201022100011110210011110232
Swanscombe              010?????????????????????????????????????????????????????????
La Chapelle aux Saints  00000???100112011000110000000010212011111020111100111111 0432
La Quina H5             010000??110122001000001100000102121111210201210001111110 0432
Spy 1                   ???0011010012001?00110110000210011011?21?002121211112102142 0
La Ferrassie 1          110000011101100111000001001001020201122000021210211121121432
Amud 1                  010?00100100200011201011001001021101112 0?2021210111121??0432
Shanidar 5              ?000?00110???0001000101100000100 0?????21?20111002?1122100432
Skhull V                0000?01001112001111011110010011112010120??131212211111 2121022
WHL 50                  ?????0?000112000??????11????????1???????????????2???????????
Gibraltar 2°            ?????0??00?02?00111000000000200201002120?00100101110111?1433
Teshik Tash°            000000??000??2001000??0000100102?????0????0200120?1032021432
Engis 2°                ?0???0??0000??00110011000000011212111221121110002?10110?1432
Velika Pecina           ????????????????????????????????????????????????????????????
Trinil                  ????????????????????????????????????????????????????????????
Zuttiyeh                ????????????????????????????????????????????????????????????
modern humans           000000??010120101020111100000110121111211023121221111 2121422
Flores                  0?0101100101210001111?00101001001101010??2122001210101?10310

                 Character 111111111111111111111111111111111111111111111111111111111111
                        222222222333333333344444444445555555555666666666677777777778
Taxon                   123456789012345678901234567890123456789012345678901234567890
----------------------------------------------------------------------------------------
Gorilla gorilla         013051?2?000031102001??02???22??????001334230222022200432032
Pan troglodytes         013221120021020101100341035512123334112334130001144522355300
Stk5                    113?3?100110011020130450245511334345002111121332101121422032
Knmer 406               110?61?0?12001101000 1??03???00??????003000231443210021455141
Knmer 1813              110??01?00?001031?220112022223412122333223011100043433122312
Knmer 1470              0102411110000?10212113420344341233443431231023200233333 55312
Knmer 3883              110140000010110010101221122223312232222521212232001222123331 1
Knmer 3733              112??1100001011220300431023334212233340323123030022321343312
Knmwt 15000°            121??010110??1??????????????????????????????????????????????
Eliye Springs           110???0?0110??12??????0?00?0?44?50?0?5?3?4?01?00?4?5?33?5?00
Laetoli H18             010??00?01?00?1?1??201020?????401?????????????????????355401
OH 9                    110??????????3?11?20?????????????????????22????0???21????22?
Bodo 1                  ???0??000?100?01?01?1001111111400110153112121100022633354310
Saldanha                ?????????0?1????????????????????????????????????????????????
Ndtu                    203?????????????????????????????????????????????????????????
Kabwe                   2003200000100?10?02010012000224000001123221212100322212 13210
Omo Kibish 2            2?0??00????1????????????????????????????????????????????????
Jebel Irhoud 1          210?20000120011520020002000022400000220123001000022333355400
Salé                    ????????10??????????????????????????????????????????????????
Sambungmacan 1          110???000110??12?????????3????4??0?????1????2??0?4????3????0
Modjokerto°             ?10?????00???2????????0????0????0??0?????4????0????6????5???
Ngandong  1             1?02????00??0?1?1??100020?????400?????????????????????155401
Ngandong  2°            ??????0?001001152002000200003440000045323400100004663335540 0
Ngandong  3             ????????????????????????????????????????????????????????????
Ngandong  5             ????000?00010014121??1?1?00?3?30?01?4??33?0?11?0?55?2?13?3?1
Ngandong  6             1122??000011101610310002011134401000453122003320044533244302
Ngandong  7             1?01?00001000??220320002100022400000333222021010034433355410
Ngandong 10             ????00?00?00?12?????0?2???????40?0??????????????????????25?3?1
```

```
            Character 111111111111111111111111111111111111111111111111111111111111
                      222222222333333333344444444445555555555666666666677777777778
Taxon                 123456789012345678901234567890123456789012345678901234567890
-----------------------------------------------------------------------------------
Ngandong 11           112??00000100?1220230001111222302011333121021220033333013312
Ngandong 12           121?200?0111030320230002011134400000453122002100044433143402
Ngawi                 1?1?00001110011010310001101111400000343112121110034433243311
Sangiran  2           110???1?0????2?????????????????????????????????????????????
Sangiran  3           ????????????????????????????????????????????????????????????
Sangiran  4           1?1?????????????????????????????????????????????????????????
Sangiran 10°          ?10?????????????????????????????????????????????????????????
Sangiran 12           ????????????????????????????????????????????????????????????
Sangiran 17           110310101111111420310001112223412121333323121220033323233310
Sangiran 26           101?????????????????????????????????????????????????????????
Sangiran 38           ??????0?1??0??1?????????????????????????????????????????????
Sinanthropus  3       111?41100010011120110002012222400111353111122320132333355300
Sinanthropus 10       ?????????1?00??????3???????????????????????????????????35530?
Sinanthropus 11       11???01?00000?12202??1?2?22?3?40?12?4??32?1?11?0?44?3?35?3?1
Sinanthropus 12       ????00001111?0?10?10001100022400000222322121000033333355310
Dali                  1101110?0000021210211002000034400000353223002200033433255400
Maba                  ????????00???210??????????0????4??0?????????????????????3????0
Narmada               110??????02?????????????????????????????????????????????????
Arago XXI et XLVII    ???23010111121111210001001022301111223324121000033422355500
Petralona             110040000000010110100001112211301121121132121210022233133300
Monte Circeo          1103??0?0010011010120001100022400000333334121000055533355400
Swanscombe            ????????????????????????????????????????????????????????????
La Chapelle aux Saints 010?200000?011112?120002000022400000232223011000033333133300
La Quina H5           110?00??00001211101??0?1?00?2?42?00?2??32?1?10?0?33?3?13?4?0
Spy 1                 012?2???00?103110?1200011????400000333?????????????????144301
La Ferrassie 1        110??0?000000111??1??0?1?00?3?40?00?3??23?1?12?0?33?2?35?4?0
Amud 1                110??0??00?01?1??????0020????400000453?????????????????255400
Shanidar 5            ?10?200?00?011102?1200011????400000333?????????????????355300
Skhull V              010??00?0010?11????31?0?00?0?44?0??????3?4?31?00???????2?5?01
WHL 50                ?????00?00?0????????????????????????????????????????????????
Gibraltar 2°          010?200?0000011120030002000034400000352334001000046633255400
Teshik Tash           0103?00?0000011020031002000023400000352334001000045633355400
Engis 2°              010?????00?00?100?1310??0????????000453?????????????????355400
Velika Pecina°        ?????????????31020030??20???34??????453???22????0???33???40?
Trinil                ??????1?11?1??0?????????????????3??????????????????????3????0
Zuttiyeh              ???330?000111?1110131002000034400000353334011000055633355410
modern humans         010340000000011020030002000034400000453334001000056633355400
Flores                1100??02110??2?????????????3????????????????????????????????

            Character 111111111111111111112222222222222222222222222222222222222222
                      888888888999999999900000000001111111111222222222233333333334
Taxon                 123456789012345678901234567890123456789012345678901234567890
-----------------------------------------------------------------------------------
Gorilla gorilla       442211100004550111001202531433324?????2000110552123422453446
Pan troglodytes       121122411322331320000201230233213102210212222443123710003114
Sterkfontein 5        4422000000045510120001011204333230?11??44???111433553630446
Knmer  406            442210100104550111122211000433324????332?00?222333632302436
Knmer  1813           330012223213213101222210010100000?4??????????001322322203122
Knmer  1470           221112422222322233100211330233212112?114?22?1111122331003222
Knmer  3883           220022423212223112111201220123001333320532322000000101252112
Knmer  3733           130122322212222223101011330133101 23???0??????111011111152323
Knmwt 15000°          ?????????????????????????????????34341053132 0????4???1222???
Eliye Springs         ?0?03?5?36?0?0?3?41?0??1?3?1?3?0?????1???????0?0?0?0???030?0
Laetoli H18           1100?????????????????????????????3??????????????????????????
OH 9                  ???????12???223?????????????0???001???????????????????0??0????
Bodo 1                110022322320203115100100010132000 3?4?1?5?1???000211202101111
```

```
                Character 111111111111111111112222222222222222222222222222222222222222
                          888888888899999999990000000000111111111122222222223333333334
Taxon                     123456789012345678901234567890123456789012345678901234567890
------------------------------------------------------------------------------------------
Saldanha                  ??????????????????????????????????4??????????????????00????0
Ndutu                     ????????????????????????????????????????????????????????????
Kabwe                     220022333312223101212011110101000445530542422000100102002001
Omo Kibish 2              ??????????????????????????????????4?????????????1????1003???
Jebel Irhoud 1            000034533321113112100200220132100445 5?054?54?000000001240111
Salé                      ????????????????????????????????????????????????0???????????
Sambungmacan 1            ????2????6?????4??2????0???0?????3?????????????01????0003??0
Modjokerto°               ?0????5????0?????5??0????2???2???????????????0??00???00030??
Ngandong  1               1100????????????????????????????????????????????????????????
Ngandong  2°              000034533630003455100210330033000????????????000100001001000
Ngandong  3               ????????????????????????????????????????????????????????????
Ngandong  5               2?0?34?3?52?1??43?11?2?13??01?0??????????????010?00?????????
Ngandong  6               2200224335211033352222101100110004???3???????000011111002111
Ngandong  7               01003343342110323310121011013100044??20??1??2000111101002000
Ngandong 10               2?0??????4?????4??11?2?????13?0???4???????????????????04??11
Ngandong 11               23001223332210323210221011000000044???0??????000100102450112
Ngandong 12               12001233342210333420021011003200004455?054?54?000111101102111
Ngawi                     12001232242110333410021001013200004455?054?54?000211201200111
Sangiran  2               ????????????????????????????????????????????????????????????
Sangiran  3               ????????????????????????????????????????????????????????????
Sangiran  4               ????????????????????????????????????????????????????????????
Sangiran 10°              ????????????????????????????????????????????????????????????
Sangiran 12               ????????????????????????????????????????????????????????????
Sangiran 17               2200334334221132231112112201110003344?054?53?000000001023011
Sangiran 26               ????????????????????????????????????????????????????????????
Sangiran 38               ????????????????????????????????????????????????????????????
Sinanthropus  3           111132411521223324111210110133001344410141320000011200002011
Sinanthropus 10           ?????????621213434?????????133000????????????????????????000
Sinanthropus 11           1?0?33?2?52?1??33?11?2?12??13?0???4??1??????1?000?11?0003?11
Sinanthropus 12           000033422521222333100211120133000044?5?0?4??2?000000001012000
Dali                      110033533521103223110210220012000?3552???5420000000000003000
Maba                      ????2????5?????2???????0???0?????????????????0?021?2?00020?0
Narmada                   ?????????????????????????????????????????????????????0003???
Arago XXI et XLVII        000034533521103223100201220012000424420532542000000000003000
Petralona                 110022333322103101100100010001000????????????000100100003000
Monte Circeo              000034533631103333101211220012000????3???????000000000003000
Swanscombe                ????????????????????????????????????????????????????????????
La Chapelle aux Saints    100033433422103112100200110000000445530542542000310141002000
La Quina H5               0?0?33?3?42?1??12?10?1?11??00?0???????????54??001?00?1003?00
Spy 1                     2200?????32211311122221????001000????????????????????10??001
La Ferrassie 1            1?0?34?3?52?1??22?12?2?12??01?0???????????????000?00?0003?00
Amud 1                    1000?????52110322222020????031000?????????54?????????00??000
Shanidar 5                1100?????63110333410121????133000?455?????54?????????00??000
Skhull V                  ?0?03?5?3?????????2?0??1?3?0?3????4????????????01??1?0003??0
WHL 50                    ????????????????????????????????????????????????????????????
Gibraltar 2°              000034533630003233100100220011000?454?????54?000200000003000
Teshik Tash°              000034533630003345100011330133000?4??3??????2000300000002000
Engis 2°                  0000?????63000334510021????033000????????????????????00??000
Velika Pecina°            ??00???33???003??????21???0???000???????????????0???20003???
Trinil                    ????????????????????????????????????????????????????????????
Zuttiyeh                  000034533630103344510021133013200 0?????0542542000100120003000
modern humans             000034533630103455100211330133000445530542542000000000003000
Flores                    ?????????????????????????????????003?330?22?2????????0003???
```

```
                  Character 222222222222222222222222222222222222222222222222222222222223
                            444444444555555555566666666667777777777888888888899999999990
Taxon                       123456789012345678901234567890123456789012345678901234567890
--------------------------------------------------------------------------------------
Gorilla gorilla             241103412431445413443332212221141230100010001200001111303151
Pan troglodytes             102204213632337044234351303121222230222013201100001122012252
Sterkfontein 5              251003101331234411333342310222010000001112022001112300222010
Knmer   406                 1311031125314464333333624003220100311121102 0??011??2111?????
Knmer  1813                 112104200032101233000303133111120131112113022001122201121130
Knmer  1470                 232214112232223534232304032111100031222013031001112300322110
Knmer  3883                 101214000032001034112323223111100031220113022001122221121121
Knmer  3733                 232?14012332001244222303033020100031220111011101012221112142
Knmwt 15000°                ??1?133????2????30?????????????00031221113032101120300320200
Eliye Springs               1???????????0?0?????????03????????????2?12032021120300320200
Laetoli H18                 ???????0003???????????040330002?????????13031???1?0?????02??
OH 9                        10?????????????0??????????2?2?1???????21??????0?????????????
Bodo 1                      10210300003211104400012323313120?0????2?????????????????????
Saldanha                    ??22?????0??????44??0???????????????????1??3??????0??????2??
Ndutu                       ??????????????????????????????????????0??????????????00??0??
Kabwe                       004114000002000042000052322223100021010113033001120300320200
Omo Kibish 2                ??221?3?????????44?????????????0?0????2?03?21?0111030032?210
Jebel Irhoud 1              102213000032111244000313132121100031220103032001120300320200
Salé                        ??????0????????????????????????0?????????3??1???1???????????
Sambungmacan 1              ??22120??0?1??1?44??0??????????0?0????2?12031?010?1310120200
Modjokerto °                ??221400???21???440????????????0003122211???30?1???????2????
Ngandong  1                 ???????0003???????????040330002?????????1302??0?010??03?????
Ngandong  2°                1025010000300000440003040330202000312222????????????????????
Ngandong  3                 ???????????????????????????????????????????13??1???????2????
Ngandong  5                 ??1?142?0032?00?3??00303???1???200312???????????????????????
Ngandong  6                 112214000032111344000304033000100031221112011001121310320200
Ngandong  7                 003204000032001044000103223100200031222113011001121300120200
Ngandong 10                 ??01???001??????44?12313???1??????1022?113013???120?????02??
Ngandong 11                 101112000030112013000313223101200011120102011001121300220200
Ngandong 12                 201202000011121334000304033000200031222112001?011?0300320200
Ngawi                       1012010000301232340003132331111000312221?201200112030032 0200
Sangiran  2                 ???????????????????????????????00030222113031?1111030032?240
Sangiran  3                 ?????????????????????????????????????????3?2000????????2?1?0
Sangiran  4                 ????????????????????????????????????????????????????????????
Sangiran 10°                ????????????????????????????????????????13032?0?113??01?10??
Sangiran 12                 ????????????????????????????????????????03031?2?112??03?11??
Sangiran 17                 102214000032000044000303133111200031221103020?01121210320220
Sangiran 26                 ????????????????????????????????????????????????????????????
Sangiran 38                 ????????????????????????????????????????1?031?????0?????02??
Sinanthropus  3             102214000032112444000303032010100031222113021?01111210321220
Sinanthropus 10             12????????????????????030300200?????????1?02????????????????
Sinanthropus 11             ??22140?1032?11?44?00?002??0???00030222113022101112310321210
Sinanthropus 12             102212000031112144000203133111200031221101 11?01102131012020 0
Dali                        1022140000320000440002031331111000312221130320?1120300320200
Maba                        0?320300?0?11?2?440?0???23?????000312221????????????????????
Narmada                     ??32140????2????44?????????????000312221??????0?????????????
Arago XXI et XLVII          00221400003200014400031313212110003122211?032?????3?????02??
Petralona                   102214000032101044000223233111100031222113032001120300320200
Monte Circeo                0032140000320003440001422202210000312221?3032001121310320200
Swanscombe                  ????????????????????????????????????????1?0321?1???30??20200
La Chapelle aux Saints      11220300002200054400012312012100003122211303 2?01120??03202??
LaQuina H5                  ??22120?0030?01?40?00142???2???0003122211?03???1?2030??20200
Spy 1                       0042???0023?????440001423122111???3122?11?032?????0?????02??
La Ferrassie 1              ??32140?0132?00?44?00?002??2???0003122211?0320?1???30??20?10
Amud 1                      0032???0003?????440003031330002???3122?11?0320?1??033??23263
Shanidar 5                  0022???0003?????440002232201210???3122?1????????????????????
Skhull V                    ??22140??0?2??0?44??0??????????00031222113033011120300320200
```

```
             Character 222222222222222222222222222222222222222222222222222222222223
                       444444444455555555556666666666777777777788888888889999999990
Taxon                  123456789012345678901234567890123456789012345678901234567890
-----------------------------------------------------------------------------------
WHL 50                 ????????????????????????????????????????1?022?????0?????02??
Gibraltar 2°           003513000021000044000113232111100031221211??13??1??0????202??
Teshik Tash            002501000030000044000204233000200031221013033311120300320200
Engis 2°               0025???0003?????44000304033011 2???3122?21?033?2?120??03?02??
Velika Pecina°         1035140??32???444???3?????2?1?1000312210????????????????????
Trinil                 ???????????????????????????????????????????31???????????????
Zuttiyeh               1022140000320000440003030330102000312221????????????????????
modern humans          102214000032000044000304033000200031222113033001120300320200
Flores                 ??22140????2????40?????????????0002122221312?101124300000200

             Character 333333333333333333333333333333333333333333333333333333333333
                       000000000111111111122222222223333333333444444444455555555556
Taxon                  123456789012345678901234567890123456789012345678901234567890
-----------------------------------------------------------------------------------
Gorilla gorilla        42222221200?211220000000010031022212134121142010121034222120 1
Pan troglodytes        402232211020231211110000100002033033011221530001112112122100
Sterkfontein 5         1211100000200100033322102320022 00?0?01000?001221??0000?002?1
Knmer   406            ??????????1022??1333????422?1?2?00??12??01?012?111?22?0?0211
Knmer  1813            320000000100002113312321423200 30???00000?00?011???0000?0???0
Knmer  1470            12111101010000000321110022110 13?11???1??0??0?1?121??3?2?0?1?
Knmer  3883            120000000000000003312321222200300211010001000101210110100210
Knmer  3733            31110100000011010331121122120 03??1??01??????01??21?11?1???10
Knmwt 15000°           040000000000000003312320333000300000010002000101210000000210
Eliye Springs          0400000000000000023123214330003?12??22??02?100?011?21?1?0210
Laetoli H18            ??????????????????31??21???000?00003010001000101211001000210
OH 9                   ????????????????2???????????????10??12??01?101?111?22?0?0100
Bodo 1                 ????????????????????????????????????????????????????????????
Saldanha               ???????????????????????1????????????????????????????????????
Ndutu                  ????????????????????????????1?30?2?00110120???????0002001???
Kabwe                  04000000000?000003312321433100300 0?001000100010121 1000000210
Omo Kibish 2           14????????0000?20331?2?143300?3?????????????01?121?????????0
Jebel Irhoud 1         040000000000000003312321433000300000010001000101210000000210
Salé                   ????????????????????????????????????????????????????????????
Sambungmacan 1         12????????0?00000331??21322000310?10011001300101??1002?102?0
Modjokerto°            ????????????????????????????????????????????????????????????
Ngandong  1            ??????????0?00000?21?221?33?00?0?0?00100?10?010?21001000??00
Ngandong  2°           ????????????????????????????????????????????????????????????
Ngandong  3            ????0????????????3??2??????0????????????????????????????????
Ngandong  5            ????????????????????????????????????????????????????????????
Ngandong  6            0400000000000000033023212330003???01000?????000???1000???2?0
Ngandong  7            040000000?0?000003312321432000300000010201000101210000020210
Ngandong 10            ??????????????????21?221???00???????????????????????????????
Ngandong 11            030000000000000003312321432000 3?????????????????????????????
Ngandong 12            04????????0000000331??214331003?00??01??01?001?121?00?0?0210
Ngawi                  040000000?0?000003312321433100300000010002000101210000000210
Sangiran  2            04????????0000000331?221433000331022031012401121212132011200
Sangiran  3            ?40?0?0?000??????33?23?14?32????????????????????????????????
Sangiran  4            ????????????????????????????????????????????????????????????
Sangiran 10°           01????????1?00000?31?220?32100??????00??0???00?1???00???00?0
Sangiran 12            04????????0?00000?32?221?33210??????????????????????????????
Sangiran 17            14????????0000000331?320222000201011111002210101210111010210
Sangiran 26            ???????????????????????????????0????0110??0???????1012?0????
Sangiran 38            ???????????????????1??21???100??????????????????????????????
Sinanthropus  3        220000000000000003312221223200300010010001100101211001000210
Sinanthropus 10        ????????????????????????????0???????01?????????????00???????
Sinanthropus 11        130000000000000003312321433200311011100000011100002012211012
```

	Character
	333333333333333333333333333333333333333333333333333333333333
	000000000111111111122222222223333333333444444444455555555556
Taxon	123456789012345678901234567890123456789012345678901234567890
Sinanthropus 12	040000000?0?0010033123214230003?????????????????????????????
Dali	040000000?0?001002102321422000311011100000011100102112221012
Maba	??
Narmada	?????10001?0??000???2321433????1?0?02311?14?020?21022101?210
Arago XXI et XLVII	???????????????????0??21???300??????????????????????????????
Petralona	040000000?0?000003302321432000311?1???00???01?00???0?222??12
Monte Circeo	040000000000000003312321233000300011110000011100002012221012
Swanscombe	???00000?????????3?02?214??2?03?????????????????????????????
La Chapelle aux Saints	0400000?0?0?000003302321?33000?11011100000011100002012221012
La Quina H5	???00000?????????3?12?214??1003??????0010?11????????????????
Spy 1	??????????????????????21???000???????0000?0?????????????????
La Ferrassie 1	???000?0?????????3?12?2?3??2003??????0002?11????????????????
Amud 1	???00000?????????3?02?214??2003??????0???0?1????????????????
Shanidar 5	??
Skhull V	040000000000000003312321433100311011100000011100002012221012
WHL 50	???????????????????2??21???000??????????????????????????????
Gibraltar 2°	?????????????????3?0??21???000??????????????????????????????
Teshik Tash°	0400000000000000033123214330003?0???00??0??000?1???00???02?0
Engis 2°	04????????0?00001?30?321?33000?2002?01002?40??????1001011???
Velika Pecina°	??
Trinil	???????????????????????????0????????????????????????????????
Zuttiyeh	??
modern humans	040000000000000003312321433000311011100000011100002012221012
Flores	000001000100000003312321333110300000010002002200210000000211

	Character
	333333333333333333333333333333333333333344444444444444444444
	666666666777777777788888888889999999999000000000011111111112
Taxon	123456789012345678901234567890123456789012345678901234567890
Gorilla gorilla	0002010011000000000110112113222?101?1110100020220?3101031?10
Pan troglodytes	000111000100140221224122010210000010023042003200001213021021
Sterkfontein 5	121?2?222??1221?31??2????1?3211?000??00?00??0??????????21?1?
Knmer 406	1??1012222?????????????2????????00?01?10?000??22?02??00?1010
Knmer 1813	010????22??????????????23??210?0???????????????????????3?0?1
Knmer 1470	000201221201231321222122???????????????????????0??0??0??????
Knmer 3883	0??20112?2??????????????31132011000110100003000000010000 21021
Knmer 3733	0??2?122?2012??3???2??2???????????????????????00??0?????????
Knmwt 15000°	000201221201231321223221311320100000002200101000000000120021
Eliye Springs	0??2010000111?12?1222121????????00?00?10?001??00?01??00?1021
Laetoli H18	000201222201261341222221110220000020002120200300100000421021
OH 9	0??1011100011?01?1222120????????00?31?31?110??11?320?22?1221
Bodo 1	??
Saldanha	??
Ndutu	???????????????????????231?321??000?00100?01?0?????????2??21
Kabwe	00020122221126134122314231121000001000100001?000000000011021
Omo Kibish 2	0??201????1??????1?22?2????????????????????????????????1?22?
Jebel Irhoud 1	000201222201231321222121311320100000001000000000000000121021
Salé	??
Sambungmacan 1	011?0?222??1251?412?21?231?310100000?010000?00?????????11021
Modjokerto°	??
Ngandong 1	0001?122?20122?32?22?1223?1320?0??????????????0000000?02?0?1
Ngandong 2°	??
Ngandong 3	??
Ngandong 5	??
Ngandong 6	00????221??1251?2?2??1??11?100000000???1??0?12?????????110?1
Ngandong 7	000201221211221321222131311320100000 1?111002100000030002102 1

Taxon	Character
	333333333333333333333333333333333333333344444444444444444444
	666666666677777777778888888888999999999900000000001111111112
	123456789012345678901234567890123456789012345678901234567890
Ngandong 10	??
Ngandong 11	??
Ngandong 12	0??2012222012?13?1222132????????00?00?10?002??00?00??00?1021
Ngawi	000201222201221321222121311320100000001000000000000000011021
Sangiran 2	020101112101241131222320110222000030003140102101202112331021
Sangiran 3	??
Sangiran 4	???????????????????????????320?1???????????????????????3?0?1
Sangiran 10°	0?????220??22?0??1??????????????00????3?????????????????2?2?
Sangiran 12	?????????????11?2????????1???21???1?????????????????????????
Sangiran 17	000201211201221221222121111320100011002010001100001001021021
Sangiran 26	???????????????????????????221?????????????????????????3????
Sangiran 38	??
Sinanthropus 3	000201221201241321222122311320000021101020011000000000011021
Sinanthropus 10	??
Sinanthropus 11	413212221200000010001002311220000000001000111000000000021021
Sinanthropus 12	??
Dali	31321?22?20000?01000?00??11320100000001?00?00?00000000?1102?
Maba	??
Narmada	0012?12222012213 2?12?02231132010002012?03?01?00000110?0200?0
Arago XXI et XLVII	??
Petralona	?1??12?21??0???00??01?02???????00?0??11?00??????????????1021
Monte Circeo	51311?22?4000??00000?00??110000?000?001?00?1??000?0000?21?2?
Swanscombe	??
La Chapelle aux Saints	313212223200000000001001311320100000000120000000000000421021
La Quina H5	??????????00?0?12100?30??1?3211?000??01?00??0??????????21?2?
Spy 1	??????????0????0000??0?????100?????????????????????????1????
La Ferrassie 1	??????????01?1?01100?00??1?3211?000??01?00??0??????????11?2?
Amud 1	??????????00???????0??0??11???1?????0?1??0?0??????0??0??????
Shanidar 5	??
Skhul V	21311?22?31000?01000?30??113201?000?001?10?00?000?0000?21?2?
WHL 50	??
Gibraltar 2°	??
Teshik Tash	0???0?222??22?1??1??2???????????00????1??0??????????????1?2?
Engis 2°	??????2???12121221?22?2??102001?000?001?20?00?000?0110?11?2?
Velika Pecina°	??
Trinil	??
Zuttiyeh	??
modern humans	213212221200000010001002311320100000000100000000000000021021
Flores	100201202202251320211321200320101000101300100310003000200001

Taxon	Character
	444444444444444444444444444444444444444444444444
	222222222333333333344444444445555555555666666666
	123456789012345678901234567890123456789012345678
Gorilla gorilla	01231012??30?0101201?010???????0301?314010133010
Pan troglodytes	132221121332002121103100000000031211120423042411
Sterkfontein 5	???2???10030402403221222231313264523111212031310
Knmer 406	00??1002003032?3033??3??1203???325??02?3?5?1???0
Knmer 1813	1?1021??113?502100330132231314226545023005300001
Knmer 1470	??????11??2?4??????????????????????????????????1
Knmer 3883	132221114131421303330??2231314?325230??????????0
Knmer 3733	?3????1131??50?2222?????2313???535??1??????????1
Knmwt 15000°	132221111130302212230022221313243534001010100001
Eliye Springs	13??211?222151?4023??0??2313???555??00?0?0?0???1
Laetoli H18	1202211??2213?14?3?33??22?1?122?????0?0???1???01
OH 9	12??211???

```
           Character 444444444444444444444444444444444444444444444444
                     222222222233333333334444444444555555555566666666
Taxon                123456789012345678901234567890123456789012345678
---------------------------------------------------------------------
Bodo 1               ????????????????????????????????????????????????
Saldanha             ??????????1???????3?????????????????????????????
Ndtu                 1320213?20315114033304?2231314?6654502?1?3?3?2??
Kabwe                1321211???2??21?03???12?????????4?????012???????
Omo Kibish 2         ?3????1022??30?40?3?????23?3???665??????????????
Jebel Irhoud 1       13222110?02?4?14?3?3??322?1?1?2?????0?0???1????1
Salé                 ???????1????????????????????????????????????????
Sambungmacan 1       1?2121?1?1224?24?3?3??222?1?1?2?????0?0???0???01
Modjokerto°          ?????????000?0?????????2????1?2???????????????01
Ngandong  1          1323?11?2031422303330232231314222524021114200100
Ngandong  2°         ????????????????????????????????????????????????
Ngandong  3          ????????????????????????????????????????????????
Ngandong  5          ????????????????????????????????????????????????
Ngandong  6          1?2121???3215?24?3?3???22?1?1???????0??????????1
Ngandong  7          13222111233?512403330232231314244534011103100000
Ngandong 10          ?????????23013?2023???0?2313??2335??0?0?0?0?0??1
Ngandong 11          ????????43304223033302?2231314?3252201?1?2?0?0?1
Ngandong 12          13??211?42??5123033302?2231314?3353401?1?3?0?1?0
Ngawi                132221113030412303230232231314233535311121121000
Sangiran  2          12002111?01?2?04???3???21???1???????????????????
Sangiran  3          ????????????????????????????????????????????????
Sangiran  4          1??321?2343?31230?230??222?314?33524????????????
Sangiran 10°         ???0????????????????????????????????????????????
Sangiran 12          ????????203?50?2221??0??1203???323??00?10??2????
Sangiran 17          132221113230412303330122231314243534001111100000
Sangiran 26          ????????????????????????????????????????????????
Sangiran 38          ????????303?22?20?2?????12?2???2????????????????
Sinanthropus  3      13222112?023313?3?3????21?1?1??4??3?0??????????1
Sinanthropus 10      ?????????0223??3?3??????2??2???????????????????0
Sinanthropus 11      1322211??13?401303230?222313142445342?0?0?0?0?01
Sinanthropus 12      ????????313251?4033??2??2313???445??00?0?0?0???1
Dali                 ?3?2??111120421403330132231314245534000002100001
Maba                 ????????????????????????????????????????????????
Narmada              0322211?????????????????????????????????????????
Arago XXI et XLVII   ????????????????????????????????????????????????
Petralona            1???2??0232050?4033??0??2313???555??00?0?0?0???1
Monte Circeo         ?3?2??4??3205?24?3?3??122?1?1?2?????0?0???0????1
Swanscombe           ????????00?0????????????2313???555??00?0?0?0????
La Chapelle aux Saints 131221102210411403330022231314256545000000100001
La Quina H5          ???2???0?0103?14?2?3??222?1?1?2?????0?0???2???01
Spy 1                ???2???0?1215?24?3?3??222?1?1?2?????0?0???0???10
La Ferrassie 1       ???1???0111?41140333002223131425554400000010000 1
Amud 1               ?????????1?0???????????2??1?1?2????????????????1
Shanidar 5           ????????????????????????????????????????????????
Skhul V              ???2??20001031?4032??0??2313???555??00?00?0????1
WHL 50               ????????????????????????????????????????????????
Gibraltar 2°         ????????????????????????????????????????????????
Teshik Tash          ???????011?050??003??0???303????45?????0?2?0???1
Engis 2°             ?3?1??2?401?42?4??3??1??2??3?????5???0?????0???1
Velika Pecina°       ????????????????????????????????????????????????
Trinil               ????????????????????????????????????????????????
Zuttiyeh             ????????????????????????????????????????????????
modern humans        132221100030502403330022231314266545000000000001
Flores               1014211?200?21021?233??201?103?31224????????????
```

Appendix 3

Listing of the changes of the states of the characters at different nodes.

Character
number of the character

Step
number of step between the states

CI
Consistence index

Change
Change between the described states

=> non ambiguous or fix character

–> ambiguous or homoplastic character

Character	Step	CI	Change		
Node 66 to Node 68					
4	1	0.133	1	->	0
44	1	0.125	0	=>	1
69	1	0.333	1	=>	0
144	1	0.222	1	=>	0
145	2	0.273	0	=>	2
157	1	0.400	3	->	0
158	1	0.455	4	->	0
168	1	1.000	0	->	2
175	1	0.250	3	=>	4
178	1	0.556	3	=>	0
179	2	0.400	1	=>	3
181	1	0.400	2	->	4
182	2	0.444	2	=>	4
183	1	0.667	1	=>	2
184	1	0.667	1	=>	2
185	1	0.333	2	->	1
187	1	0.625	4	=>	1
188	1	0.375	2	->	0
189	1	0.429	2	->	0
190	1	0.500	2	->	0
191	1	0.500	2	=>	0
192	2	0.400	2	=>	4
193	2	0.571	3	=>	5
194	2	0.625	3	=>	5
195	1	0.750	2	->	0
197	1	0.333	2	=>	1
198	1	0.385	3	->	1
208	1	0.429	2	=>	4
211	1	1.000	2	=>	3
212	1	1.000	1	=>	2
213	1	0.571	1	->	4
228	1	0.600	1	->	2
234	1	0.333	1	->	2
238	1	0.667	1	=>	4
239	1	0.444	2	->	4
240	2	0.600	4	=>	6
243	1	0.333	2	=>	1
244	1	0.400	2	=>	1
246	1	0.333	4	=>	3
252	1	0.500	2	=>	1
253	1	0.500	0	->	4
254	1	0.500	0	->	4
256	1	0.500	0	=>	4
257	1	0.286	3	->	1
258	1	0.800	4	=>	3
259	1	1.000	2	=>	3
264	1	0.444	3	->	2
266	1	0.429	3	->	1
268	1	0.300	1	=>	2
277	1	0.500	2	=>	1
278	1	0.500	2	=>	0
282	1	0.600	3	=>	0
214	1	0.750	0	->	1
331	1	0.667	3	->	2
355	1	0.400	1	->	2
360	1	1.000	0	=>	1
376	1	0.600	3	->	0
407	1	1.000	0	=>	2
408	1	1.000	0	=>	2

Character	Step	CI	Change		
412	1	0.750	0	->	1
419	1	1.000	2	=>	1
420	1	0.500	1	=>	0
421	1	0.500	1	=>	0
425	1	1.000	2	=>	1
426	1	1.000	1	=>	0
464	1	0.667	0	->	3
468	1	0.250	1	=>	0

Character	Step	CI	Change		
Node 66 to *Gorilla gorilla*					
1	1	0.167	0	->	1
6	1	0.200	1	->	0
7	1	0.200	1	->	0
8	1	0.167	1	->	0
12	1	0.333	1	=>	0
16	1	0.500	0	->	1
25	1	0.111	0	=>	1
26	1	0.500	0	->	1
51	1	0.182	2	->	0
66	1	0.154	1	=>	0
70	1	0.125	0	->	1
72	1	0.143	1	->	0
95	1	0.167	0	=>	1
104	1	0.300	2	->	0
124	1	0.429	2	->	0
125	1	0.333	4	->	5
134	1	0.286	1	=>	3
136	1	0.267	0	->	1
137	1	0.182	1	->	0
138	1	0.500	0	->	2
139	1	0.250	1	->	0
141	1	0.125	0	->	1
159	1	0.500	3	=>	1
163	1	0.250	1	->	2
165	1	0.333	1	->	0
172	1	0.462	3	->	2
173	1	0.333	2	=>	0
174	1	0.500	1	=>	0
176	1	0.400	5	=>	3
177	1	0.300	5	->	2
186	1	0.500	2	->	1
204	1	0.250	1	=>	2
205	1	0.333	2	->	5
207	1	1.000	0	=>	1
219	1	1.000	0	->	2
220	1	1.000	4	=>	0
221	1	0.667	4	=>	0
222	1	0.500	2	->	0
223	1	0.833	2	->	1
224	1	0.800	2	->	1
225	1	0.400	2	->	0
226	3	0.625	2	=>	5
227	3	0.625	2	=>	5
232	1	0.538	1	->	4
233	1	0.714	3	->	2
235	1	0.500	0	->	4
236	1	0.600	0	=>	5
242	1	0.667	3	=>	4
247	1	0.667	1	=>	4

Character	Step	CI	Change		
250	1	1.000	3	=>	4
255	1	0.700	1	->	5
259	1	1.000	3	=>	4
260	1	1.000	3	=>	4
263	1	0.545	0	->	3
265	1	0.308	3	=>	2
267	1	0.250	3	->	2
272	1	0.750	0	->	4
273	1	1.000	0	=>	1
274	1	0.667	0	->	2
279	1	0.250	2	=>	0
286	2	0.333	0	=>	2
288	1	0.500	1	->	0
289	1	0.250	1	->	0
290	1	0.286	1	->	0
291	1	0.214	2	->	1
295	1	0.375	1	->	3
296	1	0.667	2	=>	0
297	1	0.333	2	=>	3
301	3	0.267	1	->	4
303	1	0.667	1	->	2
304	1	0.667	1	->	2
305	1	0.750	1	->	2
306	1	0.400	1	->	2
307	2	0.500	0	->	2
309	2	0.667	0	=>	2
315	1	0.500	0	->	1
316	1	0.400	0	->	2
317	1	0.400	0	=>	2
318	1	1.000	3	=>	0
319	1	0.750	3	=>	0
320	1	0.500	1	->	0
325	1	0.333	2	->	1
326	1	0.250	2	->	0
328	1	0.250	0	=>	3
329	1	0.250	0	->	1
330	1	0.500	1	=>	0
332	1	0.500	0	=>	2
333	1	0.429	0	=>	2
335	1	0.600	1	->	2
337	1	0.500	0	=>	3
338	1	0.571	1	=>	4
339	1	0.200	0	->	1
341	1	0.667	0	=>	1
343	1	0.833	0	=>	4
344	1	0.500	0	=>	2
352	1	0.375	1	->	3
353	1	0.444	1	->	4
357	1	0.500	0	=>	1
359	1	0.333	1	->	0
368	1	0.500	2	->	0
370	1	0.800	2	->	1
372	1	0.333	1	->	0
373	1	0.500	2	=>	0
374	1	0.667	2	=>	0
375	1	0.250	1	->	0
377	1	0.571	2	->	0
378	1	0.500	1	=>	0
379	1	0.667	2	->	0
380	1	1.000	2	->	1
381	1	0.600	2	=>	1
382	1	0.500	1	->	0

Character	Step	CI	Change		
383	1	0.800	2	->	1
384	1	0.286	2	=>	1
390	1	0.500	0	->	2
391	1	0.333	1	=>	2
393	1	1.000	0	=>	1
398	1	0.667	0	=>	1
401	1	0.500	0	=>	1
405	1	0.500	0	=>	2
411	1	0.429	0	->	3
414	1	0.750	0	=>	1
416	1	0.250	2	=>	3
422	1	1.000	3	->	1
424	1	0.429	2	=>	3
435	1	0.222	2	=>	1
436	1	0.400	3	=>	0
438	1	0.600	3	->	2
439	1	0.300	3	=>	0
440	1	1.000	3	->	1
442	1	0.500	1	->	0
443	1	0.500	2	=>	1
444	1	0.500	2	->	0
452	1	0.417	3	=>	0
454	1	1.000	5	=>	0
457	1	0.600	1	=>	3
458	1	0.400	2	->	1
459	1	0.600	1	->	4
463	1	0.333	0	=>	1
465	1	0.750	0	->	3

Node 68 to Node 67

Character	Step	CI	Change		
45	1	0.250	1	=>	2
55	1	0.182	1	=>	0
74	1	0.286	0	->	1
82	1	0.167	1	=>	0
89	1	0.286	1	=>	0
93	1	0.500	1	=>	0
102	1	0.222	0	=>	2
108	1	0.133	0	=>	1
113	1	0.231	0	->	1
114	1	0.200	0	=>	1
121	1	0.286	0	=>	1
128	1	0.667	2	->	0
130	1	0.200	0	=>	1
149	1	0.273	2	->	1
166	1	0.235	2	=>	3
167	1	0.364	2	=>	3
186	1	0.500	2	->	0
230	1	0.444	2	=>	3
237	1	0.214	3	=>	2
267	1	0.250	3	->	0
270	1	0.429	1	=>	2
271	1	0.222	1	=>	0
272	1	0.750	0	->	1
280	2	0.800	0	=>	1
320	2	0.500	1	->	3
327	1	0.333	0	=>	2
345	1	0.500	0	=>	1
346	1	0.667	1	=>	2
361	1	1.000	0	=>	1
367	1	0.500	0	->	2
369	1	0.429	1	=>	2

Character	Step	CI	Change		
429	1	0.400	1	->	0
430	1	0.286	1	->	0
437	1	0.333	1	=>	0

Node 67 to Sterkfontein 5

Character	Step	CI	Change		
2	1	0.143	0	->	2
3	1	0.167	0	=>	1
8	1	0.167	1	->	0
12	1	0.333	1	=>	2
19	1	0.143	0	->	1
23	1	0.182	0	=>	2
27	1	0.375	0	->	1
36	1	0.167	0	=>	1
37	1	0.083	0	=>	1
42	1	0.200	0	=>	1
43	1	0.125	1	=>	0
49	1	0.167	1	=>	0
50	1	0.167	1	=>	0
52	1	0.125	0	=>	1
53	1	0.154	0	=>	1
60	1	0.100	0	->	1
68	1	0.167	0	->	1
75	1	0.091	0	=>	1
79	1	0.154	2	=>	1
87	1	0.100	0	->	1
97	1	0.286	0	->	1
105	1	0.222	0	=>	1
110	1	0.182	0	->	1
125	1	0.333	4	=>	3
131	1	0.182	0	->	1
137	1	0.182	1	=>	2
140	1	0.200	0	=>	3
142	1	0.571	3	->	4
143	1	1.000	4	->	5
146	1	0.444	3	->	4
150	1	0.400	2	->	1
151	1	0.429	1	->	3
152	1	1.000	2	->	3
153	1	0.571	3	->	4
155	1	0.500	3	->	4
156	1	0.714	4	->	5
159	1	0.500	3	->	2
160	1	0.250	3	->	1
161	1	0.300	3	->	1
162	1	0.444	4	->	1
164	1	0.333	3	=>	2
169	1	0.500	0	->	1
170	1	0.417	2	->	0
171	1	0.545	2	->	1
172	1	0.462	3	->	1
176	1	0.400	5	=>	2
177	1	0.300	5	->	2
185	1	0.333	1	=>	0
187	1	0.625	1	=>	0
195	1	0.750	0	->	1
196	1	0.286	1	=>	0
198	1	0.385	1	=>	2
201	1	0.222	1	->	0
202	1	0.333	2	=>	1
205	1	0.333	2	->	1
206	1	0.375	3	->	2

Character	Step	CI	Change		
213	1	0.571	4	->	3
214	1	0.800	1	->	0
216	1	0.667	2	->	1
217	1	0.800	4	->	1
226	1	0.625	2	=>	1
227	1	0.625	2	=>	1
228	1	0.600	2	->	1
229	1	0.400	1	->	4
232	1	0.538	1	->	5
233	1	0.714	3	->	5
234	1	0.333	2	=>	3
235	1	0.500	0	->	6
236	1	0.600	0	=>	3
237	2	0.214	2	=>	0
242	1	0.667	3	=>	5
244	1	0.400	1	=>	0
248	1	0.500	1	=>	0
249	1	0.750	2	=>	1
253	1	0.500	4	->	2
254	1	0.500	4	->	3
255	1	0.700	1	->	4
258	1	0.800	3	=>	1
263	1	0.545	0	->	4
275	1	1.000	3	=>	0
277	1	0.500	1	=>	0
279	1	0.250	2	=>	1
282	1	0.600	0	=>	2
284	1	0.300	0	=>	2
285	1	0.333	1	->	2
292	1	0.333	1	->	3
293	1	0.231	1	=>	0
294	1	0.500	1	=>	0
295	1	0.375	1	->	2
298	1	0.400	1	->	0
299	1	0.545	5	->	1
300	1	0.375	1	->	0
306	1	0.400	1	->	0
308	1	0.500	1	->	0
311	1	0.667	0	->	2
313	1	0.667	2	=>	0
321	1	0.667	0	->	2
322	1	0.429	0	->	2
323	1	0.667	0	->	1
326	1	0.250	2	->	3
330	1	0.500	1	->	2
335	1	0.600	1	->	0
340	1	0.500	2	->	0
347	1	1.000	0	->	2
352	1	0.375	1	->	0
353	1	0.444	1	->	0
354	1	0.286	2	->	0
356	1	0.250	2	->	0
362	1	0.500	0	->	2
363	1	0.500	0	->	1
365	1	0.667	0	=>	2
377	1	0.571	2	->	3
390	1	0.500	0	->	1
395	1	0.286	1	->	0
399	1	0.600	1	=>	0
428	1	0.286	2	=>	1
436	1	0.400	3	=>	4
439	1	0.300	3	->	2

Character	Step	CI	Change		
440	1	1.000	3	->	2
442	1	0.500	1	->	2
450	1	0.750	4	->	3
452	1	0.417	3	=>	6
453	1	0.385	3	->	4
455	1	0.429	1	->	2
458	1	0.400	2	->	1
460	1	0.667	0	->	2
462	1	0.500	0	->	2
465	1	0.750	0	->	1
466	1	1.000	0	->	3

Node 67 to Knmer 406

Character	Step	CI	Change		
6	1	0.200	1	->	0
7	1	0.200	1	->	0
16	1	0.500	0	->	1
26	1	0.500	0	->	1
28	1	0.167	1	=>	2
29	1	0.182	1	->	0
30	1	0.125	0	=>	1
40	1	0.167	0	=>	1
51	1	0.182	2	->	1
70	1	0.125	0	->	1
71	1	0.167	0	->	1
73	1	0.200	2	=>	1
77	1	0.111	1	=>	0
80	1	0.125	0	=>	1
83	1	0.143	0	=>	1
86	1	0.500	0	=>	1
91	1	0.167	0	=>	1
92	1	0.125	2	=>	1
94	1	0.200	1	=>	0
103	1	0.286	0	=>	1
107	1	0.200	0	=>	2
108	1	0.133	1	=>	2
109	1	0.154	0	=>	1
110	1	0.182	0	->	2
111	2	0.125	0	=>	1
117	1	0.100	0	=>	1
118	1	0.333	0	=>	4
123	1	0.429	3	=>	0
125	2	0.333	4	=>	6
131	1	0.182	0	->	2
139	1	0.250	1	->	0
141	1	0.125	0	->	1
145	1	0.273	2	=>	3
149	1	0.273	1	=>	0
150	1	0.400	2	->	0
160	1	0.250	3	->	0
161	1	0.300	3	->	0
162	1	0.444	4	->	0
163	1	0.250	1	->	2
166	1	0.235	3	=>	4
167	1	0.364	3	=>	4
168	1	1.000	2	=>	3
169	1	0.500	0	=>	2
170	1	0.417	2	->	1
171	1	0.545	2	->	0
172	1	0.462	3	->	0
178	1	0.556	0	=>	1
179	1	0.400	3	=>	4

Character	Step	CI	Change		
180	1	0.222	2	=>	1
190	1	0.500	0	=>	1
199	1	0.250	0	=>	1
200	2	0.222	0	=>	2
201	1	0.222	1	=>	2
203	1	0.143	0	=>	1
205	1	0.333	2	->	0
206	1	0.375	3	->	0
219	1	1.000	0	->	3
220	1	1.000	4	=>	3
221	1	0.667	4	=>	2
223	1	0.833	2	->	0
224	1	0.800	2	->	0
229	1	0.400	1	->	3
232	1	0.538	1	->	6
235	1	0.500	0	->	3
239	1	0.444	4	=>	3
241	1	0.222	2	=>	1
250	1	1.000	3	=>	5
255	1	0.700	1	->	6
257	1	0.286	1	->	3
263	1	0.545	0	->	6
265	1	0.308	3	=>	4
266	1	0.429	1	->	0
268	1	0.300	2	=>	3
276	1	0.250	0	=>	1
278	1	0.500	0	=>	1
283	2	0.500	0	=>	2
292	1	0.333	1	->	2
311	1	0.667	0	->	1
314	1	0.750	1	=>	2
316	1	0.400	0	->	1
325	1	0.333	2	->	4
328	1	0.250	0	->	1
334	1	0.400	1	->	0
337	1	0.500	0	=>	1
338	1	0.571	1	=>	2
349	1	0.400	2	->	1
352	1	0.375	1	->	2
353	1	0.444	1	->	2
355	1	0.400	2	->	0
364	1	0.200	2	->	1
411	1	0.429	0	->	2
422	1	1.000	3	->	0
427	1	1.000	1	->	0
433	1	0.250	4	=>	3
434	1	0.300	0	=>	2
442	1	0.500	1	->	3
445	1	0.500	2	=>	1
446	1	0.500	3	=>	2
447	1	0.333	1	->	0
453	1	0.385	3	->	2
457	1	0.600	1	=>	0
460	1	0.667	0	->	3
462	1	0.500	0	->	5
464	1	0.667	3	=>	1

Node 66 to *Pan troglodytes*

Character	Step	CI	Change		
1	1	0.167	0	->	1
4	1	0.133	1	=>	2
20	1	0.200	1	=>	0

Character	Step	CI	Change		
22	1	0.200	0	=>	1
27	1	0.375	0	->	3
28	1	0.167	1	=>	0
29	1	0.182	1	->	0
48	1	0.500	0	=>	1
60	1	0.100	0	->	1
68	1	0.167	0	->	1
72	1	0.143	1	->	0
73	1	0.200	2	=>	1
78	1	0.111	0	=>	1
84	1	0.125	1	->	0
87	1	0.100	0	->	1
94	1	0.200	1	=>	0
104	1	0.300	2	->	1
120	1	0.231	2	=>	1
125	2	0.333	4	=>	2
131	1	0.182	0	->	2
132	1	0.167	0	=>	1
134	1	0.286	1	=>	2
135	1	0.167	1	=>	0
136	1	0.267	0	->	1
137	1	0.182	1	->	0
138	1	0.500	0	->	1
149	1	0.273	2	->	1
157	1	0.400	3	->	1
158	1	0.455	4	->	1
159	1	0.500	3	->	2
165	1	0.333	1	->	0
166	2	0.235	2	=>	0
167	2	0.364	2	=>	0
168	1	1.000	0	->	1
169	1	0.500	0	->	1
170	1	0.417	2	=>	4
171	1	0.545	2	=>	4
172	1	0.462	3	->	5
174	1	0.500	1	=>	2
179	1	0.400	1	=>	0
180	1	0.222	2	=>	0
181	1	0.400	2	->	1
188	1	0.375	2	->	1
189	1	0.429	2	->	1
190	1	0.500	2	->	3
195	1	0.750	2	->	1
196	1	0.286	1	->	3
198	1	0.385	3	->	0
201	1	0.222	1	->	0
213	1	0.571	1	->	3
217	1	0.800	4	->	2
220	1	1.000	4	=>	2
221	1	0.667	4	=>	1
226	2	0.625	2	->	4
227	2	0.625	2	->	4
228	1	0.600	1	->	3
232	1	0.538	1	->	7
233	1	0.714	3	->	1
234	1	0.333	1	->	0
239	1	0.444	2	->	1
241	1	0.222	2	=>	1
242	1	0.667	3	=>	0
247	1	0.667	1	=>	2
249	1	0.750	2	=>	3
250	1	1.000	3	=>	6

Character	Step	CI	Change		
253	1	0.500	0	->	3
254	1	0.500	0	->	3
255	1	0.700	1	->	7
257	1	0.286	3	->	4
261	1	1.000	3	=>	4
263	1	0.545	0	->	5
264	1	0.444	3	->	1
266	1	0.429	3	->	0
271	1	0.222	1	=>	2
272	1	0.750	0	->	2
273	1	1.000	0	=>	2
274	1	0.667	0	->	2
283	2	0.500	0	=>	2
286	1	0.333	0	->	1
288	1	0.500	1	->	0
289	1	0.250	1	->	0
290	1	0.286	1	->	0
291	1	0.214	2	->	1
293	1	0.231	1	->	2
294	1	0.500	1	=>	2
295	1	0.375	1	=>	0
296	1	0.667	2	=>	1
298	1	0.400	1	=>	2
300	1	0.375	1	=>	2
301	3	0.267	1	->	4
302	1	0.667	2	=>	0
303	1	0.667	1	->	2
304	1	0.667	1	->	2
305	2	0.750	1	=>	3
306	1	0.400	1	->	2
307	2	0.500	0	->	2
309	1	0.667	0	->	1
311	1	0.667	0	->	2
314	1	0.750	0	->	3
315	1	0.500	0	->	1
316	1	0.400	0	->	2
317	1	0.400	0	->	1
318	1	1.000	3	=>	1
319	1	0.750	3	=>	1
325	1	0.333	2	->	1
326	1	0.250	2	->	0
330	1	0.500	1	->	2
331	1	0.667	3	->	0
332	1	0.500	0	=>	3
333	1	0.429	0	=>	3
334	1	0.400	1	->	0
335	1	0.600	1	->	3
336	1	0.400	1	=>	3
339	1	0.200	0	->	1
341	1	0.667	0	=>	2
343	1	0.833	0	=>	5
344	1	0.500	0	=>	3
346	1	0.667	1	=>	0
349	1	0.400	2	->	1
351	1	0.400	0	=>	2
357	1	0.500	0	=>	2
358	1	0.667	2	=>	1
359	1	0.333	1	->	0
364	1	0.200	2	->	1
365	1	0.667	0	=>	1
368	1	0.500	2	->	0
369	1	0.429	1	=>	0

Character	Step	CI	Change		
370	1	0.800	2	->	1
372	1	0.333	1	->	0
373	1	0.500	2	=>	1
374	1	0.667	2	=>	4
375	1	0.250	1	->	0
376	1	0.600	3	->	2
381	1	0.600	2	=>	4
385	2	0.429	2	=>	0
387	1	0.500	1	=>	0
388	1	0.429	3	=>	2
389	1	0.333	2	=>	1
391	1	0.333	1	=>	0
397	1	0.200	1	=>	0
398	1	0.667	0	=>	2
399	1	0.600	1	=>	3
401	1	0.500	0	=>	4
402	1	1.000	0	=>	2
405	1	0.500	0	=>	3
406	1	1.000	0	=>	2
411	1	0.429	0	->	1
412	1	0.750	0	->	2
413	1	1.000	0	=>	1
414	1	0.750	0	=>	3
430	1	0.286	1	->	3
432	1	0.500	0	=>	2
433	1	0.250	4	=>	0
436	1	0.400	3	=>	1
437	1	0.333	1	->	2
438	1	0.600	3	=>	1
439	1	0.300	3	=>	1
440	1	1.000	3	->	0
441	2	0.500	1	=>	3
443	1	0.500	2	=>	0
444	1	0.500	2	->	0
445	1	0.500	2	=>	0
446	1	0.500	3	=>	0
447	1	0.333	1	->	0
448	1	1.000	3	=>	0
449	1	1.000	1	=>	0
450	1	0.750	4	->	0
451	1	1.000	2	=>	0
453	1	0.385	3	=>	1
454	1	1.000	5	=>	2
456	1	0.500	3	=>	1
459	1	0.600	1	->	0
460	1	0.667	0	=>	4
461	1	0.333	1	->	2
462	1	0.500	0	=>	3
464	1	0.667	0	->	4
465	1	0.750	0	->	2
466	1	1.000	0	=>	4

Node 66 to Node 65

12	1	0.333	1	->	3
31	1	0.333	0	=>	1
32	1	0.154	2	->	1
38	1	0.182	0	=>	2
42	1	0.200	0	=>	1
43	1	0.125	1	=>	0
50	1	0.167	1	->	0
56	1	0.250	1	=>	0

Character	Step	CI	Change		
66	1	0.154	1	=>	2
74	1	0.286	0	->	1
79	1	0.154	2	->	0
81	1	0.333	0	->	1
85	1	0.167	1	=>	0
99	1	0.333	0	->	1
105	1	0.222	0	->	2
109	1	0.154	0	->	1
110	1	0.182	0	->	1
113	1	0.231	0	->	1
123	1	0.429	3	=>	0
128	1	0.667	2	->	0
137	1	0.182	1	->	2
147	1	0.556	5	=>	4
148	1	0.625	5	=>	4
150	1	0.400	2	=>	4
161	1	0.300	3	=>	2
162	1	0.444	4	=>	3
164	1	0.333	3	->	2
165	1	0.333	1	=>	2
194	1	0.625	3	=>	2
199	1	0.250	0	=>	1
203	1	0.143	0	=>	1
215	1	0.571	0	->	3
226	1	0.625	2	=>	1
227	1	0.625	2	=>	1
231	1	0.375	3	->	0
240	2	0.600	4	=>	2
245	1	0.200	0	=>	1
261	1	1.000	3	=>	2
265	1	0.308	3	=>	0
269	1	0.300	2	->	1
276	1	0.250	0	=>	1
284	1	0.300	0	->	2
292	1	0.333	1	->	2
299	1	0.545	5	->	2
313	1	0.667	2	=>	0
321	1	0.667	0	=>	1
322	1	0.429	0	->	3
327	1	0.333	0	=>	1
328	1	0.250	0	->	2
340	1	0.500	2	->	0
354	1	0.286	2	->	0
356	1	0.250	2	->	0
367	1	0.500	0	->	2
385	1	0.429	2	->	3
428	1	0.286	2	=>	1
441	1	0.500	1	->	0
455	1	0.429	1	->	3
467	1	1.000	1	->	0

Node 65 to Node 64

2	1	0.143	0	->	1
3	1	0.167	0	=>	1
5	1	0.111	0	->	1
62	1	0.111	1	=>	0
107	1	0.200	0	->	1
121	1	0.286	0	=>	1
124	1	0.429	2	->	1
143	1	1.000	4	=>	3
146	1	0.444	3	=>	2

Character	Step	CI	Change		
147	1	0.556	4	=>	3
148	1	0.625	4	=>	3
151	1	0.429	1	->	4
152	1	1.000	2	=>	1
153	1	0.571	3	=>	2
154	1	0.750	3	=>	2
156	1	0.714	4	=>	3
176	1	0.400	5	->	3
177	1	0.300	5	=>	3
183	1	0.667	1	=>	0
191	1	0.500	2	=>	1
193	1	0.571	3	=>	2
208	1	0.429	2	=>	1
211	1	1.000	2	->	0
212	1	1.000	1	=>	0
214	1	0.800	1	->	3
216	1	0.667	2	->	3
220	1	1.000	4	->	5
223	1	0.833	2	->	3
229	1	0.400	1	=>	0
230	1	0.444	2	->	0
233	1	0.714	3	->	0
235	1	0.500	0	->	2
247	1	0.667	1	=>	0
260	1	1.000	3	=>	2
280	1	0.500	0	=>	1
305	1	0.750	1	=>	0
308	1	0.500	1	=>	0
323	1	0.667	0	->	2
324	1	0.333	0	=>	1
330	1	0.500	1	=>	0
433	1	0.250	4	->	5

Node 64 to Node 63

Character	Step	CI	Change		
25	1	0.111	0	->	1
52	1	0.125	0	=>	1
97	1	0.286	0	->	1
99	1	0.333	1	->	2
102	1	0.222	0	=>	1
103	1	0.286	0	=>	1
104	1	0.300	2	->	1
117	1	0.100	0	->	1
126	1	0.333	1	=>	0
131	1	0.182	0	=>	1
142	1	0.571	3	=>	2
143	1	1.000	3	=>	2
147	1	0.556	3	=>	2
148	1	0/625	3	=>	2
150	1	0.400	4	=>	3
156	1	0.714	3	=>	2
158	1	0.455	4	->	3
170	1	0.417	2	->	3
184	1	0.667	1	=>	0
189	1	0.429	2	=>	3
195	1	0.750	2	=>	3
200	1	0.222	0	=>	1
206	1	0.375	3	->	2
226	1	0.625	1	=>	0
227	1	0.625	1	=>	0
228	1	0.600	1	->	0
239	1	0.444	2	->	1
241	1	0.222	2	=>	1
242	1	0.667	3	=>	0
248	1	0.500	1	=>	0
249	1	0.750	2	=>	0
250	1	1.000	3	=>	0
259	1	1.000	2	=>	1
260	1	1.000	2	=>	1
265	1	0.308	0	->	1
290	1	0.286	1	=>	2
297	1	0.333	2	=>	1
303	1	0.667	1	=>	0
304	1	0.667	1	=>	0
306	1	0.400	1	=>	0
321	1	0.667	1	=>	2
327	1	0.333	1	->	3
334	1	0.400	1	->	0
437	1	0.333	1	=>	0
457	1	0.600	1	=>	0

Node 63 to Node 62

Character	Step	CI	Change		
39	1	0.143	0	=>	1
64	1	0.125	1	->	0
73	1	0.200	2	=>	1
107	1	0.200	1	->	0
114	1	0.200	0	->	1
116	1	0.333	0	->	1
124	1	0.429	1	->	3
125	1	0.333	4	->	3
139	1	0.250	1	->	2
140	1	0.200	0	=>	1
142	1	0.571	2	=>	1
143	1	1.000	2	=>	1
154	1	0.750	2	=>	1
155	1	0.500	3	=>	2
165	1	0.333	2	=>	1
171	1	0.545	2	->	3
174	1	0.500	1	=>	3
185	1	0.333	2	->	3
194	1	0.625	2	=>	1
213	1	0.571	1	=>	0
215	1	0.571	3	=>	4
222	1	0.500	2	->	1
259	1	1.000	1	=>	0
260	1	1.000	1	=>	0
261	2	1.000	2	=>	0
300	1	0.375	1	=>	0
325	1	0.333	2	=>	4
336	1	0.400	1	=>	0
352	1	0.375	1	=>	0
353	1	0.444	1	=>	0
355	1	0.400	1	->	0
405	1	0.500	0	->	1
452	1	0.417	3	->	4
456	1	0.500	3	->	4

Node 62 to Knmer 1813

Character	Step	CI	Change		
5	1	0.111	1	->	0
11	1	0.333	2	=>	1
12	1	0.333	3	=>	1
18	1	0.250	2	->	1

Character	Step	CI	Change		
19	1	0.143	0	->	1
37	1	0.083	0	->	1
38	1	0.182	2	->	0
44	1	0.125	0	=>	1
51	1	0.182	2	=>	1
55	1	0.182	1	->	2
62	1	0.111	0	=>	1
63	1	0.167	0	->	1
75	1	0.091	0	=>	1
92	2	0.125	2	=>	0
98	1	0.222	0	=>	2
99	2	0.333	2	=>	0
106	2	0.154	0	=>	2
107	1	0.200	0	->	2
117	1	0.100	1	->	0
120	1	0.231	2	->	3
135	1	0.167	1	->	0
136	1	0.267	0	->	3
137	1	0.182	2	->	1
140	1	0.200	1	=>	2
144	1	0.222	1	->	2
160	1	0.250	3	->	2
163	1	0.250	1	=>	0
164	1	0.333	2	=>	1
166	1	0.235	2	=>	1
167	2	0.364	2	=>	0
170	1	0.417	3	->	4
172	1	0.462	3	=>	4
173	1	0.333	2	->	3
175	1	0.250	3	=>	1
176	1	0.400	3	=>	2
177	1	0.300	3	=>	2
181	1	0.400	2	=>	3
182	1	0.444	2	=>	3
185	1	0.333	3	->	1
187	1	0.625	4	=>	2
192	1	0.400	2	=>	3
197	1	0.333	2	=>	0
198	1	0.385	3	=>	1
199	1	0.250	1	=>	2
200	1	0.222	1	=>	2
201	1	0.222	1	=>	2
204	1	0.250	1	->	0
205	1	0.333	2	=>	0
206	1	0.375	2	->	1
209	1	0.375	3	=>	3
210	1	0.333	3	->	2
228	1	0.600	0	->	2
229	1	0.400	0	=>	3
230	1	0.444	0	->	2
231	1	0.375	0	->	2
232	1	0.538	1	=>	3
233	1	0.714	0	=>	2
234	1	0.333	1	=>	2
239	1	0.444	1	->	2
242	1	0.667	0	=>	1
244	1	0.400	2	=>	1
245	1	0.200	1	=>	0
247	1	0.667	0	=>	2
253	1	0.500	0	=>	1
256	1	0.500	0	=>	2
258	1	0.800	4	=>	3

Character	Step	CI	Change		
272	1	0.750	0	=>	2
274	1	0.667	0	=>	1
277	1	0.500	2	=>	1
278	1	0.500	2	=>	1
285	1	0.333	1	->	2
293	1	0.231	1	=>	0
299	1	0.545	2	=>	3
301	1	0.267	1	->	3
310	1	0.333	0	=>	1
315	1	0.500	0	=>	2
316	1	0.400	0	=>	1
317	1	0.400	0	=>	1
338	1	0.571	1	=>	0
342	1	0.400	1	=>	0
347	1	1.000	0	=>	1
362	1	0.500	0	=>	1
388	1	0.429	3	=>	2
389	1	0.333	2	=>	1
416	1	0.250	2	=>	3
423	1	0.500	2	=>	1
424	1	0.429	2	=>	0
436	1	0.500	3	=>	1
438	1	0.600	3	=>	0
443	1	0.500	2	=>	3
452	1	0.417	4	->	2
453	1	0.385	3	=>	6
455	1	0.429	3	->	4
456	1	0.500	4	->	5
459	1	0.600	1	=>	3
461	1	0.333	1	->	0
462	1	0.500	0	->	5
463	1	0.333	0	=>	3

Node 62 to Node 61

Character	Step	CI	Change		
21	1	0.250	0	=>	1
27	1	0.375	0	=>	3
33	1	0.111	1	=>	0
45	1	0.250	1	=>	2
70	1	0.125	0	->	1
72	1	0.143	1	=>	0
74	1	0.286	1	=>	0
80	1	0.125	0	->	1
87	1	0.100	0	->	1
112	1	0.200	0	=>	1
115	1	0.250	0	=>	1
119	1	0.182	3	=>	2
142	1	0.571	1	=>	0
143	1	1.000	1	=>	0
146	1	0.444	2	=>	1
156	1	0.714	2	=>	1
180	1	0.222	2	=>	0
188	1	0.375	2	=>	3
190	1	0.500	2	->	4
191	1	0.500	1	=>	2
193	1	0.571	2	->	1
196	1	0.286	1	->	2
235	1	0.500	2	->	0
238	1	0.667	1	=>	0
240	1	0.600	2	=>	1
257	1	0.286	3	->	4
291	1	0.214	2	=>	1

Character	Step	CI	Change		
294	1	0.500	1	=>	0
295	1	0.375	1	=>	3
297	1	0.333	1	->	0
298	1	0.400	1	=>	2
354	1	0.286	0	->	1

Node 61 to Node 39

Character	Step	CI	Change		
13	1	0.125	0	->	1
25	1	0.111	1	->	0
50	1	0.167	0	->	1
88	1	0.167	0	=>	1
120	1	0.231	2	=>	1
130	1	0.200	0	=>	1
132	1	0.167	0	=>	1
236	1	0.600	0	=>	2
302	1	0.667	2	->	4
328	1	0.250	2	->	0
342	1	0.400	1	=>	2
384	1	0.286	2	=>	1
397	1	0.200	1	=>	0
399	1	0.600	1	=>	2
458	1	0.400	2	->	0
460	1	0.667	0	->	1
463	1	0.333	0	=>	1

Node 39 to Node 38

Character	Step	CI	Change		
31	1	0.333	1	=>	2
62	1	0.111	0	=>	1
77	1	0.111	1	->	0
89	1	0.286	1	->	2
114	1	0.200	1	->	0
120	1	0.231	1	=>	0
225	1	0.400	2	->	0
235	1	0.500	0	->	2
257	1	0.286	4	->	3
284	1	0.300	2	=>	3
292	1	0.333	2	=>	3
293	1	0.231	1	=>	0
301	1	0.267	1	=>	0
326	1	0.250	2	->	3
354	1	0.286	1	->	0
436	1	0.400	3	=>	2
437	1	0.333	0	=>	1
438	1	0.600	3	=>	2
442	1	0.500	1	=>	0
446	1	0.500	3	=>	2

Node 38 to Node 36

Character	Step	CI	Change		
33	1	0.111	0	=>	1
34	1	0.167	0	=>	1
47	1	0.200	1	=>	0
129	1	0.200	0	->	1

Node 36 to Knmer 15000

Character	Step	CI	Change		
21	1	0.250	1	=>	0
24	1	0.300	0	=>	2
28	1	0.167	1	=>	0
32	1	0.154	1	=>	0

Character	Step	CI	Change		
39	1	0.143	1	->	0
46	1	0.125	1	=>	0
51	1	0.182	2	->	0
52	1	0.125	1	->	0
54	1	0.250	0	->	1
55	1	0.182	1	->	0
58	1	0.500	0	->	1
64	1	0.125	0	->	1
67	1	0.333	1	->	0
68	1	0.167	0	->	1
79	1	0.154	0	->	1
84	1	0.125	1	->	0
92	1	0.125	2	->	0
97	1	0.286	1	->	0
102	1	0.222	1	->	0
104	1	0.300	1	->	0
107	1	0.200	0	->	1
109	1	0.154	1	->	0
112	1	0.200	1	->	0
117	1	0.100	1	->	0
122	1	0.500	1	->	2
123	1	0.429	0	->	1
131	1	0.182	1	->	0
221	1	0.667	4	->	3
230	1	0.444	0	->	4
237	1	0.214	3	->	2
243	1	0.333	2	->	1
246	1	0.333	4	->	3
247	1	0.667	0	->	3
258	1	0.800	4	->	0
279	1	0.250	2	->	1
285	1	0.333	1	=>	2
286	1	0.333	0	->	1
291	1	0.214	1	->	0
299	1	0.545	2	->	0
324	1	0.333	1	->	0
325	1	0.333	4	->	3
335	1	0.600	1	->	0
374	1	0.667	2	->	3
381	1	0.600	2	->	3
382	1	0.500	1	->	2
395	1	0.286	1	->	0
400	1	0.400	0	->	2
403	1	0.500	0	->	1
415	1	0.750	0	->	1
417	1	0.500	1	->	0
433	1	0.250	5	->	3
439	1	0.300	3	->	2
450	1	0.750	4	->	3
460	1	0.667	1	->	0

Node 36 to Trinil

Character	Step	CI	Change		
6	1	0.200	1	->	0
22	1	0.200	0	->	1
25	1	0.111	0	=>	1
30	1	0.125	0	=>	1
32	1	0.154	1	=>	2
53	1	0.154	0	->	1
135	1	0.167	1	->	0
151	1	0.429	4	->	3

Character	Step	CI	Change		

Node 38 to Node 37

Character	Step	CI	Change		
27	1	0.375	3	=>	1
29	1	0.182	1	->	2
290	1	0.286	2	=>	1
322	1	0.429	3	=>	2
431	1	0.286	1	=>	0
445	1	0.500	2	=>	1

Node 37 to Sangiran 2

Character	Step	CI	Change		
20	1	0.200	1	->	0
22	1	0.200	0	->	1
39	1	0.143	1	->	0
44	1	0.125	0	=>	1
55	1	0.182	1	->	0
57	1	0.333	1	=>	0
66	1	0.154	2	->	1
73	1	0.200	1	->	0
78	1	0.111	0	->	1
84	1	0.125	1	->	0
110	1	0.182	1	->	2
113	1	0.231	1	->	3
116	1	0.333	1	->	0
119	1	0.182	2	->	1
134	1	0.286	1	->	2
276	1	0.250	1	->	0
287	1	0.667	0	->	1
291	1	0.214	1	->	0
299	1	0.545	2	->	4
431	1	0.222	3	=>	1
433	1	0.250	5	=>	2
434	1	0.222	2	->	0
435	1	0.400	2	=>	4

Node 37 to Sangiran 12

Character	Step	CI	Change		
22	1	0.200	0	->	2
23	1	0.182	0	=>	2
36	1	0.167	0	=>	1
37	1	0.083	0	=>	1
48	1	0.500	0	=>	1
61	1	0.167	0	=>	1
75	1	0.091	0	=>	1
281	1	0.333	1	=>	0
287	1	0.667	0	->	2
291	1	0.214	1	=>	2
297	1	0.333	0	->	1
298	1	0.400	2	=>	1
320	1	0.500	1	=>	2
328	1	0.250	0	=>	2
329	1	0.250	0	=>	1
374	1	0.667	2	->	1
390	1	0.500	0	->	2
429	1	0.400	1	->	2
437	1	0.333	1	->	2
439	1	0.300	3	->	1
447	1	0.333	1	->	0
452	1	0.417	4	->	3
453	1	0.385	3	->	2
454	1	1.000	5	->	3

Character	Step	CI	Change		
461	1	0.333	1	->	0
464	1	0.667	0	->	2

Node 39 to Sangiran 17

Character	Step	CI	Change		
1	1	0.167	0	->	1
2	1	0.143	1	->	0
4	1	0.133	1	->	0
5	1	0.111	1	->	0
20	1	0.200	1	->	0
23	1	0.182	0	=>	2
28	1	0.167	1	=>	0
29	1	0.182	1	=>	0
32	1	0.154	1	=>	2
36	1	0.167	0	=>	1
38	1	0.182	2	->	1
40	1	0.167	0	=>	1
46	1	0.125	1	=>	0
52	1	0.125	1	->	0
61	1	0.167	0	=>	1
63	1	0.167	0	->	1
64	1	0.125	0	->	1
70	1	0.125	1	->	0
79	1	0.154	0	=>	2
87	1	0.100	1	->	0
89	1	0.286	1	->	0
92	1	0.125	2	->	1
94	1	0.200	1	->	2
105	1	0.222	2	->	0
107	1	0.200	0	->	2
108	1	0.133	0	=>	2
113	1	0.231	1	=>	0
116	1	0.333	1	=>	0
125	1	0.333	3	->	1
129	1	0.200	0	->	1
133	1	0.200	0	->	1
136	1	0.267	0	->	4
139	1	0.250	2	->	3
145	1	0.273	0	->	1
175	1	0.250	3	=>	2
186	1	0.500	2	->	3
209	1	0.375	3	->	1
210	1	0.333	3	->	1
215	1	0.571	4	=>	3
216	1	0.667	3	->	4
223	1	0.833	3	=>	5
224	1	0.800	2	=>	3
232	1	0.538	1	->	0
255	1	0.700	1	->	0
271	1	0.222	1	->	2
279	1	0.250	2	->	1
281	1	0.333	1	=>	0
285	1	0.333	1	=>	0
324	1	0.333	1	->	0
325	1	0.333	4	=>	2
327	1	0.333	3	=>	2
331	1	0.667	3	=>	2
333	1	0.429	0	=>	1
336	1	0.400	0	=>	1
337	1	0.500	0	=>	1
339	1	0.200	0	=>	1
343	1	0.833	0	=>	2

Character	Step	CI	Change		
344	1	0.500	0	=>	1
352	1	0.375	0	=>	2
353	1	0.444	0	=>	1
356	1	0.250	0	=>	1
368	1	0.500	2	=>	1
376	1	0.600	3	=>	2
385	1	0.429	3	=>	1
396	1	0.500	0	=>	1
401	1	0.500	0	=>	1
406	1	1.000	0	=>	1
411	1	0.429	0	=>	1
414	1	0.750	0	=>	1
429	1	0.400	1	->	3
430	1	0.286	1	=>	2
433	1	0.250	5	=>	4
434	1	0.300	0	=>	1
461	1	0.500	0	=>	1
468	1	0.250	1	=>	0

Node 61 to Node 60

Character	Step	CI	Change		
10	1	0.200	0	->	1
29	1	0.182	1	->	2
38	1	0.182	2	->	1
55	1	0.182	1	->	2
60	1	0.100	0	->	1
90	1	0.250	1	=>	0
91	1	0.167	0	=>	1
95	1	0.167	0	->	1
105	1	0.222	2	->	0
111	1	0.125	0	=>	1
144	1	0.222	1	->	2
150	1	0.400	3	->	2
152	1	1.000	1	->	0
153	2	0.571	2	->	0
155	1	0.500	2	->	1
158	1	0.455	3	->	5
162	1	0.444	3	->	2
171	1	0.545	3	->	2
173	1	0.333	2	->	3
176	1	0.400	3	->	5
181	1	0.400	2	->	1
182	1	0.444	2	->	1
192	1	0.400	2	->	1
206	1	0.375	2	->	1
216	1	0.667	3	->	4
234	1	0.333	1	=>	0
265	1	0.308	1	->	0
395	1	0.286	1	=>	2
404	1	0.429	0	=>	1
429	1	0.400	1	->	2
435	1	0.222	2	->	1
453	1	0.385	3	->	4
459	1	0.600	1	->	0

Node 60 to Node 59

Character	Step	CI	Change		
14	1	0.1111	0	=>	1
15	1	0.111	0	=>	1
22	1	0.200	0	=>	1
53	1	0.154	0	=>	2
98	1	0.222	0	=>	2

Character	Step	CI	Change		
104	1	0.300	1	=>	2
106	1	0.154	0	->	1
113	1	0.231	1	=>	2
114	1	0.200	1	=>	2

Node 59 to Node 58

Character	Step	CI	Change		
40	1	0.167	0	=>	1
66	1	0.154	2	->	1
72	1	0.143	0	->	1
107	1	0.200	0	->	1
116	1	0.333	1	=>	2
127	1	0.250	1	->	0
147	1	0.556	2	->	1
148	1	0.625	2	=>	1
156	1	0.714	1	->	0
166	1	0.235	2	->	1
167	1	0.364	2	=>	1
172	1	0.462	3	=>	4
187	1	0.625	4	=>	3
217	1	0.800	4	->	5
223	1	0.833	3	=>	0
239	1	0.444	1	->	2
269	1	0.300	1	->	3
284	1	0.300	2	->	3
292	1	0.333	2	=>	0
299	1	0.545	2	=>	4
302	1	0.667	2	->	3
326	1	0.250	2	->	0
328	1	0.250	2	->	2
354	1	0.286	1	=>	0
395	1	0.286	2	=>	0
397	1	0.200	1	=>	0
405	1	0.500	1	->	0
436	1	0.400	3	->	4

Node 58 to Node 48

Character	Step	CI	Change		
24	1	0.300	0	->	1
73	1	0.200	1	=>	2
94	1	0.200	1	->	2
130	1	0.200	0	=>	1
154	1	0.750	1	=>	0
160	1	0.250	3	->	1
197	1	0.333	2	->	0
204	1	0.250	1	->	0
205	1	0.333	2	->	1
210	1	0.333	3	->	1
229	1	0.400	0	=>	1
270	1	0.429	1	->	0
369	1	0.429	1	=>	2
377	1	0.571	2	->	4
460	1	0.667	0	->	1
461	1	0.333	1	->	2
462	1	0.500	0	=>	3

Node 48 to Node 46

Character	Step	CI	Change		
23	1	0.182	0	=>	1
51	1	0.182	2	=>	0
83	1	0.143	0	->	1
100	1	0.167	1	->	0

Character	Step	CI	Change		
119	1	0.182	2	=>	3
125	1	0.333	3	->	2
134	1	0.286	1	->	3
146	1	0.444	1	=>	0
155	1	0.500	1	->	0
214	1	0.800	3	=>	4
216	1	0.667	4	->	5
234	1	0.333	0	=>	1
293	1	0.231	1	=>	0
335	1	0.600	1	=>	0
354	1	0.286	2	=>	0
430	1	0.286	1	=>	2
463	1	0.333	0	=>	1

Node 46 to Node 42

1	1	0.167	0	=>	1
24	1	0.300	1	->	0
28	1	0.167	1	=>	2
52	1	0.125	1	=>	0
89	1	0.286	1	=>	0
90	1	0.250	0	=>	1
108	1	0.133	0	->	1
113	1	0.231	2	->	1
172	1	0.462	4	->	2
173	1	0.333	3	=>	2
174	1	0.500	3	=>	1
178	1	0.556	3	->	2
185	1	0.333	3	->	2
193	1	0.571	1	->	2
194	1	0.625	1	=>	2
196	1	0.286	2	->	1
218	1	0.400	1	->	3
268	1	0.300	1	->	0
291	1	0.214	1	->	0
351	1	0.400	0	->	1
374	1	0.667	2	=>	6
388	1	0.429	3	=>	2
391	1	0.333	1	=>	0
431	1	0.222	3	->	2
432	1	0.500	0	->	1
433	1	0.250	5	=>	3
452	1	0.417	4	->	6

Node 42 to Node 40

6	1	0.200	1	=>	0
14	1	0.111	1	=>	0
15	1	0.111	1	=>	0
72	1	0.143	1	=>	0
92	2	0.125	2	=>	0
108	1	0.133	1	=>	2
111	1	0.125	1	->	0
117	1	0.100	1	=>	0
137	1	0.182	2	=>	1
384	1	0.286	2	=>	1
400	1	0.400	0	=>	1
404	1	0.429	1	=>	0
421	1	1.000	3	=>	2

Character	Step	CI	Change		

Node 40 to Laetoli H18

2	1	0.143	1	->	2
5	1	0.111	1	->	0
9	1	0.167	2	=>	0
10	1	0.200	1	=>	0
23	1	0.182	1	->	2
24	1	0.300	0	=>	2
25	1	0.111	1	->	0
29	1	0.182	2	->	1
32	1	0.154	1	=>	0
39	1	0.143	1	=>	0
45	1	0.250	2	=>	0
47	1	0.200	1	=>	0
53	1	0.154	2	->	1
66	1	0.154	1	->	0
70	1	0.125	1	=>	0
73	1	0.200	2	=>	0
77	1	0.111	1	=>	0
78	1	0.111	0	=>	1
79	1	0.154	0	=>	1
82	1	0.167	1	=>	0
91	1	0.167	1	=>	0
105	1	0.222	0	=>	1
110	1	0.182	1	->	2
114	1	0.200	2	->	1
118	1	0.333	0	=>	2
121	1	0.286	1	=>	0
140	1	0.200	1	=>	2
142	1	0.571	0	->	1
153	1	0.571	0	->	1
177	1	0.300	3	->	5
178	1	0.556	2	->	4
179	1	0.400	1	=>	0
180	1	0.222	0	->	1
215	1	0.571	4	->	3
264	1	0.444	3	->	4
269	1	0.300	2	->	0
271	1	0.222	1	=>	2
336	1	0.400	0	->	3
354	1	0.286	0	->	1
382	1	0.500	1	=>	2
385	2	0.286	3	->	1
387	1	0.500	1	->	0
395	1	0.286	0	->	2
399	1	0.600	1	->	2
401	1	0.500	0	->	2
403	1	0.500	0	->	2
406	1	1.000	0	->	3
409	1	1.000	0	->	1
415	1	0.750	0	->	4
423	2	0.500	2	->	0
441	3	0.500	0	->	3
450	1	0.750	4	->	2

Node 40 to OH 9

37	1	0.083	0	->	1
38	1	0.182	1	=>	0
49	1	0.167	1	->	0
60	1	0.100	1	=>	0

Character	Step	CI	Change		
62	1	0.111	0	=>	1
64	1	0.125	0	=>	1
71	1	0.167	0	->	1
75	1	0.091	0	->	1
83	1	0.143	1	=>	0
85	1	0.167	0	=>	1
98	1	0.222	2	=>	0
99	1	0.333	2	=>	1
113	1	0.231	1	->	3
116	1	0.333	2	->	1
136	1	0.267	0	->	1
140	1	0.200	1	->	0
163	1	0.250	1	->	2
179	1	0.400	1	=>	2
188	1	0.375	3	->	1
189	1	0.429	3	->	2
210	1	0.571	0	->	1
264	1	0.250	3	->	2
317	1	0.400	0	->	2
333	1	0.429	0	=>	1
337	1	0.500	0	=>	1
338	1	0.571	1	=>	2
344	1	0.500	0	=>	1
349	1	0.400	2	=>	1
352	1	0.375	0	=>	2
353	1	0.444	0	=>	2
358	1	0.667	2	=>	1
359	1	0.333	1	=>	0
364	1	0.200	2	=>	1
367	1	0.500	2	=>	1
368	1	0.500	2	=>	1
369	1	0.429	2	=>	0
370	1	0.800	2	=>	0
373	1	0.500	2	=>	1
375	1	0.250	1	=>	0
376	1	0.600	3	=>	1
384	1	0.286	1	=>	0
396	1	0.500	0	=>	3
397	1	0.200	0	=>	1
399	1	0.600	1	->	3
402	1	1.000	0	=>	1
403	1	0.500	0	=>	1
407	1	1.000	0	=>	1
408	1	1.000	0	=>	1
410	1	1.000	0	=>	3
411	1	0.429	0	=>	2
414	1	0.750	0	=>	2
415	1	0.750	0	->	2
418	1	1.000	0	=>	2

Node 42 to Node 41

Character	Step	CI	Change		
38	1	0.182	1	=>	2
45	1	0.250	2	=>	3
65	1	0.200	1	=>	0
66	1	0.154	1	->	0
79	1	0.154	0	=>	2
87	1	0.100	1	=>	0
121	1	0.286	1	=>	2
371	1	0.333	0	=>	1

Character	Step	CI	Change		

Node 41 to Kabwe

Character	Step	CI	Change		
2	1	0.143	1	->	2
4	1	0.133	1	->	0
5	1	0.111	1	=>	0
12	1	0.333	3	=>	2
19	1	0.143	0	=>	1
22	1	0.200	1	=>	0
32	1	0.154	1	->	2
33	1	0.111	0	->	1
34	1	0.167	0	->	1
36	1	0.167	0	->	1
40	1	0.167	1	=>	0
44	1	0.125	0	=>	1
50	1	0.167	0	->	1
53	1	0.154	2	->	1
54	1	0.250	0	=>	1
71	1	0.167	0	->	1
73	1	0.154	2	=>	1
75	1	0.091	0	->	1
77	1	0.111	0	=>	0
80	1	0.125	2	=>	0
98	1	0.222	1	->	1
102	1	0.222	1	->	0
104	1	0.300	2	=>	3
106	1	0.154	1	->	2
110	1	0.182	1	->	2
114	1	0.200	2	->	1
115	1	0.250	1	->	0
122	1	0.500	1	->	0
130	1	0.200	1	->	0
140	1	0.200	1	->	0
141	1	0.125	0	->	1
144	1	0.222	2	->	1
145	2	0.273	0	->	2
147	1	0.556	1	->	0
148	1	0.625	1	->	0
157	1	0.400	3	->	1
158	1	0.455	5	->	1
159	1	0.500	3	->	2
160	1	0.250	1	->	3
166	1	0.235	1	->	2
175	1	0.250	3	->	2
176	1	0.400	5	->	1
181	1	0.400	1	->	2
182	1	0.444	1	->	2
190	1	0.500	4	->	3
191	1	0.500	2	->	1
192	1	0.400	1	->	2
198	1	0.385	3	->	1
199	1	0.250	1	->	2
201	1	0.222	1	->	2
202	2	0.333	2	->	0
204	1	0.250	0	->	1
209	1	0.375	3	->	0
222	1	0.500	1	->	2
223	1	0.833	5	->	4
234	1	0.333	1	=>	2
237	1	0.214	3	->	2
241	1	0.222	1	->	0
243	1	0.333	2	=>	4

Character	Step	CI	Change		
244	1	0.400	2	=>	1
251	1	1.000	3	->	0
255	1	0.700	1	->	0
258	1	0.800	4	=>	2
262	1	0.500	3	->	0
263	1	0.545	0	->	5
264	1	0.444	3	->	2
265	1	0.308	0	->	3
266	1	0.429	3	->	2
267	1	0.250	3	->	2
268	1	0.300	0	->	2
270	1	0.429	0	->	3
275	1	1.000	3	->	2
277	1	0.500	2	->	0
278	1	0.500	2	->	1
279	1	0.250	2	=>	0
285	1	0.333	1	=>	3
301	1	0.267	1	->	0
328	1	0.250	0	=>	1
381	1	0.600	2	=>	3
386	1	0.800	2	=>	4
389	1	0.333	2	->	1
395	1	0.286	0	->	1
416	1	0.250	2	->	1
424	1	0.429	2	->	1
434	1	0.300	0	->	2

Node 41 to Omo Kibish 2

Character	Step	CI	Change		
18	1	0.250	2	=>	1
25	1	0.111	1	->	0
28	1	0.167	2	=>	0
29	1	0.182	2	=>	0
37	1	0.083	0	->	1
42	1	0.200	1	=>	0
74	1	0.286	0	=>	1
85	1	0.167	0	=>	1
88	1	0.167	0	->	1
106	1	0.154	1	=>	0
111	1	0.125	1	->	0
117	1	0.100	1	->	0
132	1	0.167	0	=>	1
247	1	0.667	0	=>	3
281	1	0.333	1	=>	0
284	1	0.300	3	=>	2
290	1	0.286	2	=>	1
299	1	0.545	0	=>	1
316	1	0.400	0	=>	2
323	1	0.429	3	=>	2
429	1	0.286	1	->	0
454	1	0.385	4	=>	6

Node 46 to Node 45

Character	Step	CI	Change		
3	1	0.167	1	->	0
33	1	0.111	0	->	1
62	1	0.111	0	=>	1
106	1	0.154	1	->	2
109	1	0.154	1	->	2
118	1	0.333	0	->	4
124	1	0.429	3	->	1
171	1	0.545	2	=>	4
178	1	0.556	3	->	4
194	1	0.625	1	->	0
197	1	0.333	0	->	3
200	1	0.222	1	=>	0
218	1	0.400	1	->	2
230	1	0.444	0	=>	1
231	1	0.375	0	->	1
237	1	0.214	3	->	2
245	1	0.200	1	=>	0
265	1	0.308	0	->	2
269	1	0.300	2	->	0
284	1	0.300	3	=>	1
301	1	0.267	1	->	0
377	1	0.571	4	->	2
404	1	0.429	1	=>	2
434	1	0.300	0	->	1
435	1	0.222	1	=>	2
442	1	0.500	1	=>	2
443	1	0.500	2	=>	3
458	1	0.400	2	->	1
459	1	0.600	0	=>	1
468	1	0.250	1	=>	0

Node 45 to Ngandong 7

Character	Step	CI	Change		
2	1	0.143	1	->	0
4	1	0.133	1	=>	2
9	1	0.167	2	->	0
10	1	0.200	1	=>	0
12	1	0.333	3	->	0
18	2	0.250	2	->	4
20	1	0.200	1	->	0
31	1	0.333	1	=>	0
35	1	0.167	0	=>	1
42	1	0.200	1	=>	0
46	1	0.125	1	=>	0
47	1	0.200	1	=>	0
56	1	0.250	0	->	1
58	1	0.500	0	->	2
59	1	0.667	2	->	0
60	1	0.100	1	=>	0
61	1	0.167	0	->	1
70	1	0.125	1	=>	0
74	1	0.286	2	=>	1
88	1	0.167	0	->	1
96	1	0.250	1	=>	0
98	1	0.222	2	=>	0
101	1	0.333	0	=>	1
107	1	0.200	1	=>	0
110	1	0.182	2	->	2
113	1	0.231	2	->	3
131	1	0.182	1	=>	0
136	1	0.267	0	->	2
139	1	0.250	2	->	3
140	1	0.200	1	->	2
145	1	0.273	0	->	1
147	1	0.556	1	->	0
148	1	0.625	1	->	0
158	1	0.455	5	=>	3
160	1	0.250	1	->	2
163	1	0.250	1	->	0
166	1	0.235	1	=>	0

Character	Step	CI	Change		
177	1	0.300	3	->	5
181	1	0.400	1	=>	0
186	1	0.500	2	=>	3
187	1	0.625	3	=>	4
240	1	0.600	1	=>	0
241	1	0.222	1	=>	0
243	1	0.333	2	->	3
262	1	0.500	3	=>	1
266	1	0.429	3	=>	2
271	1	0.222	1	->	2
295	1	0.375	3	=>	1
327	1	0.333	2	=>	2
340	1	0.500	0	=>	2
356	1	0.250	0	=>	2
369	1	0.429	2	=>	1
374	1	0.667	2	=>	3
377	1	0.200	0	=>	1
400	1	0.400	0	=>	1
401	1	0.500	0	=>	1
405	1	0.500	0	=>	1
412	1	0.750	0	=>	3
430	1	0.286	2	=>	3
461	1	0.333	2	->	0

Node 45 to Node 44

Character	Step	CI	Change		
37	1	0.083	0	->	1
38	1	0.182	1	=>	2
45	1	0.250	2	=>	1
123	1	0.429	0	=>	1
429	1	0.400	2	=>	3
436	1	0.400	4	=>	3
452	1	0.417	4	->	3
453	1	0.385	4	=>	3

Node 44 to Node 43

Character	Step	CI	Change		
8	1	0.167	1	->	0
12	1	0.333	3	->	1
13	1	0.125	0	->	1
14	1	0.111	1	->	0
15	1	0.111	1	->	0
22	1	0.200	1	->	2
31	1	0.333	1	=>	2
56	1	0.250	0	->	1
79	1	0.154	0	=>	2
103	1	0.286	1	=>	0
115	1	0.250	1	=>	0
119	2	0.182	3	=>	1
120	2	0.231	2	=>	0
176	1	0.400	5	->	4
182	1	0.444	1	->	2
185	1	0.333	3	->	1
196	1	0.286	2	->	3
198	1	0.385	3	->	4
201	1	0.222	1	->	0
210	1	0.333	1	->	2
224	1	0.800	2	->	4
238	1	0.667	0	->	1
239	1	0.444	0	->	1
243	1	0.333	2	->	1
246	2	0.333	4	->	2

Character	Step	CI	Change		
253	1	0.500	0	->	1
254	1	0.500	0	->	2
257	1	0.286	4	->	3
282	1	0.600	3	->	2
291	1	0.214	1	->	0
328	1	0.250	0	->	1

Node 43 to Ngandong 12

Character	Step	CI	Change		
4	1	0.133	1	=>	0
11	1	0.333	2	=>	1
18	2	0.250	2	->	4
20	1	0.200	1	->	0
34	1	0.167	0	=>	1
38	1	0.182	2	=>	0
43	1	0.125	0	->	1
49	1	0.167	1	=>	0
51	1	0.182	0	=>	2
58	1	0.500	0	->	1
63	1	0.167	0	=>	1
66	1	0.154	1	->	2
73	1	0.200	2	=>	1
76	1	0.333	0	=>	1
88	1	0.167	0	->	1
96	1	0.250	1	=>	0
102	1	0.222	1	->	0
106	1	0.154	2	=>	1
111	1	0.125	1	=>	0
122	1	0.500	1	->	2
132	1	0.167	0	=>	1
135	1	0.167	1	=>	0
136	1	0.267	0	->	3
140	2	0.200	1	=>	3
146	1	0.444	0	=>	1
149	1	0.273	2	=>	3
150	1	0.400	2	->	4
157	1	0.400	3	=>	4
163	1	0.250	1	->	0
164	1	0.333	2	=>	0
165	1	0.333	1	=>	2
167	1	0.364	1	=>	0
170	1	0.417	3	=>	4
175	1	0.250	3	->	1
179	1	0.400	1	=>	0
180	1	0.222	0	->	2
192	1	0.400	1	=>	2
199	1	0.250	1	->	2
208	1	0.429	1	=>	0
235	1	0.500	0	->	1
241	1	0.222	1	=>	2
251	1	1.000	3	=>	1
252	1	0.500	2	->	1
256	1	0.500	0	->	3
264	1	0.444	3	=>	4
265	1	0.308	2	->	0
268	1	0.300	1	->	0
271	1	0.222	1	->	2
284	1	0.300	1	=>	0
383	1	0.800	2	=>	3
429	1	0.400	3	=>	4
466	1	1.000	0	=>	1

Node 43 to Ngawi

Character	Step	CI	Change		
2	1	0.143	1	->	0
9	1	0.167	2	->	0
30	1	0.125	0	->	1
32	1	0.154	1	=>	2
33	1	0.111	1	=>	0
35	1	0.167	0	=>	1
52	1	0.125	1	=>	0
53	1	0.154	2	=>	1
55	1	0.182	2	=>	1
58	1	0.500	0	->	2
62	1	0.111	1	=>	0
80	1	0.125	1	=>	0
92	2	0.125	2	=>	0
98	1	0.222	2	=>	1
125	2	0.333	2	=>	0
129	1	0.200	0	=>	1
134	1	0.286	3	->	1
137	1	0.182	2	=>	1
139	1	0.250	2	->	3
144	1	0.222	2	=>	1
145	1	0.273	0	->	1
149	1	0.273	2	=>	1
150	1	0.400	2	->	1
158	1	0.455	5	=>	4
161	1	0.300	2	=>	1
175	1	0.250	3	->	2
178	1	0.556	4	->	3
180	1	0.222	0	->	1
188	1	0.375	3	=>	2
189	1	0.429	3	=>	2
205	1	0.333	1	=>	0
229	1	0.400	1	=>	2
232	1	0.538	1	=>	2
235	1	0.500	0	->	2
237	2	0.214	2	=>	0
246	1	0.333	2	=>	1
252	1	0.500	2	->	0
255	1	0.700	1	=>	3
256	1	0.500	0	->	2
263	1	0.545	0	=>	1
269	1	0.300	0	->	1
270	1	0.429	0	=>	1
285	1	0.333	1	=>	2
342	1	0.400	1	=>	2
384	1	0.286	2	=>	1
403	1	0.429	2	=>	0
416	1	0.250	2	->	1
430	1	0.286	2	=>	0
433	1	0.250	5	=>	4
439	1	0.300	3	->	2
456	1	0.500	4	=>	5
457	1	0.600	0	=>	3
462	1	0.500	3	=>	1
464	1	0.667	0	=>	2
465	1	0.750	0	->	1

Node 44 to Sangiran 4

Character	Step	CI	Change		
28	1	0.167	1	=>	0
30	1	0.125	0	->	1
40	1	0.167	1	=>	0
43	1	0.125	0	->	1
44	1	0.125	0	=>	1
53	1	0.154	2	=>	0
61	1	0.167	0	->	1
66	1	0.154	1	->	2
75	1	0.091	0	=>	1
77	1	0.111	1	=>	0
82	1	0.167	1	=>	0
85	1	0.167	0	=>	1
86	1	0.500	0	=>	1
90	1	0.250	0	=>	1
100	1	0.167	0	=>	1
108	1	0.133	0	=>	2
392	1	0.500	0	=>	1
416	1	0.250	2	=>	3
424	1	0.429	2	=>	3
428	1	0.286	1	=>	2
430	1	0.286	2	=>	4
433	1	0.250	5	=>	3
439	1	0.300	3	->	2
446	1	0.500	3	=>	2
455	1	0.429	3	=>	2

Node 48 to Node 47

Character	Step	CI	Change		
41	1	0.500	0	=>	1
43	1	0.125	0	=>	1
55	1	0.182	2	=>	1
65	1	0.200	1	=>	0
71	1	0.167	0	->	1
105	1	0.222	0	=>	1
109	1	0.154	1	->	2
111	1	0.125	1	->	0
112	1	0.200	1	=>	0
113	1	0.231	2	->	3
339	1	0.200	0	=>	1
424	1	0.429	2	=>	1

Node 47 to Ndutu

Character	Step	CI	Change		
18	2	0.250	2	=>	0
29	1	0.182	2	=>	1
37	1	0.083	0	->	1
39	1	0.143	1	=>	0
53	1	0.154	2	=>	1
54	1	0.250	0	=>	1
60	1	0.100	1	=>	0
66	1	0.154	1	->	0
74	1	0.286	0	=>	1
83	1	0.143	0	->	1
84	1	0.125	1	=>	0
87	1	0.100	1	=>	0
110	1	0.182	1	->	2
118	1	0.333	0	->	2
121	1	0.286	1	=>	2
122	1	0.500	1	=>	0

Character	Step	CI	Change		
123	1	0.429	0	=>	3
279	1	0.250	2	=>	0
295	1	0.375	3	->	0
298	1	0.400	2	=>	0
329	1	0.250	0	=>	1
334	1	0.400	0	->	2
341	1	0.667	0	=>	1
342	1	0.400	1	=>	2
357	1	0.500	0	=>	1
390	1	0.500	0	=>	1
424	1	0.429	1	=>	0
427	1	1.000	1	->	3
430	1	0.286	1	=>	0
432	1	0.500	0	->	1
434	1	0.300	0	->	1
442	1	0.500	1	->	4
452	1	0.417	4	->	6
453	1	0.385	4	->	6
455	1	0.429	3	->	4
456	1	0.500	4	->	5
464	1	0.667	0	->	3
466	1	1.000	0	->	2

Node 47 to Sambungmacan 1

Character	Step	CI	Change		
2	1	0.143	1	->	0
3	1	0.167	1	->	0
4	1	0.133	1	->	0
13	1	0.125	0	->	1
45	1	0.250	2	=>	1
46	1	0.125	1	=>	0
51	1	0.250	2	=>	1
63	1	0.167	0	=>	1
68	1	0.167	0	->	1
70	1	0.125	1	=>	0
78	1	0.111	0	->	1
79	1	0.154	0	=>	1
88	1	0.167	0	->	1
92	2	0.125	2	->	0
94	1	0.200	2	->	1
95	1	0.167	1	=>	0
96	1	0.250	1	=>	0
100	1	0.167	1	->	0
102	1	0.222	1	->	0
104	1	0.300	2	=>	0
108	1	0.133	0	=>	2
114	1	0.200	2	=>	1
117	1	0.100	1	=>	0
118	1	0.333	0	->	3
119	1	0.182	2	=>	1
120	2	0.231	2	->	0
136	1	0.267	0	->	2
146	2	0.444	1	->	3
165	1	0.333	1	->	2
170	1	0.417	3	->	4
185	1	0.333	3	->	2
190	1	0.500	4	->	6
196	1	0.286	2	->	4
199	1	0.500	1	->	2
208	1	0.429	1	->	0
240	1	0.600	1	->	0
246	2	0.333	4	->	2

Character	Step	CI	Change		
252	1	0.500	2	->	1
282	1	0.600	3	->	2
289	1	0.250	1	->	0
295	1	0.375	3	->	1
302	1	0.667	4	->	2
325	1	0.333	4	->	3
326	1	0.250	3	->	2
327	1	0.333	3	->	2
332	1	0.500	0	=>	1
343	1	0.833	0	=>	3
351	1	0.400	0	->	1
356	1	0.250	0	=>	1
362	1	0.500	0	->	1
363	1	0.500	0	->	1
374	1	0.667	2	->	5
389	1	0.333	2	=>	1
416	1	0.250	2	=>	1
431	1	0.222	3	->	2
432	1	0.500	0	->	2
433	1	0.250	5	=>	4
435	1	0.222	1	=>	2

Node 58 to Node 57

Character	Step	CI	Change		
49	1	0.167	1	->	0
118	1	0.333	0	->	4
190	1	0.500	4	->	5
265	1	0.308	0	->	2
285	1	0.333	1	=>	2
332	1	0.500	0	=>	1
333	1	0.429	0	=>	1
336	1	0.400	0	=>	1
337	1	0.500	0	=>	1
338	1	0.571	1	=>	0
342	1	0.400	1	=>	0
345	1	0.500	0	=>	1
348	1	1.000	1	=>	0
349	1	0.400	2	=>	0
350	1	1.000	1	=>	0
351	1	0.400	0	=>	2
353	1	0.444	0	=>	1
355	1	0.400	0	=>	2
357	1	0.500	0	=>	1
358	1	0.667	2	=>	0
360	1	1.000	0	=>	2
362	1	0.500	0	->	1
363	1	0.500	0	=>	3
365	1	0.667	0	=>	1
366	1	1.000	1	=>	2
372	1	0.333	1	=>	0
373	1	0.500	2	=>	0
374	1	0.667	2	=>	0
375	1	0.250	1	=>	0
376	1	0.600	3	=>	0
377	1	0.571	2	->	1
378	1	0.500	1	=>	0
379	1	0.667	2	=>	0
380	1	1.000	2	=>	0
381	1	0.600	2	=>	1
382	1	0.500	1	=>	0
383	1	0.800	2	=>	0
461	1	0.333	1	->	0

Node 57 to Node 56

Character	Step	CI	Change		
5	1	0.111	1	=>	0
11	1	0.333	2	=>	1
22	1	0.200	1	=>	0
25	1	0.111	1	=>	0
59	1	0.667	2	->	0
65	1	0.200	1	=>	0
82	1	0.167	1	=>	0
98	1	0.222	2	=>	0
102	1	0.222	1	->	0
136	1	0.267	0	->	1
137	1	0.182	2	=>	1
194	1	0.625	1	->	0
200	1	0.222	1	=>	0
201	1	0.222	1	->	0
210	1	0.333	3	->	2
224	1	0.800	2	->	4
301	1	0.267	1	=>	0
320	1	0.500	1	=>	0
356	1	0.250	0	->	2
361	1	1.000	0	->	3
428	1	0.286	1	->	0
431	1	0.222	3	->	2
442	1	0.500	1	->	0
453	1	0.385	4	=>	5
458	1	0.400	2	->	0

Node 56 to Node 49

Character	Step	CI	Change		
1	1	0.167	0	=>	1
124	1	0.429	3	=>	0
135	1	0.167	1	=>	0
139	1	0.250	2	=>	1
144	1	0.222	2	->	1
145	1	0.273	0	->	1
149	1	0.273	2	=>	1
150	1	0.400	2	=>	1
157	1	0.400	3	=>	1
160	1	0.250	3	->	1
170	1	0.417	3	=>	2
185	1	0.333	3	=>	2
190	1	0.500	5	=>	3
196	1	0.286	2	->	1
202	1	0.333	2	=>	1
203	1	0.143	1	->	0
204	1	0.250	1	->	0
205	1	0.333	2	=>	0
253	1	0.500	0	=>	1
263	1	0.545	0	=>	1

Node 49 to Bodo

Character	Step	CI	Change		
9	1	0.167	2	=>	0
23	1	0.182	0	=>	1
30	1	0.125	0	->	1
32	1	0.154	1	=>	0
141	1	0.125	0	=>	1
161	1	0.300	2	->	1
167	1	0.364	1	->	0
172	1	0.462	4	->	6
177	1	0.300	3	->	4
188	1	0.375	3	=>	2
189	1	0.429	3	=>	2
192	1	0.400	1	=>	0
193	1	0.571	1	=>	2
197	1	0.333	2	=>	1
198	1	0.385	3	=>	5
229	1	0.400	0	=>	2
230	1	0.444	0	=>	1
231	1	0.375	0	=>	1
232	1	0.538	1	=>	2
234	1	0.333	0	=>	2
235	1	0.500	0	=>	1
237	2	0.214	3	=>	1
238	1	0.667	0	=>	1
239	1	0.444	0	->	1
244	1	0.400	2	=>	1
245	1	0.200	1	=>	0
246	1	0.333	4	=>	3
254	1	0.500	0	=>	1
262	1	0.500	3	->	1
269	1	0.300	2	->	3
271	1	0.222	1	=>	2

Node 49 to Petralona

Character	Step	CI	Change		
2	1	0.143	1	=>	2
4	1	0.133	1	=>	0
13	1	0.125	0	->	1
21	1	0.250	1	->	0
24	1	0.300	0	->	1
28	1	0.167	1	->	2
35	1	0.167	0	->	1
42	1	0.200	1	->	0
43	1	0.125	0	->	1
45	1	0.250	2	->	3
46	1	0.125	1	->	0
47	1	0.200	1	->	0
52	1	0.125	1	->	0
75	1	0.091	0	->	1
81	1	0.333	1	->	0
87	1	0.100	1	->	0
95	1	0.167	1	->	0
97	1	0.286	1	->	0
106	1	0.154	1	->	0
107	1	0.200	1	->	0
117	1	0.100	1	->	0
118	1	0.333	4	->	1
120	1	0.231	2	->	1
125	1	0.333	3	->	4
131	1	0.182	1	=>	0
140	1	0.200	1	->	0
147	1	0.556	1	->	2
148	1	0.625	1	=>	2
151	1	0.429	4	=>	3
153	1	0.571	0	=>	1
155	1	0.500	1	=>	2
156	1	0.714	0	->	1
158	1	0.455	5	=>	2
159	1	0.500	3	=>	1
161	1	0.300	2	->	3
166	1	0.235	1	=>	2

Character	Step	CI	Change		
172	1	0.462	4	->	2
175	1	0.250	3	=>	1
176	1	0.400	5	=>	3
179	1	0.400	1	->	0
192	1	0.400	1	=>	2
197	1	0.333	2	->	0
198	1	0.385	3	->	1
208	1	0.429	1	->	0
209	1	0.375	3	=>	0
210	1	0.333	2	->	1
229	1	0.400	0	->	1
240	1	0.600	1	->	0
262	1	0.500	3	->	2
269	1	0.300	2	->	1
291	1	0.214	1	->	0
293	1	0.231	1	->	0
327	1	0.333	3	->	2
377	1	0.571	1	->	0
400	1	0.400	0	->	1
430	1	0.286	1	->	3
452	1	0.417	4	->	5

Node 56 to Node 55

Character	Step	CI	Change		
33	1	0.111	0	=>	1
35	1	0.167	0	->	1
146	1	0.444	1	=>	0
148	1	0.625	1	->	0
162	1	0.444	2	=>	4
167	1	0.364	1	->	0
171	1	0.545	2	=>	3
177	1	0.300	3	->	5
179	1	0.400	1	->	0
186	1	0.500	2	=>	4
187	1	0.625	3	=>	5
206	1	0.375	1	=>	2
208	1	0.429	1	->	0
209	1	0.375	3	=>	1
214	1	0.800	3	=>	4
218	1	0.400	1	=>	2
222	1	0.500	1	=>	2
232	1	0.538	1	=>	0
240	1	0.600	1	->	0
255	1	0.700	1	=>	0
265	1	0.308	2	=>	1

Node 55 to Node 54

Character	Step	CI	Change		
21	1	0.250	1	->	0
24	1	0.300	0	->	2
37	1	0.083	0	->	1
38	1	0.182	1	=>	2
45	1	0.250	2	->	1
79	1	0.154	0	->	2
90	1	0.250	0	->	1
94	1	0.200	1	->	2
104	1	0.300	2	->	3
114	1	0.200	2	->	1
119	1	0.182	2	->	3
147	1	0.556	1	=>	0
154	1	0.750	1	=>	0
155	1	0.500	1	=>	0

Character	Step	CI	Change		
163	1	0.250	1	=>	0
178	1	0.556	3	->	4
216	1	0.667	4	=>	5
291	1	0.214	1	->	0
344	1	0.500	0	->	1

Node 54 to Dali

Character	Step	CI	Change		
2	1	0.143	1	=>	0
12	1	0.333	3	=>	1
14	1	0.111	1	->	0
15	1	0.111	1	->	0
41	1	0.500	0	=>	1
43	1	0.125	0	->	1
44	1	0.125	0	=>	1
61	1	0.167	0	=>	1
66	1	0.154	1	->	2
68	1	0.167	0	=>	1
83	1	0.143	0	=>	1
85	1	0.167	0	=>	1
106	1	0.154	1	->	0
107	1	0.200	1	->	2
109	1	0.154	1	=>	0
110	1	0.182	1	=>	2
112	1	0.200	1	=>	0
113	1	0.231	2	->	0
115	1	0.250	1	=>	0
120	1	0.231	2	=>	3
124	1	0.429	3	->	1
125	2	0.333	3	=>	1
126	1	0.333	0	=>	1
131	1	0.182	1	=>	0
134	1	0.286	1	->	2
136	1	0.267	1	->	2
141	1	0.125	0	=>	1
149	1	0.273	2	->	3
150	1	0.400	2	->	4
160	1	0.250	3	=>	2
162	1	0.444	4	=>	3
164	1	0.333	2	->	0
165	1	0.333	1	=>	2
166	1	0.235	1	=>	2
175	1	0.250	3	->	2
186	1	0.500	4	=>	3
200	1	0.222	0	=>	1
204	1	0.250	1	->	0
215	1	0.571	4	=>	3
262	1	0.500	3	=>	2
269	1	0.300	2	->	1
293	1	0.231	1	->	0
315	1	0.500	0	=>	1
318	1	1.000	3	=>	2
319	1	0.750	3	=>	1
326	1	0.250	3	=>	2
327	1	0.333	3	->	2
349	1	0.400	0	=>	1
352	1	0.375	0	=>	1
416	1	0.250	2	=>	1
428	1	0.286	0	->	1
429	1	0.400	2	=>	1
433	1	0.250	5	=>	4
434	1	0.300	0	->	2

Character	Step	CI	Change		
442	1	0.500	0	->	1
443	1	0.500	2	=>	3
462	1	0.500	0	=>	2
463	1	0.333	0	=>	1

Node 54 to Node 53

Character	Step	CI	Change		
11	1	0.333	1	=>	2
39	1	0.143	1	=>	0
40	1	0.167	1	=>	0
51	1	0.182	2	=>	0
55	1	0.182	2	=>	0
60	1	0.100	1	=>	0
80	1	0.125	1	->	0
89	1	0.286	1	=>	0
95	1	0.167	1	=>	0
98	1	0.222	0	->	1
105	1	0.222	0	=>	1
113	1	0.231	2	->	1
136	1	0.267	1	->	0
140	1	0.200	1	=>	2
166	1	0.235	1	->	0
182	1	0.444	1	->	0
218	1	0.400	2	=>	3
434	1	0.300	0	->	1
452	1	0.417	4	->	5
455	1	0.429	3	=>	4
456	1	0.500	4	=>	5

Node 53 to Node 50

Character	Step	CI	Change		
10	1	0.200	1	=>	0
18	1	0.250	2	=>	3
35	1	0.167	1	=>	0
57	1	0.333	1	->	0
58	1	0.500	0	=>	3
59	1	0.667	0	->	1
87	1	0.100	1	=>	0
91	1	0.167	1	->	0
103	1	0.286	1	->	0
104	1	0.300	3	=>	1
116	1	0.333	2	=>	1
117	1	0.100	1	->	0
139	1	0.250	2	->	1
158	1	0.455	5	=>	3
262	1	0.500	3	=>	1
266	1	0.429	3	=>	2
267	1	0.250	3	=>	0
271	1	0.222	1	->	0
377	1	0.571	1	=>	0

Node 50 to Monte Circeo

Character	Step	CI	Change		
4	1	0.133	1	=>	0
22	1	0.200	0	=>	2
27	1	0.375	3	->	2
45	1	0.250	1	->	3
46	1	0.125	1	=>	0
64	1	0.125	0	=>	1
65	1	0.200	0	=>	1
67	1	0.333	1	->	0
75	1	0.091	0	=>	1
77	1	0.111	1	=>	0
78	1	0.111	0	->	1
79	1	0.154	2	->	1
81	1	0.333	1	=>	0
93	1	0.500	1	=>	0
97	1	0.286	1	=>	0
98	1	0.222	1	=>	2
109	1	0.154	1	=>	2
111	1	0.125	1	=>	0
112	1	0.200	1	=>	0
118	1	0.333	4	=>	2
144	1	0.222	2	=>	1
145	1	0.273	0	=>	1
161	1	0.300	2	->	3
163	1	0.250	0	=>	1
170	1	0.417	3	->	5
171	1	0.545	3	=>	5
172	1	0.462	4	->	5
181	1	0.400	1	=>	0
190	1	0.500	5	->	6
191	1	0.500	2	=>	3
196	1	0.286	2	->	3
197	1	0.333	2	->	3
201	1	0.222	0	=>	1
241	1	0.222	1	=>	0
243	1	0.333	2	=>	3
256	1	0.500	0	->	3
263	1	0.545	0	->	4
264	1	0.444	3	=>	2
265	1	0.308	1	=>	2
268	1	0.300	1	=>	2
291	1	0.214	O	=>	1
320	1	0.500	O	=>	1
325	1	0.333	4	->	2
332	1	0.500	1	=>	0
333	1	0.429	1	=>	0
338	1	0.571	0	=>	1
361	1	1.000	3	=>	5
364	1	0.200	2	=>	1
370	1	0.800	2	=>	4
388	3	0.429	3	=>	0
389	2	0.333	2	=>	0
391	1	0.333	1	=>	0
404	1	0.429	0	=>	1
427	1	1.000	1	=>	4
430	1	0.286	1	->	3
435	1	0.222	1	->	2
443	1	0.500	2	=>	1

Node 50 to La Chapelle aux Saints

Character	Step	CI	Change		
2	1	0.143	1	=>	0
3	1	0.167	1	=>	0
4	1	0.133	1	=>	2
7	1	0.200	1	->	0
13	1	0.125	0	=>	1
14	1	0.111	1	->	0
15	1	0.111	1	->	0
17	1	1.000	0	=>	1
27	1	0.375	3	->	0
28	1	0.167	1	=>	0
31	1	0.333	1	=>	2

Character	Step	CI	Change		
32	1	0.154	1	->	0
33	1	0.111	1	=>	0
38	1	0.182	2	=>	0
44	1	0.125	0	=>	1
45	1	0.250	1	->	2
47	1	0.200	1	=>	0
53	1	0.154	2	->	1
55	1	0.182	0	->	1
70	1	0.125	1	=>	0
74	1	0.286	0	=>	2
76	1	0.333	0	=>	1
79	1	0.154	2	->	0
82	1	0.167	0	=>	1
84	1	0.125	1	=>	0
99	1	0.333	2	=>	1
102	1	0.222	0	=>	2
109	1	0.154	1	=>	0
121	1	0.286	1	=>	0
125	1	0.333	3	->	2
133	1	0.200	0	=>	1
136	1	0.267	0	->	1
137	1	0.182	1	->	2
157	1	0.400	3	=>	2
159	1	0.500	3	=>	2
160	1	0.250	3	=>	2
162	1	0.444	4	=>	3
164	1	0.333	2	->	1
172	1	0.462	4	->	3
175	1	0.250	3	->	1
176	1	0.400	5	=>	3
177	1	0.300	5	=>	3
178	1	0.556	4	=>	3
186	1	0.500	4	=>	3
187	1	0.625	5	=>	4
190	1	0.500	5	->	4
192	1	0.400	1	=>	2
196	1	0.286	2	->	1
197	1	0.333	2	->	1
198	1	0.385	3	->	2
203	1	0.143	1	->	0
204	1	0.250	1	->	0
205	1	0.333	2	=>	1
206	1	0.375	2	=>	1
209	1	0.375	1	=>	0
210	1	0.333	2	=>	0
229	1	0.400	0	=>	3
230	1	0.444	0	=>	1
232	1	0.538	0	=>	1
233	1	0.714	0	=>	4
234	1	0.333	0	=>	1
237	1	0.214	3	=>	2
242	1	0.667	0	=>	1
245	1	0.200	1	=>	0
246	1	0.333	4	=>	3
251	1	1.000	3	=>	2
256	1	0.500	0	->	5
263	1	0.545	0	->	2
369	1	0.429	1	->	3
384	1	0.286	2	->	1
400	1	0.400	0	->	2
415	1	0.750	0	->	4
423	1	0.500	2	->	1

Character	Step	CI	Change		
430	1	0.286	1	->	2
431	1	0.222	2	->	1
433	1	0.250	5	=>	4
453	1	0.385	5	->	6
463	1	0.333	0	=>	1

Node 53 to Node 52

Character	Step	CI	Change		
29	1	0.182	2	->	1
32	1	0.154	1	->	0
52	1	0.125	1	=>	0
53	1	0.154	2	->	1
69	1	0.333	1	=>	0
73	1	0.200	1	=>	2
83	1	0.143	0	->	1
106	1	0.154	1	=>	2
157	1	0.400	3	=>	4
175	1	0.250	3	->	2
198	1	0.385	3	->	2
209	1	0.375	1	=>	3
268	1	0.300	1	=>	0
269	1	0.300	2	->	0
270	1	0.429	1	=>	0
271	1	0.222	1	->	2

Node 52 to Amud 1

Character	Step	CI	Change		
5	1	0.111	0	->	1
7	1	0.200	1	->	0
13	1	0.125	0	=>	1
23	1	0.182	0	=>	1
24	1	0.300	2	->	3
25	1	0.111	0	=>	1
31	1	0.333	1	=>	0
36	1	0.167	0	->	1
57	1	0.333	1	->	0
58	1	0.500	0	=>	2
62	1	0.111	0	=>	1
72	1	0.143	1	=>	0
78	1	0.111	0	->	1
91	1	0.167	1	->	0
94	1	0.200	2	=>	1
100	1	0.167	1	->	0
102	1	0.222	0	=>	2
103	1	0.286	1	->	0
104	1	0.300	3	->	2
113	1	0.231	1	->	2
117	1	0.100	1	->	0
133	1	0.200	0	=>	1
199	1	0.250	1	->	2
200	2	0.222	0	=>	2
203	1	0.143	1	->	0
210	1	0.333	2	->	1
241	1	0.222	1	=>	0
243	1	0.143	2	=>	3
293	2	0.231	1	=>	3
297	3	0.333	0	=>	3
299	1	0.545	0	=>	6
300	3	0.375	0	=>	3
328	1	0.250	0	=>	2

Character	Step	CI	Change		

Node 52 to Node 51

Character	Step	CI	Change		
4	1	0.133	1	=>	2
28	1	0.167	1	->	2
55	1	0.182	0	->	1
82	1	0.167	0	=>	1
92	1	0.125	2	=>	1
108	2	0.133	0	=>	2
109	1	0.154	1	=>	2
114	1	0.200	1	=>	2
119	1	0.182	3	=>	2
121	1	0.286	1	=>	0
140	1	0.200	2	=>	3
150	1	0.400	2	->	4
206	1	0.375	2	->	3
210	1	0.333	2	->	3
285	1	0.333	2	=>	3
293	1	0.231	1	->	0
320	1	0.500	0	=>	1
361	1	1.000	3	->	2
429	1	0.400	2	->	0
430	1	0.286	1	=>	0

Node 51 to Skhul V

Character	Step	CI	Change		
5	1	0.111	0	->	1
7	1	0.200	1	->	0
9	1	0.167	2	=>	0
18	2	0.250	2	=>	0
24	1	0.300	2	->	3
36	1	0.167	0	->	1
43	1	0.125	0	=>	1
46	1	0.125	1	->	0
60	1	0.100	0	=>	1
71	1	0.167	0	=>	1
76	1	0.333	0	=>	1
78	1	0.111	0	->	1
79	1	0.154	2	=>	1
97	1	0.286	1	=>	0
100	1	0.167	1	->	0
118	1	0.333	4	=>	0
141	1	0.125	0	=>	1
164	1	0.333	2	->	3
180	1	0.222	0	=>	1
199	1	0.250	1	->	2
229	1	0.400	0	=>	1
232	1	0.538	0	=>	1
287	1	0.667	0	=>	1
329	1	0.250	0	=>	1
366	1	0.200	2	=>	1
370	1	0.800	2	=>	3
371	1	0.333	0	=>	1
382	1	0.500	0	=>	3
401	1	0.500	0	=>	1
427	1	1.000	1	=>	2
431	1	0.222	2	->	1
433	1	0.250	5	=>	3
439	1	0.300	3	=>	2

Node 51 to Modern humans

Character	Step	CI	Change		
2	1	0.143	1	=>	2
11	1	0.333	2	=>	0
14	1	0.111	1	->	0
15	1	0.111	1	->	0
33	1	0.111	1	=>	0
35	1	0.167	1	=>	0
37	1	0.083	1	=>	0
38	1	0.182	2	=>	1
45	1	0.250	1	=>	0
53	1	0.154	1	=>	0
54	1	0.250	0	=>	1
56	1	0.250	0	=>	1
66	1	0.154	1	=>	0
75	1	0.091	0	=>	1
87	1	0.100	1	=>	0
92	1	0.125	1	=>	0
95	1	0.167	0	=>	1
101	1	0.333	0	->	1
103	1	0.286	1	=>	2
125	1	0.333	3	->	4
131	1	0.182	1	=>	0
137	1	0.182	1	->	2
139	1	0.250	2	->	0
149	1	0.273	2	->	3
161	1	0.300	2	->	3
164	1	0.333	2	->	0
170	1	0.417	3	->	5
171	1	0.545	3	->	6
172	1	0.462	4	->	6
175	1	0.250	2	->	3
181	1	0.400	1	->	0
190	1	0.500	5	->	6
191	1	0.500	2	->	3
192	1	0.400	1	->	0
196	1	0.286	2	->	4
197	1	0.333	2	->	5
198	1	0.385	2	->	5
205	1	0.333	2	->	3
208	1	0.429	0	=>	1
264	1	0.444	3	->	4
265	1	0.308	1	->	0
431	1	0.222	2	->	3
434	1	0.300	1	->	0
435	1	0.222	1	->	2
452	1	0.417	5	=>	6
453	1	0.385	5	->	6

Node 55 to Arago

Character	Step	CI	Change		
14	1	0.111	1	->	0
15	1	0.111	1	->	0
19	1	0.143	0	=>	1
27	1	0.375	3	=>	0
29	1	0.182	2	->	1
31	1	0.333	1	=>	0
124	1	0.429	3	=>	2
127	1	0.250	0	->	1
129	1	0.200	0	=>	1
130	1	0.200	0	=>	1

Character	Step	CI	Change		
132	1	0.167	0	=>	1
133	1	0.200	0	=>	1
134	1	0.286	1	->	2
138	1	0.500	0	=>	1
144	1	0.222	2	->	1
151	1	0.429	4	=>	3
153	1	0.571	0	=>	1
156	1	0.714	0	->	1
157	1	0.400	3	=>	2
158	1	0.455	5	=>	2
166	1	0.235	1	->	0
173	1	0.333	3	=>	2
174	1	0.500	3	=>	2
178	1	0.556	3	->	5
181	1	0.400	1	=>	0
182	1	0.444	1	->	0
203	1	0.143	1	->	0
215	1	0.571	4	=>	2
217	1	0.800	5	->	4
221	1	0.667	4	=>	3
241	1	0.222	1	=>	0
256	1	0.500	0	=>	1
263	1	0.545	0	=>	1
267	1	0.250	3	=>	2
275	1	0.500	2	=>	3
291	2	0.214	1	=>	3
328	1	0.250	0	=>	3

Node 57 to Sinanthropus 11

Character	Step	CI	Change		
3	1	0.167	1	->	0
4	1	0.133	1	=>	2
6	1	0.200	1	=>	0
8	1	0.167	1	=>	0
10	1	0.200	1	=>	0
12	1	0.333	3	=>	1
13	1	0.125	0	->	1
19	1	0.143	0	=>	1
23	1	0.182	0	=>	2
24	1	0.300	0	->	2
28	1	0.167	1	=>	0
30	1	0.125	0	->	1
32	1	0.154	1	=>	0
50	1	0.167	0	->	1
51	1	0.182	2	=>	1
53	1	0.154	2	=>	1
55	1	0.182	2	=>	0
60	1	0.100	1	=>	0
61	1	0.167	0	=>	1
62	1	0.111	0	=>	1
63	1	0.167	0	=>	1
64	1	0.125	0	->	1
66	1	0.154	1	->	2
72	1	0.143	1	->	0
77	1	0.111	1	=>	0
78	1	0.111	0	->	1
80	1	0.125	1	->	0
100	1	0.167	1	->	0
103	1	0.286	1	=>	2
108	1	0.133	0	->	1
109	1	0.154	1	->	2
119	1	0.182	2	=>	1
127	1	0.250	0	->	1
131	1	0.182	1	=>	0
136	1	0.267	0	->	2
142	1	0.571	0	=>	1
146	1	0.444	1	=>	2
147	1	0.556	1	->	2
149	1	0.273	2	=>	3
155	1	0.500	1	=>	2
157	1	0.400	3	=>	4
170	1	0.417	3	=>	4
171	1	0.545	2	=>	4
180	1	0.222	0	=>	1
186	1	0.500	2	=>	3
188	1	0.375	3	=>	2
196	1	0.286	2	->	3
197	1	0.333	2	->	3
225	1	0.400	2	=>	1
231	1	0.375	0	->	1
239	1	0.444	0	->	1
249	1	0.750	0	=>	1
254	1	0.500	0	=>	1
264	1	0.444	3	=>	0
268	1	0.300	1	->	0
276	1	0.250	1	=>	0
284	1	0.300	3	->	2
286	1	0.333	0	=>	1
290	1	0.545	2	=>	1
291	1	0.214	1	=>	2
297	1	0.333	0	->	1
299	1	0.545	0	=>	1
302	1	0.667	4	->	3
328	1	0.250	0	->	2
344	1	0.500	0	->	1
356	1	0.250	0	->	1
361	1	1.000	0	->	4
388	1	0.429	3	=>	2
391	1	0.333	1	=>	0
403	1	0.500	0	=>	1
405	1	0.500	0	->	1
433	1	0.250	5	=>	4
436	1	0.400	4	->	3
439	1	0.300	3	=>	2
457	1	0.600	0	=>	2

Node 59 to Sinanthropus 3

Character	Step	CI	Change		
4	1	0.133	1	=>	2
8	1	0.167	1	=>	0
13	1	0.125	0	->	1
19	1	0.143	0	=>	1
23	1	0.182	0	=>	2
24	1	0.300	0	->	1
30	1	0.125	0	->	1
37	1	0.083	0	->	1
44	1	0.125	0	=>	1
49	1	0.167	0	->	0
50	1	0.167	0	->	1
62	1	0.111	0	=>	1
64	1	0.125	0	->	1
68	1	0.167	0	=>	1
75	1	0.091	0	=>	1
78	1	0.111	0	->	1

Character	Step	CI	Change		
84	1	0.125	1	->	0
94	1	0.200	1	->	2
101	1	0.333	0	=>	1
106	1	0.154	1	=>	2
108	2	0.133	0	=>	2
123	1	0.429	0	=>	1
125	1	0.333	3	->	4
126	1	0.333	0	=>	1
136	1	0.267	0	->	1
139	1	0.250	2	->	1
160	1	0.250	3	->	1
161	1	0.300	2	=>	1
162	1	0.444	2	->	1
165	1	0.333	1	=>	2
166	1	0.235	2	=>	3
169	1	0.500	0	=>	1
177	1	0.300	3	->	5
179	1	0.400	1	->	0
183	1	0.667	0	=>	1
184	1	0.667	0	=>	1
188	1	0.375	3	=>	1
189	1	0.429	3	=>	1
190	1	0.500	4	->	5
193	1	0.571	1	->	2
194	1	0.625	1	=>	2
196	1	0.286	2	->	3
198	1	0.385	3	=>	4
204	1	0.250	1	->	0
205	1	0.333	2	->	1
213	1	0.571	0	=>	1
220	1	1.000	5	=>	1
225	1	0.400	2	->	0
230	1	0.444	0	=>	1
231	1	0.375	0	->	1
232	1	0.538	1	=>	2
237	1	0.214	3	=>	2
253	1	0.500	0	=>	1
254	1	0.500	0	=>	1
255	1	0.700	1	=>	2
256	1	0.500	0	=>	4
267	1	0.250	3	=>	2
268	1	0.300	1	->	0
270	1	0.429	1	->	0
290	1	0.286	2	=>	1
297	1	0.333	0	->	1
301	1	0.267	1	->	2
322	1	0.429	3	=>	2
325	1	0.333	4	=>	2
343	1	0.833	0	=>	1
351	1	0.400	0	->	1
374	1	0.667	2	=>	4
391	1	0.333	1	=>	0
396	1	0.500	0	=>	1
401	1	0.500	0	=>	2
416	1	0.250	2	=>	1
428	1	0.286	1	=>	2
430	1	0.286	1	=>	0
431	1	0.222	3	->	2
432	1	0.500	0	=>	3
433	1	0.250	5	=>	3
434	1	0.300	0	=>	3
445	1	0.500	2	=>	1

Character	Step	CI	Change		
Node 60 to Narmada					
2	1	0.143	1	->	0
5	1	0.111	1	->	0
23	1	0.182	0	=>	1
25	1	0.111	1	->	0
30	1	0.125	0	->	1
34	1	0.167	0	=>	1
73	1	0.200	1	=>	2
79	1	0.154	0	=>	1
80	1	0.125	1	->	0
84	1	0.125	1	->	0
89	1	0.286	1	->	0
94	1	0.200	1	->	0
105	1	0.222	0	->	1
109	1	0.154	1	->	2
110	1	0.182	1	=>	2
114	1	0.200	1	->	0
118	1	0.333	0	=>	2
120	1	0.231	2	->	3
131	1	0.182	1	=>	2
243	1	0.333	2	=>	3
306	1	0.400	0	=>	1
310	1	0.333	0	=>	1
326	1	0.250	2	->	3
332	1	0.500	0	=>	1
337	1	0.500	0	=>	2
338	1	0.571	1	=>	3
339	1	0.200	0	=>	1
340	1	0.500	0	=>	1
343	1	0.833	0	=>	4
346	1	0.667	1	=>	2
352	1	0.375	0	=>	2
353	1	0.444	0	=>	2
356	1	0.250	0	=>	1
363	1	0.500	0	=>	1
369	1	0.429	1	=>	2
379	1	0.667	2	=>	1
382	1	0.500	1	=>	0
398	1	0.667	0	=>	2
401	1	0.500	0	=>	3
411	1	0.429	0	->	1
412	1	0.750	0	=>	1
417	1	0.500	1	=>	0
420	1	0.500	1	=>	0
421	1	0.500	1	=>	0
Node 63 to Knmer 3883					
1	1	0.167	0	=>	1
14	1	0.111	0	->	1
15	1	0.111	0	->	1
18	1	0.250	2	->	1
19	1	0.143	0	->	1
27	1	0.375	0	=>	2
28	1	0.167	1	=>	0
29	1	0.182	1	->	2
31	1	0.333	1	=>	2
34	1	0.167	0	->	1
37	1	0.083	0	->	1
49	1	0.167	1	=>	0

Character	Step	CI	Change		
53	1	0.154	0	=>	1
63	1	0.167	0	->	1
66	1	0.154	2	=>	1
69	1	0.333	1	=>	0
77	1	0.111	1	->	0
83	1	0.143	0	=>	1
85	1	0.167	0	=>	1
88	1	0.167	0	=>	1
89	1	0.286	1	->	0
91	1	0.167	0	=>	1
96	1	0.250	1	=>	0
97	1	0.286	1	->	2
110	1	0.182	1	=>	2
111	1	0.125	0	=>	1
118	1	0.333	0	=>	4
120	1	0.231	2	->	3
127	1	0.250	1	=>	0
133	1	0.200	0	=>	1
135	1	0.167	1	->	0
137	1	0.182	2	->	1
141	1	0.125	0	->	1
145	1	0.273	0	->	1
151	1	0.429	4	->	3
157	1	0.400	3	=>	2
158	1	0.455	3	->	2
159	1	0.500	3	=>	5
160	1	0.250	3	->	2
161	1	0.300	2	=>	1
162	1	0.444	3	=>	2
166	1	0.235	2	=>	3
170	1	0.417	3	->	1
172	1	0.462	3	=>	2
175	1	0.250	3	=>	2
180	1	0.222	2	=>	1
197	1	0.333	2	=>	1
198	1	0.385	3	=>	2
203	1	0.143	1	=>	0
209	1	0.375	3	=>	2
217	1	0.800	4	->	3
218	1	0.400	1	=>	2
221	1	0.667	4	=>	3
236	1	0.600	0	->	5
237	1	0.214	3	->	2
243	1	0.333	2	=>	1
263	1	0.545	0	=>	2
265	1	0.308	1	->	2
266	1	0.429	3	=>	2
279	1	0.250	2	->	0
285	1	0.333	1	->	2
293	1	0.231	1	->	2
327	1	0.333	3	->	2
334	1	0.400	0	->	2
371	1	0.333	0	=>	1
383	1	0.800	2	=>	3
384	1	0.286	2	=>	1
392	1	0.500	0	=>	1
395	1	0.286	1	=>	0
396	1	0.500	0	=>	1
404	1	0.429	0	=>	3
411	1	0.429	0	->	1
429	1	0.400	1	=>	4
432	1	0.500	0	=>	1

Character	Step	CI	Change		
433	1	0.250	5	->	4
434	1	0.300	0	=>	2
435	1	0.222	2	=>	1
453	1	0.385	3	=>	2
455	1	0.429	3	->	2
468	1	0.250	1	=>	0

Node 64 to Knmer 3733

Character	Step	CI	Change		
4	1	0.133	1	=>	2
8	1	0.167	1	=>	0
9	1	0.167	2	=>	0
14	1	0.111	0	->	1
15	1	0.111	0	->	1
18	2	0.250	2	=>	4
22	1	0.200	0	=>	2
23	1	0.182	0	=>	2
31	1	0.333	1	=>	3
32	1	0.154	1	->	2
34	1	0.167	0	->	1
46	1	0.125	1	=>	0
51	1	0.182	2	->	0
60	1	0.100	0	->	1
67	1	0.333	1	=>	0
71	1	0.167	0	=>	1
75	1	0.091	0	=>	1
77	1	0.111	1	->	0
80	1	0.125	0	=>	1
84	1	0.125	1	->	0
94	1	0.200	1	=>	2
109	1	0.154	1	->	0
119	2	0.182	3	=>	1
123	1	0.429	0	=>	2
132	1	0.167	0	=>	1
136	1	0.267	0	->	2
139	1	0.250	1	=>	3
142	1	0.571	3	=>	4
149	1	0.273	2	->	3
151	1	0.429	4	->	2
159	1	0.500	3	=>	0
165	1	0.333	2	=>	3
166	2	0.235	2	=>	0
167	1	0.364	2	=>	3
176	1	0.400	3	->	4
181	1	0.400	2	->	1
182	1	0.444	2	=>	3
187	1	0.625	4	=>	3
196	1	0.286	1	->	2
202	2	0.333	2	=>	0
205	1	0.333	2	->	3
211	1	1.000	0	->	1
214	1	0.800	3	->	2
230	1	0.444	0	->	1
231	1	0.375	0	->	1
233	1	0.714	0	->	1
235	1	0.500	2	->	1
236	1	0.600	0	->	5
237	1	0.214	3	->	2
238	1	0.667	1	=>	3
240	1	0.600	2	->	3
256	1	0.500	0	=>	2
257	1	0.286	3	->	4

Character	Step	CI	Change		
268	1	0.300	1	=>	0
269	1	0.300	1	->	2
270	1	0.429	1	=>	0
279	1	0.250	2	->	0
282	1	0.600	3	=>	1
284	1	0.300	2	->	1
286	1	0.333	0	->	1
289	1	0.250	1	->	0
293	1	0.231	1	->	2
296	1	0.667	2	=>	1
299	1	0.545	2	->	4
300	1	0.375	1	=>	2
301	1	0.267	1	->	3
302	2	0.667	2	=>	1
313	1	0.667	0	=>	1
314	1	0.750	0	->	1
316	1	0.400	0	=>	1
322	1	0.429	3	->	2
323	1	0.667	2	->	1
429	1	0.400	1	=>	3
436	1	0.400	3	=>	2
437	1	0.333	1	->	2
438	1	0.600	3	->	2
439	1	0.300	3	->	2
452	1	0.417	3	=>	5

Node 65 to Knmer 1470

4	1	0.133	1	->	0
11	1	0.333	2	=>	0
18	2	0.250	2	=>	0
32	1	0.154	1	=>	0
33	1	0.111	1	=>	0
37	1	0.083	0	->	1
39	1	0.143	0	=>	1
45	1	0.250	1	=>	0
51	1	0.182	2	->	0
64	1	0.125	1	=>	0
65	1	0.250	1	=>	0
87	1	0.100	0	->	1
92	2	0.125	2	=>	0
105	1	0.222	2	->	1
106	2	0.154	0	=>	2
109	1	0.154	1	=>	2
113	1	0.231	1	->	2
118	1	0.333	0	=>	1
120	1	0.231	2	->	3
128	1	0.667	0	->	1
129	1	0.200	0	=>	1
138	1	0.500	0	->	1
139	1	0.250	1	->	2
140	1	0.200	0	=>	1
141	1	0.125	0	->	1
144	1	0.222	1	=>	2
149	1	0.273	2	->	3
155	1	0.500	3	->	4
160	1	0.250	3	=>	1
164	1	0.333	2	->	0
166	1	0.235	2	=>	3
171	1	0.545	2	=>	3
173	1	0.333	2	=>	3
174	1	0.500	1	=>	3

Character	Step	CI	Change		
185	1	0.333	2	->	1
196	1	0.286	1	->	2
197	1	0.333	2	=>	3
201	1	0.222	1	->	0
205	1	0.333	2	->	3
213	1	0.571	1	->	2
215	1	0.571	3	->	1
219	1	1.000	0	=>	1
225	1	0.400	2	=>	1
231	1	0.375	0	->	2
232	1	0/538	1	->	3
238	1	0.667	1	=>	2
250	1	1.000	3	=>	2
253	1	0.500	0	->	2
254	1	0.500	0	->	2
255	1	0.700	1	->	3
256	1	0.500	0	=>	5
264	1	0.444	3	->	4
267	1	0.250	3	->	2
284	1	0.300	2	->	3
292	1	0.333	2	->	3
293	1	0.231	1	=>	0
294	1	0.500	1	=>	0
295	1	0.375	1	->	3
299	1	0.545	2	->	1
300	1	0.375	1	=>	0
310	1	0.333	0	=>	1
319	1	0.750	3	=>	2
322	1	0.429	3	->	1
328	1	0.250	2	->	1
333	1	0.429	0	=>	1
353	1	0.444	1	=>	3
355	1	0.400	1	->	2
367	1	0.500	2	->	1
431	1	0.222	3	=>	2

Appendix 4

Diagnoses of the different species

Diagnosis of *Homo rudolfensis*

Character	Step	CI	Change		
4	1	0.133	1	->	0
11	1	0.333	2	=>	0
18	2	0.250	2	=>	0
32	1	0.154	1	=>	0
33	1	0.111	1	=>	0
37	1	0.083	0	->	1
39	1	0.143	0	=>	1
45	1	0.250	1	=>	0
51	1	0.182	2	->	0
64	1	0.125	1	=>	0
65	1	0.250	1	=>	0
87	1	0.100	0	->	1
92	2	0.125	2	=>	0
105	1	0.222	2	->	1
106	2	0.154	0	=>	2
109	1	0.154	1	=>	2
113	1	0.231	1	->	2
118	1	0.333	0	=>	1
120	1	0.231	2	->	3
128	1	0.667	0	->	1
129	1	0.200	0	=>	1
138	1	0.500	0	->	1
139	1	0.250	1	->	2
140	1	0.200	0	=>	1
141	1	0.125	0	->	1
144	1	0.222	1	=>	2
149	1	0.273	2	->	3
155	1	0.500	3	->	4
160	1	0.250	3	=>	1
164	1	0.333	2	->	0
166	1	0.235	2	=>	3
171	1	0.545	2	=>	3
173	1	0.333	2	=>	3
174	1	0.500	1	=>	3
185	1	0.333	2	->	1
196	1	0.286	1	->	2
197	1	0.333	2	=>	3
201	1	0.222	1	->	0
205	1	0.333	2	->	3
213	1	0.571	1	->	2
215	1	0.571	3	->	1
219	1	1.000	0	=>	1
225	1	0.400	2	=>	1
231	1	0.375	0	->	2
232	1	0/538	1	->	3
238	1	0.667	1	=>	2
250	1	1.000	3	=>	2
253	1	0.500	0	->	2
254	1	0.500	0	->	2
255	1	0.700	1	->	3
256	1	0.500	0	=>	5
264	1	0.444	3	->	4
267	1	0.250	3	->	2
284	1	0.300	2	->	3
292	1	0.333	2	->	3
293	1	0.231	1	=>	0
294	1	0.500	1	=>	0
295	1	0.375	1	->	3
299	1	0.545	2	->	1
300	1	0.375	1	=>	0
310	1	0.333	0	=>	1
319	1	0.750	3	=>	2
322	1	0.429	3	->	1
328	1	0.250	2	->	1
333	1	0.429	0	=>	1
353	1	0.444	1	=>	3
355	1	0.400	1	->	2
367	1	0.500	2	->	1
431	1	0.222	3	=>	2
300	1	0.375	1	=>	0
310	1	0.333	0	=>	1
319	1	0.750	3	=>	2
322	1	0.429	3	->	1
328	1	0.250	2	->	1
333	1	0.429	0	=>	1
353	1	0.444	1	=>	3
355	1	0.400	1	->	2
367	1	0.500	2	->	1
431	1	0.222	3	=>	2

Diagnosis of *Homo kenyaensis nov. sp*

Character	Step	CI	Change		
4	1	0.133	1	=>	2
8	1	0.167	1	=>	0
9	1	0.167	2	=>	0
14	1	0.111	0	->	1
15	1	0.111	0	->	1
18	2	0.250	2	=>	4
22	1	0.200	0	=>	2
23	1	0.182	0	=>	2
31	1	0.333	1	=>	3
32	1	0.154	1	->	2
34	1	0.167	0	->	1
46	1	0.125	1	=>	0
51	1	0.182	2	->	0
60	1	0.100	0	->	1
67	1	0.333	1	=>	0
71	1	0.167	0	=>	1
75	1	0.091	0	=>	1
77	1	0.111	1	->	0
80	1	0.125	0	=>	1
84	1	0.125	1	->	0
94	1	0.200	1	=>	2
109	1	0.154	1	->	0
119	2	0.182	3	=>	1
123	1	0.429	0	=>	2
132	1	0.167	0	=>	1
136	1	0.267	0	->	2
139	1	0.250	1	=>	3
142	1	0.571	3	=>	4
149	1	0.273	2	->	3
151	1	0.429	4	->	2
159	1	0.500	3	=>	0
165	1	0.333	2	=>	3
166	2	0.235	2	=>	0
167	1	0.364	2	=>	3
176	1	0.400	3	->	4
181	1	0.400	2	->	1
182	1	0.444	2	=>	3
187	1	0.625	4	=>	3
196	1	0.286	1	->	2
202	2	0.333	2	=>	0
205	1	0.333	2	->	3
211	1	1.000	0	->	1
214	1	0.800	3	->	2
230	1	0.444	0	->	1
231	1	0.375	0	->	1
233	1	0.714	0	->	1
235	1	0.500	2	->	1
236	1	0.600	0	->	5
237	1	0.214	3	->	2
238	1	0.667	1	=>	3
240	1	0.600	2	->	3
256	1	0.500	0	=>	2
257	1	0.286	3	->	4
268	1	0.300	1	=>	0
269	1	0.300	1	->	2
270	1	0.429	1	=>	0
279	1	0.250	2	->	0
282	1	0.600	3	=>	1
284	1	0.300	2	->	1
286	1	0.333	0	->	1
289	1	0.250	1	->	0
293	1	0.231	1	->	2
296	1	0.667	2	=>	1
299	1	0.545	2	->	4
300	1	0.375	1	=>	2
301	1	0.267	1	->	3
302	2	0.667	2	=>	1
313	1	0.667	0	=>	1
314	1	0.750	0	->	1
316	1	0.400	0	=>	1
322	1	0.429	3	->	2
323	1	0.667	2	->	1
429	1	0.400	1	=>	3
436	1	0.400	3	=>	2
437	1	0.333	1	->	2
438	1	0.600	3	->	2
439	1	0.300	3	->	2
452	1	0.417	3	=>	5

Character	Step	CI	Change		
Diagnosis of *Homo okotensis nov. sp*					
1	1	0.167	0	=>	1
14	1	0.111	0	->	1
15	1	0.111	0	->	1
18	1	0.250	2	->	1
19	1	0.143	0	->	1
27	1	0.375	0	=>	2
28	1	0.167	1	=>	0
29	1	0.182	1	->	2
31	1	0.333	1	=>	2
34	1	0.167	0	->	1
37	1	0.083	0	->	1
49	1	0.167	1	=>	0
53	1	0.154	0	=>	1
63	1	0.167	0	->	1
66	1	0.154	2	=>	1
69	1	0.333	1	=>	0
77	1	0.111	1	->	0
83	1	0.143	0	=>	1
85	1	0.167	0	=>	1
88	1	0.167	0	=>	1
89	1	0.286	1	->	0
91	1	0.167	0	=>	1
96	1	0.250	1	=>	0
97	1	0.286	1	->	2
110	1	0.182	1	=>	2
111	1	0.125	0	=>	1
118	1	0.333	0	=>	4
120	1	0.231	2	->	3
127	1	0.250	1	=>	0
133	1	0.200	0	=>	1
135	1	0.167	1	->	0
137	1	0.182	2	->	1
141	1	0.125	0	->	1
145	1	0.273	0	->	1
151	1	0.429	4	->	3
157	1	0.400	3	=>	2
158	1	0.455	3	->	2
159	1	0.500	3	=>	5
160	1	0.250	3	->	2
161	1	0.300	2	=>	1
162	1	0.444	3	=>	2
166	1	0.235	2	=>	3
170	1	0.417	3	->	1
172	1	0.462	3	=>	2
175	1	0.250	3	=>	2
180	1	0.222	2	=>	1
197	1	0.333	2	=>	1
198	1	0.385	3	=>	2
203	1	0.143	1	=>	0
209	1	0.375	3	=>	2
217	1	0.800	4	->	3
218	1	0.400	1	=>	2
221	1	0.667	4	=>	3
236	1	0.600	0	->	5
237	1	0.214	3	->	2
243	1	0.333	2	=>	1
263	1	0.545	0	=>	2
265	1	0.308	1	->	2
266	1	0.429	3	=>	2
279	1	0.250	2	->	0
285	1	0.333	1	->	2
293	1	0.231	1	->	2
327	1	0.333	3	->	2
334	1	0.400	0	->	2
371	1	0.333	0	=>	1
383	1	0.800	2	=>	3
384	1	0.286	2	=>	1
392	1	0.500	0	=>	1
395	1	0.286	1	=>	0
396	1	0.500	0	=>	1
404	1	0.429	0	=>	3
411	1	0.429	0	->	1
429	1	0.400	1	=>	4
432	1	0.500	0	=>	1
433	1	0.250	5	->	4
434	1	0.300	0	=>	2
435	1	0.222	2	=>	1
453	1	0.385	3	=>	2
455	1	0.429	3	->	2
468	1	0.250	1	=>	0

Character	Step	CI	Change		

Diagnosis of *Homo ergaster*

Character	Step	CI	Change		
5	1	0.111	1	->	0
11	1	0.333	2	=>	1
12	1	0.333	3	=>	1
18	1	0.250	2	->	1
19	1	0.143	0	->	1
37	1	0.083	0	->	1
38	1	0.182	2	->	0
44	1	0.125	0	=>	1
51	1	0.182	2	=>	1
55	1	0.182	1	->	2
62	1	0.111	0	=>	1
63	1	0.167	0	->	1
75	1	0.091	0	=>	1
92	2	0.125	2	=>	0
98	1	0.222	0	=>	2
99	2	0.333	2	=>	0
106	2	0.154	0	=>	2
107	1	0.200	0	->	2
117	1	0.100	1	->	0
120	1	0.231	2	->	3
135	1	0.167	1	->	0
136	1	0.267	0	->	3
137	1	0.182	2	->	1
140	1	0.200	1	=>	2
144	1	0.222	1	->	2
160	1	0.250	3	->	2
163	1	0.250	1	=>	0
164	1	0.333	2	=>	1
166	1	0.235	2	=>	1
167	2	0.364	2	=>	0
170	1	0.417	3	->	4
172	1	0.462	3	=>	4
173	1	0.333	2	->	3
175	1	0.250	3	=>	1
176	1	0.400	3	=>	2
177	1	0.300	3	=>	2
181	1	0.400	2	=>	3
182	1	0.444	2	=>	3
185	1	0.333	3	->	1
187	1	0.625	4	=>	2
192	1	0.400	2	=>	3
197	1	0.333	2	=>	0
198	1	0.385	3	=>	1
199	1	0.250	1	=>	2
200	1	0.222	1	=>	2
201	1	0.222	1	=>	2
204	1	0.250	1	->	0
205	1	0.333	2	=>	0
206	1	0.375	2	->	1
209	1	0.375	3	=>	3
210	1	0.333	3	->	2
228	1	0.600	0	->	2
229	1	0.400	0	=>	3
230	1	0.444	0	->	2
231	1	0.375	0	->	2
232	1	0.538	1	=>	3
233	1	0.714	0	=>	2
234	1	0.333	1	=>	2
239	1	0.444	1	->	2
242	1	0.667	0	=>	1
244	1	0.400	2	=>	1
245	1	0.200	1	=>	0
247	1	0.667	0	=>	2
253	1	0.500	0	=>	1
256	1	0.500	0	=>	2
258	1	0.800	4	=>	3
272	1	0.750	0	=>	2
274	1	0.667	0	=>	1
277	1	0.500	2	=>	1
278	1	0.500	2	=>	1
285	1	0.333	1	->	2
293	1	0.231	1	=>	0
299	1	0.545	2	=>	3
301	1	0.267	1	->	3
310	1	0.333	0	=>	1
315	1	0.500	0	=>	2
316	1	0.400	0	=>	1
317	1	0.400	0	=>	1
338	1	0.571	1	=>	0
342	1	0.400	1	=>	0
347	1	1.000	0	=>	1
362	1	0.500	0	=>	1
388	1	0.429	3	=>	2
389	1	0.333	2	=>	1
416	1	0.250	2	=>	3
423	1	0.500	2	=>	1
424	1	0.429	2	=>	0
436	1	0.500	3	=>	1
438	1	0.600	3	=>	0
443	1	0.500	2	=>	3
452	1	0.417	4	->	2
453	1	0.385	3	=>	6
455	1	0.429	3	->	4
456	1	0.500	4	->	5
459	1	0.600	1	=>	3
461	1	0.333	1	->	0
462	1	0.500	0	->	5
463	1	0.333	0	=>	3

Character	Step	CI	Change		
Diagnosis of *Homo erectus*					
13	1	0.125	0	->	1
25	1	0.111	1	->	0
50	1	0.167	0	->	1
88	1	0.167	0	=>	1
120	1	0.231	2	=>	1
130	1	0.200	0	=>	1
132	1	0.167	0	=>	1
236	1	0.600	0	=>	2
302	1	0.667	2	->	4
328	1	0.250	2	->	0
342	1	0.400	1	=>	2
384	1	0.286	2	=>	1
397	1	0.200	1	=>	0
399	1	0.600	1	=>	2
458	1	0.400	2	->	0
460	1	0.667	0	->	1
463	1	0.333	0	=>	1

BIBLIOGRAPHY

Abbate E., Albianelli A., Azzaroli A., Benvenuti M., Tesfamariam B., Bruni P., Cipriani N., Clarke R.J., Ficcarelli G., Macchiarelli R., Napoleone G., Papini M., Rook L., Sagri M., Tecle T.M., Torre D., Villa I., 1998 – A one million year old Homo cranium from the Danakil (Afar) Depression of Eritrea, *Nature*, 393: 458-460.

Acharyya S., Basu P., 1993 – Toba ash on the Indian subcontinent and its implications for correlation of Late Pleistocene alluvium. *Quaternary Research,* 40: 10-19.

Albrecht G.H., 1978 – Some comments on the use of ratios. *Systematic Zoology*, 27: 67-71.

Ambrose S., 1998 – Late Pleistocene human population bottlenecks, volcanic winter, and differentiation of modern humans. *Journal of Human Evolution,* 34: 623-651.

Andrews P., 1984 – An alternative interpretation of characters used to define *Homo erectus*, *Courier Forschungsinstitut Senckenberg*, 69: 167-175.

Arambourg C., 1957 – *Les Pithecanthropiens. In* Chatrusse et Cie (Ed.), Melanges Pittard, Brive, pp. 33-41.

Arambourg C., 1958 – Les stades évolutifs de l'humanité. *The Leech* (Raymond A. Dart commemorative Number), 28: 106-111.

Argue D., Donlon D., Groves C., Wright R., 2006 – *Homo floresiensis*: Microcephalic, pygmoid, Australopithecus, or Homo? *Journal of Human Evolution,* 51: 360-374.

Ascenzi A., Mallegni F., Manzi G., Segre A. and Segre Naldini E., 2000 – A re-appraisal of Ceprano calvaria affinities with *Homo erectus*, after the new reconstruction. *Journal of Human Evolution,* 39: 443-450.

Asfaw B., Gilbert W., Beyene Y., Hart W., Renne P., WoldeGabriel G., Vrba E., White T,. 2002 – Remains of *Homo erectus* from Bouri, Middle Awash, Ethiopia. *Nature,* 416: 317-320.

Atchley W.R., Anderson D., 1978 – Ratios and the statistical analysis of biological data. *Systematic Zoology*, 27: 71-78.

Atchley W.R., Gaskins C.T., Anderson D., 1976 – Statistical properties of ratios. I. Empirical results. *Systematic Zoology*, 25: 137-148.

Azzaroli A., 1995 – The "Elephant-Equus" and the "End Villafranchian" events in Eurasia. *In* E. Vrba,, G. Denton, T. Partridge and L. Burckle (Eds.), Paleoclimate and evolution, with emphasis on human origins, pp. 311-318.

Baba H., Aziz F., Kaifu Y., Suwa G., Kono R.T., Jacob T., 2003 – *Homo erectus* calvarium from the Pleistocene of Java. *Science*, 299: 1384-1388.

Baer K.E. von, 1828 – *Uber Entwicklungsgesschichte der Tiere. Beobachtung und Reflexion*. Bornträger, Königsberg.

Balzeau A., Grimaud-Hervé D., 2006 – Cranial base morphology and temporal bone pneumatization in Asian *Homo erectus*. *Journal of Human Evolution,* 51: 350-359.

Balzeau A., 2005 – Spécificités des caractères morphologiques internes du squelette céphalique chez *Homo erectus*. Thèse du Muséum National d'Histoire Naturelle, Paris.

Balzeau A., 2006 – Are thickened cranial bones and equal participation of the three structural bone layers autapomorphic traits of *Homo erectus*? *Bulletins et mémoires de la société d'Anthropologie de Paris*, 18: 145-163.

Balzeau A., Badawi-Fayad J., 2005 – La morphologie externe et interne de la région sus-orbitaire est elle corrélée à des contraintes biomécaniques ? Analyses structurelles des populations d'*Homo sapiens* d'Afalou Bou Rhummel (Algérie) et de Taforalt (Maroc). *Bulletins et mémoires de la société d'Anthropologie de Paris*, 17: 185-197.

Balzeau A., Grimaud-Hervé D., Sémah F., 2008 – Characteristics and variation of the temporal bone pneumatizationin asian *Homo erectus*. *In* Pautreau, J.P., Coupey A.S., Zeitoun V. and Rambault E. (Eds.), From *Homo erectus* to the living tradition Archaeology in southeast Asia 11th International Conference of EURASEAA, pp 21-27.

Barriel V., 1991 – Caractères ostéologiques et odontologiques chez les Hominoidea (Primates, mammalia): analyse de parcimonie. *Bulletins et mémoires de la société d'Anthropologie de Paris*, 3: 45-72.

Barriel V., 1994 – *Les relations de parenté au sein des hominoidea et la place de Pan paniscus : comparaison et analyse méthodologique des phylogénies morphologique et moléculaire*. Thèse de doctorat de l'Université Paris 6.

Barriel V., Tassy P., 1994 – Characters observations and steps: comment on Lipscomb's "parsimony, homology and te analysis of multistate characters". *Cladistics,* 9: 223-232.

Barstra G., 1994 – Indonesia in the period of *Homo sapiens neanderthalensis* and contempo-

raries. *In* De Laet S. (Ed.), History of humanity, Paris: Unesco, pp. 167-171.
BARSTRA G.J., SOEGONDHO S., VAN DER WIJK A., 1988 – Ngandong man: age and artefacts. *Journal of Human Evolution*, 17: 325-337.
BAR-YOSEF O., BELFER-COHEN A., 2001 – From Africa to Eurasia – early dispersals. *Quaternary International,* 75: 19-28.
BAR-YOSEF O., VANDERMEERSCH B., 1993 – Modern humans in the Levant. *Scientific American*, 286: 64-70.
BEALS K., SMITH C., DODD S., 1984 – Brain size, cranial morphology, climate, and time machines. *Current Anthopology,* 25:301- 330.
BEDEN M., 1987 – Les éléphantidés (mammalia, Proboscidea). In Coppens Y. and Howell F.C. (Eds.), Les faunes plio-pleistocènes de la vallée de l'Omo (Ethiopie), Paris: CNRS. *Cahiers de Paléontologie* 2: 9-161.
BERMUDEZ DE CASTRO J.M., ARSUAGA J.L., CARBONELL E., ROSAS A., MARTINEZ I., MOSQUERA M., 1997 – A Hominid from the Lower Pleistocene of Atapuerca, Spain: Possible ancestor to Neandertals and Modern Humans. *Science*, 276: 1392-1395.
BERNARD C., 1865 – *Introduction à l'étude de la médecine expérimentale*. Univers des Lettres, Bordas (édition 1966).
BILSBOROUGH A., 1983 – The pattern of evolution within the genus *Homo*. *In* Navaratnam V. and Harrison R.J. (Eds.), Progress in anatomy, vol. 3, Cambridge University Press, Cambridge, pp.143-168.
BILSBOROUGH A., WOOD B., 1986 – The Nature, origin and face of *Homo erectus*. *In* Wood B., Martin L. and Andrews P. (Eds.), *Major topics in primate and human evolution*, Cambridge University Press, Cambridge, pp.295-316.
BLACK D., 1927 – On the lower molar hominid tooth from the Chou Kou Tien deposit. *Paleontologica Sinica,* serie D, 7: 1-28.
BLACK D., 1930 – Interim report on the skull of Sinanthropus. *Bulletin of the Geological Society of China*, 9: 7-23.
BLACK D., 1931 – On an adolescent skull of *Sinanthropus pekinensis* in comparison with an adult skull of the same species and with other hominid skulls, recent and fossil. *Palaeontologica Sinica*, serie D, 7: 1-142.
BLACK D., 1934 – On the discovery, morphology and environment of *Sinanthropus pekinensis*. *Philosophical transactions of the Royal Society of London*, series B, 223: 57-120.
BLANC A.C., 1939 – L'Uomo fossile del Monte Circeo: un cranio neandertaliano delle grotta Guattari a San Felice Ciceo. *Rendiconti della R. Accademia Nazionale dei Lincei. Classe di Scienze Fisiche, Matematiche e naturali*., 29: 205-210.
BONDE N., 1977 – Cladistic classification as applied to vertebrates. *In* Hecht M.K., Goody P.C. and Hecht B.M. (Eds.), *Major patterns in vertebrate evolution,* Plenum Press, pp. 741-804.
BONDE N., 1981 – Problems of species concepts in palaeontology. *In* Martinell J. (Ed.), *Internatinal Symposium on Concept and Method in Paleontology*, University of Barcelona Press, Barcelona, pp.19-34.
BONDE N., 1989 – *Erectus* and *neanderthalensis* as species or subspecies of *Homo* with a model of speciation in hominids. *In* Giacobini J. (Ed.), *Hominidae. Proceding 2nd International Congress of Human Paleontology* Jaca Book, Milan, pp. 205-208.
BONIS L. DE, MELENTIS J., 1982 – L'Homme de Pétralona : comparaisons avec l'Homme de Tautavel. *In* Lumley de H. (Ed.), *l'Homo erectus et la place de l'Homme de Tautavel parmi les Hominidés fossiles.* 1er Congrès international de Paléontologie humaine, prétirage, pp. 847-873.
BONIS L. DE, MELENTIS J., 1991 – Age et position phylétique du crâne de Pétralona. In Bonifay E. and Vandermeersch B. (Eds.), *Les premiers européens.* Actes du 114ème congrès des sociétés savantes de Paris. Editions du C.T.H.S., pp. 285-290.
BONNEFILLE R., 1995 – A reassessment of the plio-pleistocene pollen record of east Africa. *In* Vrba, E., G. Denton, T. Partridge and L. Burckle (Eds.), Paleoclimate and evolution, with emphasis on human origins, pp. 299-310.
BOOKSTEIN F.L., 1978 – The measurement of biological shape and shape change. *Lecture notes in biomathematics* n° 24 Levin S. (Ed.), Springer-Verlag, Berlin.
BORDES, F. 1955 – L'industrie moustérienne de Teshik-Tash. Affinités et âge probable. *L'Anthropologie*, 59: 354-356.
BORDES F., 1959 – Le contexte archéologique des Hommes du Moustier et de Spy. *L'Anthropologie*, 63: 154-157.
BOULE M., 1911-1913 – *L'homme fossile de La Chapelle-aux-Saints*. Masson, Paris.
BOUTIÉ P., 1979 – Les gisements moustériens de Palestine. *Paléorient*, 5: 17-65.
BRACE L.C., 1981 – Tales of the phylogenetic woods: the evolution and significance of evolu-

tionary trees. *American Jounal of Physical Anthropology*, 56: 411-429.
Bräuer G., 1984a – A craniological approach to the origin of anatomically modern *Homo sapiens* in Africa and implications for the appearance of modern Europeans. *In* Smith F.H. and Spencer F. (Eds.), *The origins of modern humans: a world of the fossil evidence*, New York, A.R. Liss, pp. 327-410.
Bräuer G., 1984b – the Afro-european sapiens-hypothesis, and hominid evolution in east Asia during the late Middle and Upper Pleistocene. *Cour. Forsch. Inst. Senckenberg,* 69: 145-165.
Bräuer G., 1989 – The evolution of modern humans: a comparison of the african and non-african evidence. *In* Mellars P. and Stringer C.B. (Eds.), *The human revolution*, Edinburgh. Edinburgh University Press, pp. 123-154.
Braüer G., 1990 – The occurence of some controversal *Homo erectus* cranial features in the Zhoukoudian and east african hominids. *Acta Anthropologica Sinica,* 9: 352-358.
Bräuer G., 1991 – L'hypothèse africaine de l'origine des hommes modernes. *In* Hublin J.J. and Tillier A-M. (Eds.), *Aux origines d'Homo sapiens*, Paris, Presses Universitaires de France, pp. 181-215.
Bräuer G., Leakey R.E.F., 1986 – A new archaïc *Homo sapiens* cranium from Eliye Springs, West Turkana, Kenya. *Zeitschrift für Morphologie und Anthropologie*, 76: 245-252.
Bräuer G., Mbua E., 1992 – *Homo erectus* features used in cladistics and their variability in Asian and African hominids. *Journal of Human Evolution*, 22: 79-108.
Bräuer G., Rimbach K.W., 1990 – Late archaic and modern *Homo sapiens* from Europe, Africa, and Southwest Asia: craniometric comparisons and phylogenetic implications. *Journal of Human Evolution,* 19: 789-807.
Brock A., Isaac G., 1974 – Paleomagnetic stratigraphy and chronology of hominid-bearing sediments east of Lake Rudolf, Kenya. *Nature,* 247: 344-348.
Brown F. H., Harris J., Leakey R., Walker A., 1985 – Early *Homo erectus* skeleton from West Lake turkana, Kenya. *Nature,* 316: 88-792.
Brown P. 1994 – Cranial vault thickness in asian *Homo erectus* and *Homo sapiens*. *Courier Forschungsinstitut Senckenberg*, 171: 33-46.
Brown P., Sutikna T., Morwood M., Soejono Jatmiko, Wayhu Saptomo E., Awe Due R., 2004 – A new small bodied hominin fossil of the late Pleistocene of Flores, Indonesia. *Nature,* 431: 1055-61.
Brundin L., 1968 – Application of phylogenetic principles in systematics and evolutionary theory, *In* Orvig T. (Ed.), *Current problems of lower vertebrate phylogeny*, Nobel Symposium, pp. 473-495.
Bunopas S., Wasson J., Vella P., Fontaine H., Hada S., Suphajunya T., Khositanont S., 1999 – Catastrophic loess, mass mortality and forest fires suggest that a Pleistocene cometary impact in Thaïland caused the Australasian tektite field. *Journal of the Geological Society of Thailand,* 1: 1-17.
Cain A.J., 1954 – *Animal species and their evolution*, Hutchinson and Cie, London.
Campbell B.G., 1963 – Quantitative taxonomy and human evolution. *In* Washburn S.L. (Ed.) *Classification and human evolution*. Chicago, Aldine, pp. 50-74.
Campbell B.G., 1965 – *The nomenclature of the Hominidae*. Royal Anthropological Institute of Great Britain and Ireland, Londres.
Campbell, B.G. 1972 – Conceptual progress in physical anthropology: fossil man. *Annual Review of Anthropoly*, 1: 27-54.
Carbonell E., Bermudez de Castro J., Arsuaga J.L., Allue E., Bastir M., Benito A., Caceres I., Canals T., Diez J., van der Made J., Mosquera M., Olie A., Perez-Gonzalez A., Rodriguez J., Rodriguez X., Rosas A., Rosell J., Sala R., Vallverdu J., Verges J., 2005 – An early Pleistocene hominin mandible from Atapuerca TD6, Spain. *PNAS,*16 : 5674-5678.
Carbonell E., Bermudez de Castro J., Arsuaga J.L., Diez J.C., Rosas A., Cuenca-Bescos G., Sala R., Mosquera X., Rodriguez X., 1995 – Lower Pleistocene hominids and artefacts from Atapuerca TD6 (Spain). *Science*, 269: 826-829.
Chaline J., 1985 – *Histoire de l'Homme et des climats au Quaternaire*, Paris. Doin.
Chamberlain A., Wood B.A., 1987 – Early hominid phylogeny. *Journal of Human Evolution*, 16: 119-133.
Chesner C., Rose W., Deino A., Drake R., Wastegate J., 1991 – Eruptive history of earth's largest Quaternary caldera (Toba, Indonesia) clarified. *Geology,* 19: 200-203.
Chinzei K., 1970 – The Amud cave site and its deposits. *In* Suzuki H. and Takai F. (Eds.).

The Amud man and his cave site. Tokyo, University of Tokyo, pp. 21-52.

Clark J., Beyene Y., WoldeGabriel G., Hart W., Renne P., Gilbert H., Defleur A., Suwa G., Katoh S., Ludwig K., Boisserie J.R., Asfaw B., White T., 2003 – Stratigraphic, chronological and behavioural contexts of Pleistocene Homo sapiens from Middle Awash, Ethiopia. *Nature,* 423: 747-752.

Clarke R.J., 1976 – New cranium of *Homo erectus* from Lake Ndutu, Tanzania. *Nature,* 262: 485-487.

Clarke R.J., 1990 – The Ndutu cranium and the origin of *Homo sapiens*. *American Journal of Physical Anthropology*, 19: 699-736.

Conroy G., 2002 – Speciosity in the early Homo lineage: too many, too few, or just about right ? *Journal of Human Evolution,* 43: 759-766.

Conroy G., Smith R., 2007 – The size of scalable brain components in the human evolutionary lineage: With a comment on the paradox of *Homo floresiensis*. *Homo-Journal of Comparative Human Biology,* 58:1-12.

Cooke B., 1985 – Plio-Pleistocene Suidae in relation to African hominid deposits. *In* Beden M. *et al.* (Eds.), Colloque international juin 1981, Fondation Singer Polignac, Paris: Masson, pp. 101-117.

Cope D.A., Lacy M.G., 1992 – Falsification of a single hypothesis using the coefficient of variation: a simulation approach. *American Journal of Physical Anthropology*, 89: 359-378.

Coppens Y., 1983 – Les Hominidés du Pliocène et du Pléistocène d'Afrique orientale et leur environement. *In* Sakka M. (Ed.) *Morphologie évolutive, morphogenése du crâne et origine de l'homme*. Edition du C.N.R.S, pp. 185-194.

Cracraft J., 1974 – Phylogenetic models and classification. *Systematic Zoology,* 23: 71-90.

Cracraft J., Eldredge N., 1979 – *Phylogenetic analysis and paleontology*. Columbia University Press, New York.

Cronin J.E., Boaz N.T., Stringer C.B., Rak Y., 1981 – Tempo and mode in hominid evolution, *Nature,* 292: 113-122.

Cunningham D.J., 1908 – The evolution of the eyebrow region of the forehead with special reference to the excessive supraorbital development in the Neanderthal race. *Transaction of the Royal society of Edinburgh*, 46: 283-311.

Curnoe D., 2003 – Problems with the use of cladistic analysis in palaeoanthropology. *Homo*, 53: 225-234.

Darlu P., Tassy P., 1993 – *La reconstruction phylogénétique. Concepts et méthodes. Collection biologie théorique*. Masson, Paris.

Darroch J., Mosiman J., 1985 – Canonical and principal components of shape. *Biometrika,* 72: 241-252.

Darwin C., 1871 – *The descent of Man, and selection in relation to sex*, Murray J. (Ed.), London.

Day M.H., 1986 – *Guide to fossil man*. 4e édition, London, Cassel.

Day M.H., Stringer C.B., 1982 – A reconsideration of the Omo Kibish remains and erectus-sapiens transition. *In* Lumley de H. (Ed.), *l'Homo erectus et la place de l'Homme de Tautavel parmi les Hominidés fossiles.* 1er Congrès international de Paléontologie humaine, prétirage, pp. 814-846.

Day M.H., Stringer, C.B., 1991 – Les restes crâniens d'Omo-Kibbish et leur classification à l'intérieur du genre *Homo*. *L'Anthropologie*, 95: 573-594.

Day M.H., Leakey M.D., Magori C., 1980 – A new hominid fossil skull (L.H. 18) from the Ngaloba Beds, Laetoli, northern Tanzania. *Nature,* 284: 55-56.

Day M.H., Leakey R.E.F., Walker A. C., Wood B.A., 1976 – New hominids from East Turkana, Kenya II, *American Journal of Physical Anthropology*, 45: 369-436.

Day M.H., Leakey R.E.F., Walker A.C., Wood B.A., 1975 – New hominids from East Rudolf, Kenya I. *American Journal of Physical Anthropology*, 42: 461-476.

De Beer G., 1930 – *Embryology and evolution*, Clarendon Press, Oxford.

De Beer, G. 1954. Archaeopteryx and evolution. *The advancement of science*, 21: 160-170.

De Beer G., 1958 – *Embryos and ancestors*, Clarendon Press, Oxford.

De Terra H., 1943 – Pleistocene geology on early man in Java. *Transaction of the American Philantropic Society*, 32: 437-466.

De Vos J., Sartono S., Harditja-Sasmita S., Sondaar P.Y., 1982 – The fauna from Trinil, type locality of *Homo erectus*; a reinterpretation. *Geologie en Mijnbouw*, 4: 207-211.

De Vos J., Aziz F., Sondaar P.Y., 1993 – Les faunes quaternaires de Java. *Les dossiers de l'archéologie*, 184: 56-61.

Dean C., Stringer C.B., Bromage T.G., 1986 – Age at death of the neanderthal child from Devil's Tower, Gibraltar, and the implications for studies of general growth and development in

Neanderthals. *American Journal of Physical Anthropology*, 70: 301-309.
Delson E., Eldredge N., Tattersall I., 1977 – Reconstruction of hominid phylogeny: a testable framework based on cladistic analysis. *Journal of Human Evolution*, 6: 263-278.
Delson E., 1984 – Cercopithecid biochronology of the African Plio-Pleistocene: correlation among eastern and southern hominid-bearing localities. *Courier Forschung Institut Senckenberg,* 69: 199-218
Denton G.H., 1995 – The problem of pliocene paleoclimate and ice-sheet evolution in Antarctica, *In* Vrba E., Denton G., Partridge T. and Burckle L. (Eds.), Paleoclimate and evolution, with emphasis on human origins, pp. 213-229.
Denys C., 1996 – Olduvai rodent faunas: palaeoecological and palaeobiogeographical affinities. a comparison between east and south African Plio-Pleistocene faunas. *Kaupia,* 6: 247-261.
Deschamps R., Maes F., 1985 – Essai de reconstitution des climats et des végétations de la basse vallée de l'Omo au plio pléistocène à l'aide des bois fossiles. *In* M. Beden, *et al.* (Eds.), Colloque international juin 1981, Fondation Singer Polignac, Paris: Masson, pp. 175-222.
Detroit F., 2002 – Origine et evolution des *Homo sapiens* en Asie du Sud-est: descriptions et analyses morphométriques de nouveaux fossiles. PhD dissertation Museum National d'Histoire Naturelle, Paris.
Dines H.G., 1964 – The geology and archeology. *In* Ovey Cameron D. (Ed.), *The Swanscombe skull. A survey of research on a Pleistocene site*. Royal Anthropological Institute of Great Britain and Ireland, pp. 5-11.
Dobzhansky Th., 1944 – On species and races of living and fossil man. *American Journal of Physical Anthropology*, 2: 251-256.
Dodson P., 1978 – On the use of ratios in growth studies. *Systematic Zoology*, 27: 62-67.
Dubois E., 1893 – Palaeontologishe onderzoekingen op Java. *Verslag Mijnwezen (Batavia)*, 10: 10-14.
Dubois E., 1894 – Palaeontologishe onderzoekingen op Java. *Verslag Mijnw.ezen (Batavia)*, 4: 14-18.
Dubois E., 1896 – On *Pithecanthropus erectus* form between man and the apes. *The scientific transactions of the Royal Dublin Society*, 7: 1-18.
Dubois E., 1940 – The fossil humain remains discovered in Java by Dr. G.H.R. von Koenigswald and attribuated by him to *Pithecanthropus erectus*, in reality remains of *Homo sapiens soloensis*. (continuation). *Nederlandjie Akademie van Wetenschappen*, 43: 841-854.
Ducros A., 1967 – Le chignon occipital, mesure sur le squelette, *L'Anthropologie*, 7: 75-96.
Dupont L.M., Leroy, S.A.G., 1995 – Steps towards drier climatic conditions in northwestern Africa during the upper Pliocene. *In* Vrba E., Denton T., Partridge T., Burckle L. (Eds.), Paleoclimate and evolution, with emphasis on human origins, pp. 289-298.
Durband A., 2002 – The view from down under: A test of the multiregional hypothesis of modern human origins using the basicranial evidence from Australasia, Acts of the XVIIth Congress of the Indo Pacific Prehistory association, Taipei.
Duyfjes J., 1936 – Zur geologie und stratigraphie des Kendenggebietes zwischen Trinil und Soerabaja (Java). *De Ingenieur in Nederlansch Indië*, 8: 136-149.
Dzaparidze V., Bosinski G., Bugianisvili T., Gabunia L., Justus A., Klopotovskaja N., Kvavadze E., Lordkipanidze D., Majsuradze G., Mgeladze N., Nioradze M., Pavlenisvili E., Schmincke H.U., Sologasvili D., Tusabramisvili D., Tvalcrelidze M., Vekua A., 1989 – Der altpaläolithische Fundplatz Dmanisi in Georgien (Kaukasus). *Jahrbuch des Römisch-Germanischen ZentralMuseums Mainz*, 67-128.
Eck G., Jablonski N., Leakey M., 1987 – Les faunes plio-pleistocènes de la vallée de l'Omo (Ethiopie). Cercopithecidae de la formation de Shungura. *In* Coppens Y. and Howell F.C. (Eds.), Les faunes plio-pleistocènes de la vallée de l'Omo (Ethiopie). *Cahiers de Paléontologie,* 3: 1-169.
Eisenman V., 1987 – Les Equidés des gisements de la vallée de l'Omo en Ethiopie. Les Périssodactyles, les artiodactyles (les Bovidae) Expédition internationale 1967-1976. *In* Coppens Y. and Howell F.C. (Eds.), Les faunes plio-pleistocènes de la vallée de l'Omo (Ethiopie). *Cahiers de Paléontologie* 1: 13-67.
Eldredge N., Cracraft J., 1980 – *Phylogenetic patterns,and the evolutionary process*. Columbia University Press, New York.
Eldredge N., Gould S.J., 1972 – Ponctuated equilibria : an alternative to phyletic gradualism, *In* Schopf T.J.M (Ed.), *Models in paleontology*, Freeman Cooper and Cie, San Francisco, pp. 82-115.
Elyaqtine M., 1995 – Variabilité et évolution de l'os temporal chez Homo sapiens, comparaison avec *Homo erectus*. PhD dissetation, Université de Bordeaux I.
Ennouchi E., 1962 – Un Néandertalien: l'Homme du Djebel Irhoud (Maroc).

L'Anthropologie, 66: 279-299.
Falguères C., Yokoyama Y., Jacob T., Sémah F. 2001 – Advancements in the dating of Solo Man. Actes du XIVe Congrès de l'Union Internationale des sciences préhistoriques et protohistoriques, Liège.
Falk D., Hildebolt C., Smith K., Morwood M, Sutikna T., Brown P., Jatmiko, Wahyu Saptomo E., Brunsden B., Prior F., 2005a – The brain of LB1 *Homo floresiensis*. *Science,* 308:242.
Falk D., Hilderolt C., Smith K., Morwood M., Sutikna T., Jatmiko, Wahyu Saptomo E., Brunsden B., Prior F., 2005b – Response to comment on "the brain of LB1 *Homo floresiensis*". *Science,* 310: 236c.
Farris J.S., 1991 – Excess homoplasy ratios. *Cladistics*, 7: 81-91.
Feibe C.S., Brown F.H., Mac Dougall I., 1989 – Stratigraphic context of fossil hominids from the Omo Group deposits: Northern Turkana Basin, Kenya and Ethiopia. *American Journal of Physical Anthropology*, 78: 595-622.
Ferguson W., 1995 – A new species of the genus *Homo* (primates: Hominidae) from the Plio/Pleistocene of Koobi Fora, in Kenya. *Primates,* 36: 69-89.
Fitch F.J., Miller J.A., 1970 – Radioisotopic age determination of Lake Rudolf artefact site. *Nature,* 226: 226-228.
Fitch F.J., Findlater I.C., Watkins R.T., 1974 – Dating the rock succession containing fossil hominids at East Rudolf, Kenya. *Nature,* 251: 213-214.
Fitch F.J., Miller J.A., Mitchell J.G., 1996 – Dating of the KBS tuff and *Homo rudolfensis*. *Journal of Human Evolution*, 30: 277-286.
Fitch W.M., 1971 – Towards defining the course of evolution minimum change for a specific tree topology. *Systematic zoology*, 19: 172-191.
Foley R., 1993 – African terrestrial primates: the comparative evolutionary biology of Teropithecus and the Hominidae. *In* N. Jablonski (Ed.) Theropithecus the rise and fall of a primate genus, Cambridge: Cambridge University Press, pp. 245-270.
Fontaine H., 1976 – Les tektites du Vietnam méridional. *Comptes Rendus de la Société de Géologie,* 2: 37-40.
Fraipont J., Lohest M., 1887 – La race humaine de Néandertal ou de Canstadt en Belgique: recherches ethnographiques sur des ossements humains découverts dans les dépôts quaternaires d'une grotte à Spy et détermination de leur âge chronologique. *Archives de Biologie*, 7: 587-757.
Fullagar R., Price D., Head L., 1996 – Early human occupation of northern Australia and thermoluminescence dating of Jinmium rock-shelter (Northern Territory). *Antiquity*, 70: 751-773.
Gabunia L., Vekua A., 1995 – A Plio-Pleistocene hominid from Dmanisi, East Georgia, Caucasus. *Nature*, 373: 509-512.
Garrod D.A.E., 1928 – Archaeology and Geology. *In* Garrod D.A.E., Buxton L.H.D., Smith G.E. and Bate D. (Eds.), Excavation of a mousterian rock-shelter at Devil's Tower, Gibraltar. *Journal of the Royal Anthroplological Institut,* 58: 34-56.
Gauld S.C., 1989 – Patterns of cranial bone thickness in extant hominoid species. *American Journal of Physical Anthropology*, 78: 225
Geoffroy Saint-Hilaire E., 1818 – *Philosophie anatomique*, Paris.
Geraads D., 1987 – Dating the northern African cercopithecid fossil record, *Human Evolution* 2: 19-27.
Geraads D., 1994 – Evolution of bovid diversity in the Plio-Pleistocene of Africa. *Historical Biology,* 7: 221-237.
Geraads D., Coppens Y., 1995 – Evolution des faunes de mammifères dans le Plio-pleistocène de la basse vallée de l'Omo (Ethiopie): apports de l'analyse factorielle. *Comptes Rendus Académie Sciences,* 320: 625-637.
Gibbons A., 1997 – A new face or human Ancestors. *Science,* 276: 1331-1333.
Gilbert W., White T., Asfaw B., 2003 – *Homo erectus, Homo ergaster, "Homo cepranensis*", and the Daka cranium. *Journal of Human Evolution*, 45: 255-259.
Gingerich P.D., 1977 – Patterns of evolution in the Mammalian fossil record. *In* Hallam A. (Ed.), Patterns of Evolution, Elsevier Scientific Publishing Company, Amsterdam, pp. 469-500.
Gingerich P.D., 1979 – The stratophenetic approach to phylogeny reconstruction in vertebate paleontology. *In* Cracraft J., Eldredge N. (Eds.), Phylogenetic analysis and paleontology, Columbia University Press, New York, pp. 41-77.
Gould S.J., 1977 – *Ontogeny and phylogeny*, Harvard University Press, Cambridge.
Gremiatskij M.A., 1949 – Le crâne de l'enfant néandertalien de la grotte de Teshik Tash, sud de

l'Ouzbékistan. *In* Gremiatskij M.A. and Nesturkh N.F. (Eds.), Teshik-Tash, l'homme paléolithique. Moscou. *Tudy. Uzb. fil. Acad. Nauk. URSS*, pp.137-180.

Grémy F., Salmon D., 1969 – Bases statistiques pour la recherche médicale et biologique. Dunod.

Grimaud D., 1982 – *Évolution du pariétal de l'Homme fossile, position de l'Homme de Tautavel parmi les hominidés*. Thèse de 3e cycle. Muséum National d'Histoire Naturelle.

Grimaud D., Jacob T., 1983 – Les pariétaux du Pithécanthrope Sangiran 10. *L'Anthropologie*, 87: 469-474.

Grimaud-Hervé D., 1986 – The parietal bone of Indonesian *Homo erectus*. *Human Evolution*, 1: 167-182.

Grimaud-Hervé D., 1997 – L'évolution de l'encéphale chez *Homo erectus* et *Homo sapiens*, exemple de l'Asie et de l'Europe. Cahiers de Paléoanthropologie, Paris: Editions CNRS.

Groves C.P., 1989 – *A theory of human and primate evolution*. Oxford University Press, Oxford.

Groves C.P., Mazak V., 1975 – An approach to the taxonomy of the Hominidae: gracile Villafranchian hominids of Africa. *Casopis pro Mineralogii a Geologii,* 20: 225-247.

Grün R. Stringer C.B., 1991 – Electron spin resonance dating and the evolution of modern human. *Archaeometry*, 33: 153-199.

Habgood P.J., 1989a – An investigation into the usefulness of a cladistic approach to the study of the origin of anatomically modern humans. *Human Evolution*, 4: 241-252.

Habgood P.J., 1989b – The origin of anatomically-modern humans in Australasia. *In* Mellars P., Stringer C.B. (Eds.), The human revolution, Edinburgh. Edinburgh University Press, pp. 245-273.

Haeckel E., 1866 – *Generelle Morphologie der Organismen*, Reimer G. (Ed.), Berlin.

Hauser D.L., Presch W., 1991 – The effect of ordered characters on phylogenic reconstruction. *Cladistics,* 7: 243-265.

Heerkeren H.R. Van, 1972 – The stone age of Indonesia. *Verhandelingen van Het Koninklijk Instituut voor Taal, land en Volkendunde*, 61: 38-49.

Heim J.L., 1976 – Les Hommes fossiles de La Ferrassie. tome I. Le gisement, les squelettes adultes (crâne et squelette du tronc), *Archives de l'Institut de Paléontologie Humaine*, mémoire 35, Paris, Masson.

Hemmer H., 1975 – Notes on the phylogenetical position of the Petralona man. *Journal of Human Evolution*, 4: 457-459.

Henneberg M., 1987 – Hominid cranial capacity change through time: a darwinian process. *Human Evolution*, 2: 213-220.

Hennig G.J., Her W., Weber E., Xirotiris N.I., 1981 – E.S.R. dating of the fossil hominid cranium from Petralona Cave, Greece. *Nature,* 292: 533-536.

Hennig W., 1950 – *Grundzüge einer Theorie der phylogenetischen sytematik*, Deutsher Zentralverlag, Berlin.

Hennig W., 1966 – *Phylogenetic systematics*. University of Illinois Press, Urbana (édition 1979).

Hershlovitz I., Kornreich L., Laron Z., 2007 – Comparative skeletal features between *Homo floresiensis* and patients with primary growth hormone insensitivity (Laron Syndrome). *American Journal of Physical Anthropology,* 134: 198-208.

Hills M., 1978 – On ratios – A response to Atchley, Gaskins, and Anderson. *Systematic Zoology*, 27: 61-62.

Hlusko L., 2004 – Integrating the genotype and phenotype in hominid paleontology. *PNAS,* 101: 2653-2657.

Holloway R.L., 1975 – Early hominids endocasts; volumes, morphology, and significance. *In* Tuttle R. A. (Ed.), Primate functionnal morphology and evolution. The Hague, Mouton, pp. 393-416.

Holloway R.L., 1979 – Brain size, allometry and reorganisation, toward a synthesis, *In* Hahn M.E., Jensen C., Dudek B.C. (Eds.), Development and evolution of brain size, New York Academic Press, pp. 59-88.

Holloway R.L., 1981a – Volumetric and asymmetric determinations on recent hominid endocasts: Spy I and II, Djebel Ihroud I, and the Salè *Homo erectus* specimens, with some notes on Neandertal brain size. *American Journal of Physical Anthropology*, 55: 385-393.

Holloway R.L., 1981b – The indonesian *Homo erectus* brain endocasts revisited. *American Journal of Physical Anthropology*, 55: 215-230.

Holloway R.L., 1983 – Human brain evolution: a search for units, models and synthesis. *Canadian Journal of Anthropology / Revue Canadienne d'Anthropologie*, 3: 385-393.

Hooijer D., Kurten B., 1984 – Trinil and Kedung brubus: the Pithecanthropus- bearing and their relative age. *Annals Zoologici Fennici*, 21: 135-141.

Hopwood A.T., 1936 – Earth-movements, ice ages, and faunas. *Geol. Mag.,* 73: 185-188.
Hopwood A.T., 1940 – Fossil mammalas and pleistocene correlation. *Proc. Geol. Assoc.,* 51: 79-88.
Hou Y., Potts R., Baoyin Y., Zhengtang G., Deino A., Wei W., Clark J., Xie G., Huang W., 2000 – Mid-Pleistocene acheulean-like stone technology of the Bose Basin, South China. *Science,* 287: 1622-1626.
Howell F., 1999 – Paleo-demes, species clades, and extinctions in the pleistocene hominin record. *Journal of Anthropological Research* 55: 191-253.
Howell F.C., 1956 – Upper Pleistocene men of the southwest asian Mousterian. *In* Koenigswald von G.H.R. (Ed.), Hundert Jahre neanderthaler Neanderthal centenary. 1856-1956. Köln, Böhlau Verlag, pp. 185-198.
Howell F.C., 1978 – Hominidae, *In* Maglio V.J., Cooke H.B.S (Eds.), Evolution of African Mammals, Cambridge, Harvard University Press, pp. 154-248.
Howell F.C., 1991 – Préface, *In* Hublin J.J., Tillier A.M. (Eds.), *Aux origines d'Homo sapiens,* Presses Universitaires de France.
Howells W.W., 1966 – *Homo erectus. Scientific American*, 215: 46-53.
Howells W.W., 1980 – *Homo erectus* –who, when and where: a survey. *Yearbook of Physical anthropology,* 23: 1-23.
Howells W.W., 1973a – *Evolution of the genus Homo*. Adison Wesley, Reading, Masson.
Howells W.W., 1973b – *Cranial variation in Man. A study by multivariate analysis of patterns of difference among recent human populations*. Peabody Museum papers, vol. 67, Peabody Museum, Harvard University Press, Cambridge, Massachussetts.
Hublin J.J., 1978a – *Le torus occipital transverse et les structures associées : évolution dans le genre Homo*, Thèse de 3e cycle. Université P. et M. Curie, Paris VI.
Hublin J.J., 1978b – Quelques caractères apomorphes du crâne néandertalien et leur implication phylogénétique. *Compte Rendu de l'Académie des Sciences,* 287: 923-926.
Hublin J.J., 1978c – Anatomie du centre de l'écaille de l'occipital, le problème de l'inion. *Cahiers d'Anthropologie* (Paris), 2: 65-83.
Hublin J.J., 1983 – Les superstructures occipitales chez les prédécesseurs d'*Homo erectus* en Afrique : quelques remarques sur l'origine du torus occipital transverse. *Bulletins et Mémoires de la société d'anthropologie de Paris*, 10: 103-112.
Hublin J.J., 1985 – Human fossils from the North African Middle Pleistocene and the origine of *Homo sapiens*. *In* Liss A.R. (Ed.), Ancestors: the hard evidence, pp. 283-288.
Hublin J.J., 1986 – Some comments on the diagnostic features of *Homo erectus*, in Fossil Man –New facts, new ideas. *Anthropos*, (Brno), 23: 175-187.
Hublin J.J., 1989a – Les origines de l'Homme moderne: Europe occidentale et Afrique du Nord. *In* Giacobini J. (Ed.), *Hominidae, Proceedings of the 2nd International Congress of Human Paleontology*, Editoriale Jaca Book, Milan, pp. 423-430.
Hublin J.J., 1989b – Les caractères dérivés d'*Homo erectus* : relation avec l'augmentation de la masse squelettique. *In* Giacobini J. (Ed.), *Hominidae, Proceedings of the 2nd International Congress of Human Paleontology*, Editoriale Jaca Book, Milan, pp. 194-204.
Hublin J.J., 1991 – *L'émergence des Homo sapiens archaïques : Afrique du nord-ouest et Europe occidentale*. PhD dissertation, Université de Bordeaux I.
Hublin J.J., Tillier, A.M., 1991 – Gradualisme et rupture. *In* Hublin J.J. and Tillier A.M. (Eds.), Aux origines d'*Homo sapiens*, Paris, Press Universitaire de France, pp. 291-327.
Huffman F., 2001 – Geological context and age of the Perning/Modjokerto *Homo erectus*, East Java. *Journal of Human Evolution,* 40: 353-362.
Huffman F., Shipman P., Hertler C., De Vos J., Aziz F., 2005 – Historical evidence of the 1936 Mojokerto skull discovery, East Java. *Journal of Human Evolution* 48: 321-363.
Huffman F., Zaim Y., Kappelman J., Ruez D., De Vos J., Rizal Y., Aziz F., Hertler C., 2006 – Relocation of the 1936 Mojokerto skull discovery site near Perning, East Java. *Journal of Human Evolution* 50: 431-451.
Huxley J.S., 1940., *The new systematics*, Oxford University Press, London.
Huxley J.S., 1957 – The three types of evolutionary process. *Nature,*180: 454-455.
Huxley J.S., 1958 – Evolutionary processes and taxonomy with special references to grades. *In* Hedberg O. (Ed.) Systematics of today, *Uppsala Universitats Arsskrift*, 6: 21-39.
Hyodo M., Watanabe N., Sunata W., Susanto E.E., Wahyono H., 1993 –

Magnetostratigraphy of hominid fossil bearing formations in Sangiran and Modjokerto, Java. *Anthropological Science*, 101: 157-186.
ISAAC G.L., LEAKEY R.E.F., BEHRENSMEYER A.K., 1971 – Archaeological traces of early hominid activities, east of Lake Rudolf, Kenya. *Science*, 173: 1129-1134.
ITHARA M., SHIBASAKI T., SODIJONO, HAYASHI T., FURUYAMA K., 1985a – Outline of hominid fossil-bearing formations at Sonde, Ngandong and Kedungbrubus. *Quaternary Geology of the Hominid Fossil Bearing Formations in Java*, 1: 11-44.
ITHARA M., SODIJONO, KADAR D., SHIBASAKI T., KUMAI H., YOSHIKAWA F., AZIZ F., SOERADI T., WIKARNO, KADAR A.P., HASIBUAN F., KAGEMORI Y., 1985b – Geology and stratigraphy of the Sangiran area. *Quaternary Geology of the Hominid Fossil Bearing Formations in Java*, 1: 63-67.
JACOB T., 1965 – The paramastoïd crest in Indonesian skulls. *Anthropologica*, n.s. 7: 239-347.
JACOB T., 1967a. – Recent *Pithecanthropus* finds in Indonesia. *Current Anthropology*, 8: 501- 505.
JACOB T., 1967b – *The racial history of the indonesian region.* Drukkerij Neerlandia Utrecht.
JACOB T., 1976 – Early populations in the Indonesian region. *In* Kirk R.L., Thorne A.G. (Eds.), The origins of the australians. Australian Institut of Aboriginal Studies, Camberra, pp. 81-93.
JACOB T., 1977 – Evolution of Man in southeast Asia. *Journal of the Medical Sciences*, 9 : 175-186.
JACOB T., 1978 – The puzzle of Solo man. *Modern Quaternary Research in Southeast Asia*, 4: 31-40.
JACOB T., 1980 – The Pithecanthropus of Indonesia: phenotype, genetics and ecology. *In* L.K. Königsson (Ed.), Current argument on early Man. Pergamon Press, pp. 170-179.
JACOB T., 1982 – *Meganthropus, Pithecanthropus* and *Homo sapiens* in Indonesia: evidence and problems. Colloques internationaux du CNRS, pp. 81-84.
JACOB T., 1983 – Early man in Indonesia: the "defossilization" of human fossils. *Canadian Journal of Anthropology / Revue Canadienne d'Anthropologie*, 3: 191-194.
JACOB T., 1984 – The fossil skull cap from Sambungmacan and its implication to human evolution. *Berkala Bioanthropologi Indonesia*, 1: 19-27.
JACOB T., CURTIS G.H., 1971 – Preliminary potassium-argon dating of early man in Java. Contributions University of California, *Archeological Dating Facility*, 12: 50.
JACOB T., INDRIATI E., SOEJONO R., HSÜ K., FRAYER D., ECKHARDT R., KUPERAVAGE A., THORNE A., HENNEBERG M., 2006 – Pygmoid australomelanesian *Homo sapiens* skeletal remains from Liang Bua, Flores: population affinities and pathological abnormalities. *PNAS,* 103:13421-13426.
JAEGER J.J., 1973 – Un Pithécanthrope évolué. *La Recherche*, 39: 1006-1007.
JAEGER J.J., 1975 – Découverte d'un crâne d'Hominidé dans le Pléistocène moyen du Maroc. Colloque international CNRS n°218, Paris, 4-9 juin 1973. *Problèmes actuels de Paléontologie: Evolution des Vertébrés*, pp. 897-902.
JANVIER P., 1984 – Cladistics: theory, purpose and evolutionary implications. *In* Pollard J.W. (Ed.), *Evolutionary theory: paths into the future*, New York, pp. 39-75.
JANVIER P., TASSY P., THOMAS H., 1980 – Le cladisme. *La Recherche*, 117: 1396-1406.
JELINEK J., 1978 – Comparison of Mid-pleistocene evolutionnary process in Europe and in South-East Asia. *Proc. Symp. Natur. Select. Liblice. Csav*, Prague, 251-267.
JELINEK J., 1981 – Was *Homo erectus* already *Homo sapiens* ? *In* D. Ferembach (Ed.), *Les processus de l'hominisation*, CNRS, Paris, pp. 85-89.
JOHANSON D., EDEY M., 1983 – *Lucy, une jeune femme de 3 500 000 ans*, Robert Laffont Edition.
JOHANSON D.C., MASAO F.T., ECK G.G., WHITE T.D., WALTER R.C., KIMBEL W.H., ASFAW B., MANEGA P., NDESSOKIA P., SUWA G., 1987 – New partial skeleton of *Homo habilis* from Olduvai Gorge, Tanzania. *Nature,* 327: 205-209.
KENNEDY G.E., 1991 – On the autapomorphic traits of *Homo erectus*. *Journal of Human Evolution*, 20: 355-371.
KENNEDY K.A.R., 1998 – Middle and Late Pleistocene hominids of South Asia. *Humanity from African Naissance to coming millennia*, pp.167-175.
KENNEDY K.A.R., SONAKIA A., CHIMENT J., VERMA K.K., 1991 – Is the Narmada hominid an indian *Homo erectus* ? *American Journal of Physical Anthropology*, 86: 475-496.
KIDDER J., DURBAND A., 2004 – A re-evaluation of the metric diversity within *Homo erectus*. *Journal of Human Evolution,* 46: 297-313.

Kimbel W.H., 1991 – Species, species concepts and hominid evolution. *Journal of Human Evolution*, 20: 355-371.
Koenigswald G.H.R. Von, 1958 – Der Solo-Mensch von Java: ein tropischer Neanderthaler. *In* Hundert Jahre Neanderthaler 1856-1956 Böhlau-Verlag, Köln, pp: 21-26.
Koenigswald G.H.R. Von, 1936 – Ein fossiler Hominide aus dem Altpleistocän Ostjavas. *De Ingenieur in Nederlandsch-Indië*, 8: 149-157.
Koenigswald G.H.R. Von, 1940 – *Neue Pithecanthropus Funde*, 1936-1938 (Landsdrukkerij, Batavia).
Koenigswald G.H.R. Von, 1968 – Observation upon two Pithecanthropus mandibles from Sangiran. *Proceedings of the Konin. Ned. Akad. v. Weennsch.*, 71: 99-101.
Koenigswald G.H.R. Von, 1973 – *Australopithecus, Meganthropus* and *Ramapithecus. Journal of Human Evolution*, 2: 487.
Koenigswald G.H.R. Von, 1975 – Early man in Java: catalogue and problems. *In* Tuttle R.H. (Ed.), Paleontology, pp. 303-309.
Kokkoros P., Kanellis A., 1960 – Découverte d'un crâne d'Homme paléolithique dans la péninsule chalcidique. *L'Anthropologie*, 64: 438-446.
Kurten B., Poulianos A.N., 1977 – New stratigraphic and faunal material from Petralona cave, with special reference to the Carnivora. *Anthropos (Athènes),* 4: 47-130.
Laitman J., Tattersall I., 2001 – *Homo erectus newyorkensis*: An Indonesian fossil rediscovered in Manhattan sheds light on the middle phase of Human evolution. *The Anatomical Record,* 262: 341-343.
Lampl M., 1993 – Croissance saltatoire chez le nourisson. *Bulletins et Mémoires de la Société d'Anthropologie de Paris*, 5: 53-60.
Lang C., Leuenberger M., Schwande J., Johnsen S., 1999 – 16°C rapid temperature variation in Central Greenland 70 000 years ago. *Science,* 286: 934-936.
Langbroek M., Roebroeks W., 2000 – Extraterrestrial evidence on the age of the hominids from Java. *Journal of Human Evolution,* 38: 595-600.
Le Gros Clark W.E., 1964 – The fossil evidence for human evolution, 2e Edition, University of Chicago Press.
Leakey L.S.B., 1966 – *Homo habilis*, *Homo erectus* and the Australopithithecines. *Nature,* 209: 1279-1281.
Leakey M., 1993 – Evolution of Theropithecus in the Turkana Basin. *In* Jablonski N. (Ed.), Theropithecus the rise and fall of a primate genus, Cambridge: Cambridge University Press, pp. 85-124.
Leakey M., Spoor F., Brown F., Gathogo P., Leakey L., 2003 – A new hominin calvaria from Ileret (Kenya). *American Journal of Physical Anthropology,* 136.
Leakey M.D., 1971 – *Olduvai gorge*. Excavations in Beds I and II, 1960-1963. vol. 3, Cambridge University Press.
Leakey M.D., Hay R.L., 1982 – The chronological positions of the fossil hominids of Tanzaia. *In* Lumley de H. (Ed.) l'*Homo erectus* et la place de l'Homme de Tautavel parmi les Hominidés fossiles. Ier Congrès international de Paléontologie Humaine, prétirage, pp. 753-765.
Leakey R.E.F., 1973a – Evidence for an advanced Plio-Pleistocene hominid from East Rudolf, Kenya. *Nature,* 242: 447-450.
Leakey R.E.F., Walker A.C., 1976 – *Australopithecus*, *Homo erectus* and the single species hypothesis. *Nature,* 261: 572-574.
Leakey R.E.F., Walker A.C., 1985 – Further hominids from the Plio-pleistocene of Koobi Fora formation, Kenya. *American Journal of Physical Anthropology*, 67: 135-163.
Leakey R.E.F., Leakey M.G., Berhensmeyer A.K., 1978 – The hominid catalogue. *In* Leakey M.G., Leakey R.E. (Eds.) Koobi Fora research project, 1: The fossil hominids and an introduction to their context, 1968-1974. Clarendon Press, Oxford, pp. 86-182.
Lee S.H., 2005 – Brief communication: Is variation in the cranial capacity of Dmanisi sample too high to be from a single species ? *American Journal of Physical Anthropology*, 127: 263-266.
Leigh S.R., 1992 – Cranial evolution in *Homo erectus* and early *Homo sapiens. American Journal of Physical Anthropology*, 87: 1-13.
Leinders J.J.M., Aziz F., Sondaar P.Y., De Vos J., 1985 – The age of the hominid-bearing deposits of Java: state of the art. *Geologie en Mijnbouw*, 64: 167-173.
Lestrel P.E., 1975 – Hominid brain size *versus* time: revised regression estimates. *Journal of Human Evolution*, 5: 207-212.
Lestrel P.E., Read D.W., 1973 – Hominid cranial capacity *versus* time: a regression approach, *Journal of Human Evolution*, 2: 405-411.
Li X., 1983 – Le paléolithique inférieur en Chine du Nord. *L'Anthropologie*, 87: 185-199.

Lieberman D., 2005 – Further fossil finds from Flores. *Nature,* 437: 957-958.
Lieberman D.E., Wood B.A., Pilbeam D.R., 1996 – Homoplasy and early *Homo*: an analysis of the evolutionary relationships of *H. habilis sensu stricto* and *H. rudolfensis. Journal of Human Evolution*, 30: 97-120.
Lindsay, E., Opdyke N., Johnson N. 1980 – Pliocene dispersal o the horse Equus and late Cenozoic mammalian dispersal events. *Nature,* 287:135-138.
Liptak P., 1969 – On the evolutionary systematics of *Hominidae*. *Symp. Biol. Hung.*, 8: 107-111.
Liu Z.C., 1983 – Le remplissage de la grotte de l'Homme de Pékin, Choukoutien – Localité 1. *L'Anthropologie*, 87: 163-176.
Løvtrup S., 1989 – Recapitulation, epigenesis and heterochrony. *Géobios*, mém. spé. 12: 269-281.
Lumley de H., Lumley de M.A., 1971 – Découvertes de restes humains anténéanderthaliens à la Caune de l'Arago (Tautavel, Pyrénées orientales. *Compte Rendu de l'Académie des Sciences* 272: 1739-1742.
Lumley de H., Fournier A., Park Y.C, Yokoyama, Y., Demouy A., 1984 – Stratigraphie du remplissage pléistocène moyen de la Caune de l'Arago à tautavel, étude de huit carottages effectués de 1981 à 1983. *L'Anthropologie*, 88: 5-18.
Lumley de M.A., 1976 – Les Anté-Néandertaliens dans le sud. In Lumley de H. (Ed.) *La préhistoire française. Civilisations paléolithiques et mésolithiques.* t.1. Edition C.N.R.S., Paris, pp. 547-560.
Lumley de M.A., Sonokia A., 1985 – Contexte stratigraphique et archéologique de l'homme de la Narmada, Hathorna, Madhya Pradesh, Inde. *L'Anthropologie*, 89: 3-12.
Lumley de M.A., Lordkipanidze D., 2006 – L'homme de Dmanissi (*Homo georgicus*), il y a 1 810 000 ans. *C. R. Palevol,* 5: 273-281.
Lumley de M.A., Gabounia L., Vekua A., Lordkipanidze D., 2006 – Les restes humains du Pliocène final et du début du Pléistocène inférieur de Dmanissi, Géorgie (1991-2000). I – Les crânes, D2280, D2282, D2700. *L'Anthropologie* 110: 1-110.
MacIntosch N.W.G., Larnach S.L., 1972 – The persistence of *Homo erectus* traits in australian aboriginal crania. *Archaeology and Physical Anthropology in Oceania*, 7: 1-7.
Maglio V.J., 1971 – Vertebrate faunas from Kubi Algi, koobi Fora and Ileret areas, East Rudolf, Kenya. *Nature,* 231: 248-249.
Maheshwari H., Silverman B., Dupuis J., Baumann., 1988 – Phenotype and genetic analysis of a syndrome caused by an inactivating mutation in the growth hormone-releasing hormone receptor: Dwarfism of Sindh. *Journal of Clinical Endocrinology and Metabolism,* 11: 4065-4074
Maier W.O., Nkini A.T., 1985 – The phylogenetic position of Olduvai hominid 9 especially as determined from basicranial evidence. *In* Liss A.R. (Ed.), Ancestors: the hard evidence, pp.249-254.
Mallegni F., Carnieri E., Bisconti M., Tartarelli G., Ricci S., Biddittu I., Segre, A. 2003 – *Homo cepranensis* sp. nov. and the evolution of African-European Middle Pleistocene hominids. *C. R. Palevol,* 2: 153-159.
Mann A., Trinkaus E., 1974 – Neandertal and Neandertal-like fossils from the Upper Pleistocene. *Yearbook of Physical Anthropology*, 17: 169-193.
Manzi G., Bruner E., Passarello P., 2003 – The one-million-year-old Homo cranium from Bouri (Ethiopia): a reconsideration of its *H. erectus* affinities. *Journal of Human Evolution,* 44: 731-736.
Manzi G., Mallegni F., Ascenzi A., 2001 – A cranium for the earliest europeans: phylogenetic position of the hominid from Ceprano, Italy. *PNAS,* 17: 10011-10016.
Marshall L.G., 1994 – L'évolution et le contexte paléoécologique des faunes de mammifères en Amérique du sud pendant le pléistocène. *L'Anthropologie* (Paris) 98: 55-80.
Martin R., MacLarnon A., Phillips J., Dussubieux L., Williams P., Dobyns W., 2006b – Comment on "the brain of LB1, *Homo floresiensis*". *Science,* 312:999b
Martin R., Saller K., 1959 – *Lehrbuch der anthropologie in systematischer darstellung mit besonderer berücksichtigung der anthropologischen methoden.* G. Fischer Verlag, Stuttgart.
Martin R., MacLarnon A., Phillips J., Dobyns W., 2006a – Flores hominid: new species or microcephalic dwarf ? *The Anatomical Record part A,* 288A:1123-1145.
Maslin T.P., 1952 – Morphological criteria of phyletic relationships. *Systematic Zoology*, 1: 48-70.
Matile L., Tassy P., Goujet D., 1987 – *Introduction à la systématique zoologique.* Biosystema 1, Société française de systématique, Paris.
Matsu'ura S., Aziz F., Watanabe N., 1992 – Probable chronological positions of two additional

mandibule specimens of the Sangiran hominids. *Journal of Anthropological Society of Nippon*, 100: 154-187.
MAUREILLE B., 1994 – *La face chez* Homo erectus *et* Homo sapiens *: recherche sur la variabilité morphologique et métrique*. PhD dissertation, Université de Bordeaux I.
MAUREILLE B., 2007 – Portrait d'un néandertalien d'Europe de l'ouest. *In* Vandermeersch B., Maureille B. (Eds.) Les néandertalies, biologie et culture. Editions du CTHS, Paris, pp.53-67.
MAYR E., 1950 – Taxonomic categories in fossil hominids, *Cold Springs Harbor Symposia on Quantitative Biology*, 25: 109-118;
MAYR E., 1957 – Species concepts and definitions. *In* Mayr E. (Ed.), *The species problem. American Association of Advanced Science Publications,* 50: 1-22.
MAYR E., 1969 – *Principles of systematic zoology.* Mc Graw-Hill, New York.
MAYR E., 1974 – Cladistic analysis or cladistic classification? *Zeitschrift für zoologische Systematik und Evolutionforschung*, 2: 94-128.
MAYR E., 1981 – Biological classification: toward a synthesis of opposing methodologies. *Science*, 214: 510-516.
MAYR E., 1986 – La systématique évolutionniste et les quatre étapes du processus de classification. In Tassy P. (Ed.), L'ordre et la diversité du vivant. Fayard, Fondation Diderot, Paris, pp. 143-160.
MELLARS P., 1986 – A new chronology for the french Mousterian period. *Nature*, 322: 410-411.
MERCIER N., 1992 – *Apport des méthodes de datations radionucléaires à l'étude du peuplement préhistorique de l'Europe et du Proche-Orient au cours du Pléistocène supérieur*. PhD dissertation, Université de Bordeaux I.
MERCIER N., VALLADAS H., BAR-YOSSEF O.,VANDERMEERSCH B., STRINGER C.B., JORON J.L., 1993 – Thermoluminescence date for the mousterian burial site of Es-Skhùl, Mt. Carmel. *Journal of Archaelogical Science*, 20: 169-174;
MILLER J.A., 1991 – Does brain size variability provide evidence of multiple species in *Homo habilis* ? *American Journal of Physical Anthropology*, 84: 385-398.
MORWOOD M., BROWN P., JATMIKO, SUTIKNA T., WAHYU SAPTOMO E., WESTAWAY K., AWE DUE R., ROBERTS R., MAEDA T., WASISTO S.,DJUBIANTONO T. 2005 – Further evidence for small-bodied hominins from the late Pleistocene of Flores, Indonesia. *Nature,* 437: 1012-1017.
MORWOOD M., AZIZ F., O'SULLIVAN P., NASRUDDIN, HOBBS D., RAZA A., 1999 – Archaeological and palaeontological research in central Flores, East Indonesia: results of fieldwork 1997-98. *Antiquity,* 73: 273-286.
MORWOOD M., SOEJONO R., ROBERTS R., SUTIKNA T., TUMEY C., WESTAWAY K., RINK W., ZHAO J., VAN DEN BERGH G., AWE DUE R., HOBBS D., MOORE M., BIRD M., FIFIELD L., 2004 – Archaeology and age of new hominin from Flores in eastern Indonesia. *Nature,* 431: 1087-1091.
MOSIMAN J., JAMES F., 1979 – New statistical methods for allometry with application for Florida red-winged blackbirds. *Evolution,* 33:144-159.
MTURI A.A., 1976 – New hominid from lake Ndutu, Tanzania. *Nature,* 262: 484-485.
MURRAY R., MAHESHWARI H., RUSSELL E., BAUMANN G., 2000 – Pituitary hypoplasia in patients with a mutation in the growth hormone-releasing hormone receptor gene. *American Journal of Neuroradiology,* 21: 685-689.
MURRILL R.I., 1981 – *Petralona Man*. Thomas C.C. Publisher.
NARA T. 1994 – *Étude de la variabilité de certains caractères métriques et morphologiques des néandertaliens*. PhD dissertation, Université de Bordeaux I.
NELSON G., 1970 – Outline of a theory of comparative biology. *Systematic Zoology*, 19: 673-384.
NELSON G., 1972 – Phylogenetic relationship and classification. *Systematic Zoology*, 21: 227-230.
NELSON G., 1973 – The higher-level phylogeny of vertebrates. *Systematic Zoology*, 22: 87-91.
NELSON G., 1978 – Ontogeny, phylogeny and the biogenetic law. *Systematic Zoology*, 27: 324-345.
NELSON G., PLATNICK N., 1981 – *Systematics and biogeography, cladistics and vicariance*. Columbia University Press, New York.
OAKLEY P., 1960 – The dating of Kabwe (Rhodesian Man). *Hundert Jahre Neanderthaler 1856-1956*, Düsseldorf, pp. 265-266.
OBENDORF P.J., OXNARD C.E., KEFFORD B.J., 2007 – Are the small human-like fossils found on Flores human endemic cretins ? *Proc. R. Soc. B.* doi:10.109 8/rspb2007.1488.
OLIVIER G., 1974 – Les corrélations structurales de la voûte crânienne. *L'Anthropologie,* 78: 553-588.
OPPENOORTH W.F., 1932 – *Homo (javanthropus) soloensis*, en Plistocene mensch van Java.

Wetensch. Med. Dienst. Mijnb. Ned Indië, 20: 49-74.

OPPENOORTH W.F., 1937 – The place of *Homo soloensis* among fossil men. *In* Maccurdy G. (Ed.), Early Man, pp.349-360.

PARTRIDGE T.C., 1973 – Geomorphological dating of cave openings at Makapansgat, Sterkfontein, Swartkrans and Taung. *Nature,* 246: 75-79.

PARTRIDGE T.C., 1982 – The chronological positions of the fossil hominids of Southern Africa. *In* Lumley de H. (Ed.), l'*Homo erectus* et la place de l'Homme de Tautavel parmi les Hominidés fossiles. 1er Congrès international de Paléontologie humaine, prétirage, pp. 617-675.

PATTERSON C., ROSEN D.E., 1977 – Review of Ichthyodectiform and other Mesozoic Teleostfishes and the theory and practice of classifying fossils. *Bulletin of American Museum of Natural History*, 185: 81-172 .

PICQ P., 1983 – *L'articulation temporo-mandibulaire des Hominidés fossiles : anatomie comparée, biomécanique, évolution, biométrie.* Thèse de 3ème cycle de l'Université Pierre et Marie Curie, Paris VI.

PILBEAM D., GOULD S.J., 1974 – Size and scaling in human evolution, *Science*, 186: 892-901.

PIVETEAU J., 1982 – La place de l'*Homo erectus* dans le phénomène de l'hominisation. *In* Lumley de H. (Ed.), l'*Homo erectus* et la place de l'Homme de Tautavel parmi les Hominidés fossiles. 1er Congrès international de Paléontologie humaine, prétirage, pp. 5-18.

POPE G.G., 1992 – Craniofacial evidence for the origin of modern humans in China. *Yearbook of Physical Anthropology*, 35: 243-298.

POPE G.G., CRONIN J.E., 1984 – The asian hominidae, *Journal of Human Evolution*, 13: 377-396.

POPPER K.R., 1978 – *La connaissance objective.* Editions complexe, Bruxelles.

POPPER K.R.,1973 – *La logique de la découverte scientifique*. Payot, Paris.

POULIANOS A.N., 1967 – The place of the Petralonian man among Palaeoanthropoids, *Anthropos (Brno)*, 11 216-221.

RAHMAWATI N., HASTUTI J., JACOB T., 2007 – Stature and cephalometric study of Rampasasa and other East Indonesian populations (Tana Ai, Bugis and Naulu people). International seminar on Southeast Asian Paleoanthropology-july 23-25 Yogjakarta, Indonesia p. 63.

RAK Y., KIMBEL W.H., HOVERS E., 1994 – A neandertal infant from Amud cave, Israël. *Journal of Human Evolution*, 26: 313-321.

RAYNAL J.P., 1990 – Essai de datation directe. *In* Raynal J-P. and Pautrat Y. (Eds.), *La Chapelle-aux-Saints et la préhistoire en Corrèze*. Bordeaux, La Nef, pp.43-46.

READ D.W., 1990 – From multivariate to qualitative measurment: representation of shape. *Human Evolution,* 5: 417-429.

RIGHTMIRE P., 1979 – Cranial remains from *Homo erectus* from bed II and IV, Olduvai Gorge, Tanzania. *American Journal of Physical Anthropology*, 51: 99-116.

RIGHTMIRE P., 1983 – The lake Ndutu cranium and early *Homo sapiens* in Africa. *American Journal of Physical Anthropology*, 61: 245-254.

RIGHTMIRE P., 1984 – Comparisons of *Homo erectus* from Africa and southeast Asia. *Courier Forschung Institut Senckenberg*, 69: 83-98.

RIGHTMIRE P., 1986a – Problèmes dans la classification des *Homo* du Pléistocène moyen. Les restes humains de Tautavel comparés aux Hominidés provenant du lac Ndutu et de Kabwe. *L'Anthropologie*, 90: 483-493.

RIGHTMIRE P., 1986b – Species recognition and *Homo erectus. Journal of Human Evolution*, 15: 823-826.

RIGHTMIRE P., 1990 – *The evolution of Homo erectus. Comparative anatomical studies of an extinct human species*, Cambridge University Press, Cambridge.

RIGHTMIRE P., 1991 – L'évolution d'*Homo erectus*: stase ou gradualisme. *In* Hublin J.J. and A.M. Tillier (Eds.), Aux origines d'*Homo sapiens*, Presse Universitaires de France, pp.75-95.

RIGHTMIRE P., 2004 – Brain size and encephalization in early to Mid Pleistocene *Homo. American Journal of Physical Anthropology,* 124: 109-123.

RIGHTMIRE P., LORDKIPANIDZE D., VEKUA A., 2006 – Anatomical descriptions, comparative studies and evolutionary significance of the hominin skulls from Dmanisi, republic of Georgia. *Journal of Human Evolution*, 50: 115-141.

ROBINSON J.T., 1967 – Variation and the taxonomy of the early hominids. *Evolutionary Biology*, 1: 69-100.

ROCHE H., DELAGNES A., BRUGAL J.P., FEIBEL C., KIBUNJIA M., MOURRÉ V., TEXIER P.J., 1999 – Early hominid stone tool production and technical skill 2.34 Myr ago in West Turkana, Kenya. *Nature,* 399: 57-60.

Rosen D.E., 1984 – Hierarchies and history. *In* Pollard J.W. (Ed.), *Evolutionnary theory: paths into the future*. Wiley J. and sons, New York, pp.34-47.
Santa Luca A.P., 1978 – A re-examination of presumed Neanderthal-like fossils. *Journal of Human Evolution*, 4: 619-636.
Santa Luca A.P., 1980 – *The Fossil Hominids: a comparative study of a far eastern Homo erectus group*, Yale University Publications in Anthropolgy, New Haven, number 78.
Sartono S., 1964 – On a new find of another Pithecanthropus skull: an announcement. *Bulletin of the Geological Survey of Indonesia*, 1: 1-5
Sartono S., 1968 – Early Man in Java: *Pithecanthropus* skull VII, a male specimen of *Pithecanthropus erectus* (I). *Koniklijke Nederlandse Akademie van Wetenschappen*, Amsterdam, serie B, 71: 396-422.
Sartono S., 1971 – Observations on a new skull of *Pithecanthropus erectus* (Pithecanthropus VIII) from Sangiran, Central Java. *Koninklijke Nederlandse Akademie van Wetenshappen*, Amsterdam, serie B, 74: 185-194.
Sartono S., 1976 – Genesis of the Solo terraces. *Modern Quaternary Research in Southeast Asia*, 2: 1-21.
Sartono S., 1982 – Characteristics and chronology of early men in Java. *In* Lumley H. de (Ed.), l'*Homo erectus* et la place de l'Homme de Tautavel parmi les Hominidés fossiles. 1er Congrès international de Paléontologie humaine, prétirage, pp. 491-541.
Sartono S., 1991a – *Homo (Pithecanthropus) erectus*: le débat sans fin. *L'Anthropologie*, 95: 123-136.
Sartono S., 1991b – New *Homo erectus* skull from Ngawi, East Java : relation to Upper Pleistocene terraces in Java. *Indo-Pacific Prehistory Association Bulletin*, 11: 14-22.
Sartono S., Grimaud D., 1983 – Les parétaux des pithécanthropes Sangiran 12 et Sangiran 17. *L'Anthropologie*, 87: 475-482.
Schaeffer B., Hecht M., Eldredge N., 1972 – Phylogeny and paleontology, *Evolutionary biology*, 6: 31-46.
Schneider D., Kent D., Mello G., 1992 – A detailed chronology of the Australasian impact event, the Brunhes-Matyama geomagnetic reversal, and global climate change. *Earth Planet. Sci. Let.* 111: 395-405.
Schwalbe G., 1901 – Der Neanderthalschädel. *Bonner Jahrbücher, Jahrbücher desvereins von altertumsfreuden in Rheinlande,* 106: 1-71.
Schwalbe G., 1904 – *Die Vorgeschichte des Menschen*. Braunschweig (Friedrich Viewug).
Schwartz J., Tattersall I., 1999 – Morphology and diversity in fossil hominids: accepting *Homo erectus* and *Homo ergaster* as separate taxa is just the beginning. *American Journal of Physical Anthropology*, 28: 246.
Scott L., 1995 – Pollen evidence for vegetational and climatic change in southern Africa during the neogene and quaternary. *In* Vrba E., Denton G., Partridge T. and Burckle L. (Eds.), Paleoclimate and evolution, with emphasis on human origins. pp: 65-76.
Sémah A.M., Sémah F., Djubiantono T., 1993 – Les grands sites fossilifères de Java. *Les dossiers de l'archéologie*, 184: 20-25.
Sémah F., 1986 – Le peuplement ancien de Java. Ebauche d'un cadre chronologique. *L'Anthropologie*, 90: 359-400.
Sémah F., Saleki H., Falguères C., 2000 – Did early man reach Java during the Late Pliocene? *Journal Archaeological Science,* 27: 763-769.
Sémah F., Sartono S., Zaim Y., Djubiantono T. 1980 – Premiers résultats concernant l'étude paléomagnétique de la partie ouest deu dôme de Sangiran (Java, Indonésie), *Compte Rendu de l'Académie des Sciences*, 290: 477-480.
Sémah F., Sémah A.M., Djubiantono T., Simanjuntak H.T., 1992 – Did they also make tools? *Journal of Human Evolution*, 23: 439-446.
Shen G., Michel V., 2007 – Position chronologique des sites d l'homme moderne en Chin d'après la datation U-Th. *L'Anthropologie*, 111: 157-165.
Simons E.L., Pilbeam D.R., 1965 – Preliminary revision of the Dryopithecinae (Pongidae, Anthropoidea), *Folia Primatologia*, 3: 81-152.
Simpson G.G., 1951 – The species concept, *Evolution*, 5: 285-293.
Simpson G.G., 1961 – *Principles of animal taxonomy,* Columbia University Press, New York.
Singer R., 1958 – The Rhodesian, Florisbad and Saldanha skulls. *In Hundert Jahre Neanderthaler 1856-1956*. Kemink en Zoon, Utrecht, pp. 52-62.
Skinner M., Gordon A., Collard N., 2006 – Mandibular size and shape variation in the hominins at Dmanisi, republic of Georgia. *Journal of Human Evolution*, 51: 36-49

Smith F.H., 1976 – A fossil hominid frontal from Velika Pecina (Croatia) and a consideration of Upper Pleistocene hominids from Yougoslavia. *American Journal of Physical Anthropology*, 44: 127-134.

Smith F.H., 1985 – Continuity and change in the origin of modern *Homo sapiens. Z. Morph. Anthrop.*, 75: 197-222.

Smith F.H., Ranyard G.C., 1980 – Evolution of the supraorbital region in upper Pleistocene fossil hominids from south-central Europe. *American Journal of Physical Anthropology*, 53: 589-609.

Sneath P.H.A., Sokal R.R., 1973 – *Numerical taxonomy, the principles and practice of numerical classification*, Freeman, San Francisco.

Sohn S., Wolpoff M.H., 1993 – Zuttiyeh face: view from the East, *American Journal of Physical Anthropology*, 91: 325-347.

Sokal R.R., Sneath P.H.A., 1963 – *Principles of numerical taxonomy.* Freeman, San Francisco.

Solecki R.S., 1960 – Three adult Neanderthal skeletons from Shanidar, northern Iraq. *Annual Report of the Smithsonian Institution for 1959*, pp. 603-635.

Sonokia A., 1985 – Early *Homo* from Narmada valley, India. *In* Liss A.R. (Ed.), Ancestors: the hard evidence, pp. 334-338.

Spencer L.M., 1997 – Diatery adaptations of Plio-Pleistocene Bovidae: implications for hominid habitat use. *Journal of Human Evolution*, 32: 201-228.

Spitery J., 1982 – Le frontal de l'Homme de Tautavel. *In* Lumley H. de (Ed.), l'*Homo erectus* et la place de l'Homme de Tautavel parmi les Hominidés fossiles. 1er Congrès international de Paléontologie humaine, prétirage, pp. 21-61.

Spitery J., 1985 – Evolution de l'os frontal chez les hominidés fossiles. *L'Anthropologie*, 89: 63-74.

Spoor F., Leakey M., Gathogo P., Brown F., Anton S., McDougall I., Kiarie C.,Manthi F., Leakey L., 2007 – Implications of a new early Homo fossils from Ileret, east of Lake Turkana, Kenya. *Nature,* 448: 688-691.

Stewart T.D., 1959 – The restored Shanidar 1 skull. *Annual Report of the Smithsonian Institution for 1958,* pp.473-478.

Strait D., Grine F., Moniz M., 1997 – A reappraisal of early hominid phylogeny, *Journal of Human Evolution*,32: 17-82.

Stringer C.B., 1983 – Some further notes on the morphology and dating of Petralona Hominid. *Journal of Human Evolution*, 12: 731-742.

Stringer C.B., 1984 – The definition of *Homo erectus* and the existence of the species in Africa and Europe, *Courier Forschungsinstitut Senckenberg*, 69: 131-143.

Stringer C.B., 1986 – The credibility of *Homo habilis. In* Wood B. (Ed.), *Major Topics in Primate and Human Evolution*. Cambridge University Press, pp.266-294.

Stringer C.B., 1987 – A numerical cladistic analysis for the genus *Homo*. *Journal of Human Evolution*, 16: 135-146.

Stringer C.B., 1991 – *Homo erectus* et "*Homo sapiens* archaïque" peut-on définir *Homo erectus* ? *In* Hublin J.J.and Tillier A.M. (Eds.), Aux origines d'*Homo sapiens*, Presses Universitaires de France, pp. 51-74.

Stringer C.B., Grün R., Schwarcz H.P., Goldberg P., 1989 – ESR dates for the hominid burial site of Es Skhùl in Israël. *Nature,* 338: 756-758.

Stringer C.B., Hublin J.J., Vandermeersch B., 1984 – The origin of anatomically modern humans in Western Europe. *In* Smith F.H. and Spencer F. (Eds.), The origins of modern humans, a world survey of the fossil evidence, New York, pp. 51-136.

Suzuki H., 1970 – The skull of the Amud man. *In* Suzuki H. and Takai F. (Eds.), The Amud man and his cave site. University of Tokyo, pp. 123-206.

Suzuki M., Wikarno, Budisantoso, Saefudin I., Ithara M., 1985 – Fission track ages of pumice tuff, tuff layers and javites of hominids fossil bearing formations in Sangiran area, Central Java. *Quaternary Geology of the Hominid Fossil Bearing Formations in Java* 2: 309-357.

Swisher C.C., Curtis G.H., Jacob T., Getty A.G., Suprijo A., Widiasmoro., 1994 – Age of the earliest known hominids in Java, Indonesia. *Science*, 263: 1118-1121.

Swisher C.C., Rink W.J., Anton S.C., Schwarcz H.P., Curtis G.H., Suprijo A., Widiasmoro., 1996 – Latest *Homo erectus* of Java: potential contemporaneity with *Homo sapiens* in southeast Asia. *Science*, 274: 1870-1874.

Swofford D.L., 1993 – PAUP. Phylogenetic Analysis Using Parcimony, version 3.1s. Computer program distributed by the Illinois Natural History Survey. Champaign, Illinois, USA.

Tassy P., 1986 – *L'ordre et la diversité du vivant. Quel statut scientifique pour les classifications biologiques ?* Fondation Diderot, Fayard, Paris.
Tassy P., 1988 – The classification of Proboscidea: how many cladistic classification? *Cladistics*, 4: 43-57.
Tassy P., 1991 – *L'arbre à remonter le temps. Les rencontres de la systématique et de l'évolution.* C. Bourgois, Paris.
Tassy P., 1994 – Ontogénie et analyse cladistique: Haeckel ou Nelson ? *Bulletin de la Société zoologique de France*, 119: 149-158.
Tattersall I., 1986 – Species recognition in human paleontology. *Journal of Human Evolution*, 15: 165-175.
Terhune C., Kimbel W., Lockwood C., 2007 – Variation and diversity in *Homo erectus*: a 3D geometric morphometric analysis of the temporal bone. *Journal of Human Evolution,* 53: 41-60.
Thiele K., 1993 – The holy grail of the perfect character : the cladistic treatment of morphometric data. *Cladistics*, 9: 275-304.
Thoma A., 1973 – New evidence for the polycentric evolution of *Homo sapiens*. *Journal of Human Evolution*, 2: 529-536.
Thorne A., Wolpoff M., 1992 – L'évolution multirégionale de L'Homme. *Pour la science*, 176: 40-46.
Tillier A.M., 1983 – Le crâne d'enfant d'Engis 2: un exemple de distribution des caractères juvéniles, primitifs et néanderthaliens. *Bulletin de la société royale belge d'Anthropologie et de Préhistoire*, 94: 51-75.
Tillier A.M., 1988 – La place des restes de Devil's Tower (Gibraltar) dans l'ontogenèse des Néanderthaliens. *Bulletins et Mémoires de la Société d'Anthropologie de Paris*, 5: 257-266.
Tillier A.M., 1995 – Les plus anciens fossiles humains européens : le cas de la France . *In* E.Bonifay and B. Vandermeersch (Eds.) Colloque : les premiers européens Acte du 114e Congrès National des Sociétés Savantes. Edition C.T.H.S, Paris, pp. 291-297.
Tillier A.M., Vandermeersch, B. 1982 – Le problème de la radiation géographique de *Homo erectus*. *Géobios*, mém. spécial 6: 483-492.
Tobias P.V.T., 1971 – *The brain in hominid evolution*. New York, Columbia Press.
Tobias P.V.T., 1978 – The earliest Transvaal members of the genus *Homo* with another look at some problems of hominid taxonomy and systematics. *Zeitschrift für Morphologie und Anhropologie*, 69 : 225-265.
Tobias P.V.T., 1983 – Recent advances in the evolution of the hominids with especiall reference to brain and speech. *Pont. Acad. Sci.* 50: 85-140.
Tobias P.V.T., 1985 – Single characters and total morphological pattern redefined the sorting affected by a selection of morphological features of the early hominids. *In* Liss A.R. (Ed.), Ancestors: the hard evidence, pp. 94-101.
Tobias P.V.T., 1991 – *Olduvai Gorge, vol. 4.* Cambridge University Press, Cambridge.
Tobias P.V.T., von Koenigswald F., 1964 – A comparison between the Olduvai hominids and those of Java and some implications for hominid phylogeny. *Nature* 204: 515-518.
Tocheri M., Orr C., Larson S., Sutikna T., Jatmiko, Wahyu Saptomo E., Awe Due R., Djubiantono T., Morwood M., Jungers W. 2007 – The primitive wrist of *Homo floresiensis* and its implications for hominin evolution. *Science,* 317: 1743-1747.
Trinkaus E., 1983 – *The Shanidar Neandertals*. New York, Academic Press.
Trinkaus E., 1990 – Cladistics and the hominid fossil record. *American Journal of Physical Anthropology*, 83: 1-11.
Trinkaus E., 1991 – Les Hommes fossiles de la grotte de Shanidar, Irak : évolution et continuité parmi les hommes archaïques tardifs du Proche-Orient. *L'Anthropologie*, 95: 535-572.
Turner A., 1999 – Assessing earliest human settlement of Eurasia: Late Pliocene dispersions from Africa. *Antiquity,* 73: 563-570.
Turner A., Chamberlain A., 1989 – Speciation, morphological change and the status of African *Homo erectus*. *Journal of Human Evolution*, 18: 115-130.
Vandermeersch B., 1965 – Position stratigraphique et chronologique relative des restes humains du Paléolithique moyen dans le Sud-Ouest de la France. *Annales de Paléontologie*, 51: 69-126.
Vandermeersch B., 1976 – Les Néandertaliens en Charente. *In* Lumley de H. (Ed.), La Préhistoire française. tome I. civilisations paléolithiques et mésolithiques. Paris, édition du CNRS, pp.584-587.
Vandermeersch B., 1981 – *Les hommes fossiles de Qafzeh (Israël).* Cahiers de Paléontologie (Paléoanthropologie). Editions du C.N.R.S.
Vandermeersch B., 1982a – The first *Homo sapiens sapiens* in the Near East.

In Ronen A. (Ed.) The transition from lower to middle Palaeolithic and the origin of modern man. Oxford, B.A.R., 151: 297-299

VANDERMEERSCH B., 1982b – L'Homme de Biache-Saint-Vaast : comparaison avec l'Homme de Tautavel. *In* Lumley de H. (Ed.), l'*Homo erectus* et la place de l'Homme de Tautavel parmi les Hominidés fossiles. *1er Congrès international de Paléontologie humaine, prétirage*, pp. 894-900.

VANDERMEERSCH B., 1989 – L'origine de l'homme moderne : le point de vue du paléoanthropologue. *In* Giacobini J. (Ed.), *Hominidae. Proceeding 2nd International Congress of Human Paleontology*, Jaca Book, Milan, pp. 415-421.

VEKUA A., LORDKIPANIDZE D., RIGHTMIRE P., AGUSTI J., FERRING R., MAISURADZE G., MOUSKHELISHVILI A., NIORADZE M., PONCE DE LEON M., TAPPEN M., TVALCHRELIDZE M., ZOLLIKOFER C., 2002 – A new skull of early Homo from Dmanisi, Georgia. *Science*, 297: 85-89.

VLCEK E., 1986 – Les anténéandertaliens en Europe centrale et leur comparaison avec l'Homme de Tautavel. *L'Anthropologie*, 90: 503-513

VONDRA C.F., JOHNSON G.D., BOWEN B.E., BEHRENSMEYER A.K., 1971 – preliminary stratigraphical studies of the East Rudolf Basin, Kenya. *Nature,* 231: 245-248.

VRBA, E. 1980 – Evolution, species and fossils: how does life evolve ? *South African Journal of Science,* 76: 61-84.

VRBA E., 1982 – Biostratigraphy and chronology, based particulary on *Bovidae*, of southern hominid-associated assemblages: Makapansgat, Sterkfontein, Taung, Krondraai, Swartkrans; also Elandsfontein (Saldanha), Kabwe (now Kabwe) and Cave of Hearth. *In* Lumley de H. (Ed.) l'*Homo erectus* et la place de l'Homme de Tautavel parmi les Hominidés fossiles. *1er Congrès international de Paléontologie humaine, prétirage*, pp. 707-752.

VRBA E., 1996 – Climate, heterochrony and human evolution, *Journal of Anthropological Research*, 52: 1-28.

WALKER A.C., 1981 – The Koobi Fora and their bearing on the origins of the genus *Homo*. *In* Sigmond F. and Cybulski J.S. (Eds.), *Homo erectus*, Papers in honor of Davidson Black, University Toronto Press, Toronto, pp.193-215.

WALKER A.C., LEAKEY R.E.F., 1978 – The hominids of East Turkana. *Scientist American*, 239: 44-56.

WALKER A.C., LEAKEY R.E.F., 1994 – *The Nariokotome Homo erectus skeleton,* Springer Verlag.

WANG D., 1985 – Découvertes récentes de restes humains fossiles en Chine. *L'Anthropologie*, 89: 135-139.

WEBER J., CZARNETZKI A., PUSCH C., 2005 – On "the brain of LB1 *Homo Floresiensis*". *Science,* 310: 236b

WEIDENREICH F., 1940 – The torus occipitalis and related structure and their transformations in the course of human evolution. *Bulletin of the Geological Society of China*, 19: 480-588.

WEIDENREICH F., 1943 – The skull of *Sinanthropus pekinensis* a comparative study. *Palaeontologia sinica* (New series D), 10: 1-484.

WEIDENREICH F., 1945 – The paleolithic child from Teshik-Tash cave in southern Uzbekistan (central Asia). *American Journal of Physical Anthropology*, 3: 151-162.

WEIDENREICH F., 1951 – Morphology of Solo man. *Anthropological Papers of the American Museum of Natural History*, 43: 205-322.

WEINER J.S., CAMPBELL B.G., 1964 – The taxonomic status of the Swanscombe skull. *In* Ovey Cameron D. (Ed.), *The Swanscombe skull. A survey of research on a Pleistocene site.* Royal Anthropological Institute of Great Britain and Ireland, pp. 175-207.

WESSELMAN H., 1995 – Of mice and almost-men: regional paleoecology and human evolution in the Turkana Basin. *In* Vrba E., Denton G., Partridge T. and Burckle L. (Eds.), Paleoclimate and evolution, with emphasis on human origins, pp. 356-368.

WHITE T., KIMBEL W., 1981 – *Australopithecus africanus*: its phyletic position reconsidered. *South Africa Journal of Science*, 77: 445-470.

WHITE T., ASFAW B., DEGUSTA D., GILBERT H., RICHARDS G., SUWA G., HOWELL F. C., 2003 – Pleistocene *Homo sapiens* from Middle awash, Ethiopia. *Nature*, 423: 742-747.

WIDIANTO H., ZEITOUN V., 2003 – Morphological description, biometry and phylogenetic position of the skull of Ngawi 1 (East Java, Indonesia). *International Journal of Osteoarchaeology,* 13: 339-351.

WIDIANTO H., 1993 – *Unité et diversité des hominidés fossiles de Java : présentation de*

restes humains fossiles inédits. PhD dissertation, Muséum National d'Histoire Naturelle, Paris.
WILEY E.O., 1978 – The evolutionary species concept reconsidered. *Systematic Zoology*, 27: 17-26.
WILEY E.O., 1979 – An annoted Linnean hierarchy, with comments on natural taxa and competing systems. *Systematic Zoology*, 27: 308-337.
WILEY E.O., 1981 – *Phylogenetics. The theory and practice of Phylogenetic Systematics*, Wiley J. and sons, New York.
WILEY E.O., SIEGEL-CAUSEY D., BROOKS D.R., FUNK V. A., 1991 – *The compleat cladist. A primer of phylogenetic procedures*. Museum of Natural History, Dyche Hall, the Universiy of Kansas, Lawrence, Kansas.
WOLPOFF M.H., 1980 – *Paleoanthropology*, Knopf, New York.
WOLPOFF M.H., SPUHLER J.N., SMITH F. H., RADOVCIC I., POPE G., FRAYER D.W., ECKHARDT R., CLARK G., 1988 – Modern human origins. *Science*, 241: 772-773.
WOLPOFF M.H., 1991 – *Homo erectus* et les origines de la diversité humaine. In Hublin J-J. and Tillier A-M. (Eds.), Aux origines d'*Homo sapiens*, Presses Universitaires de France, pp. 98-155.
WOLPOFF M.H., THORNE A.G., JELINEK J., YINYUN Z., 1994 – The case for sinking *Homo erectus*. 100 years of Pithecanthropus is enough. *Courier Forschungsinstitut Senckenberg*, 171: 341-361.
WOOD B., 1984 – The origin of *Homo erectus*. *Courier Forschungsinstitut Senckenberg*, 69: 389-406.
WOOD B., 1991 – *Koobi Fora research project*, vol.4, Clarendon Press Oxford.
WOOD B., 1992a – Origin and evolution of the genus *Homo*. *Nature,* 355: 667-669.
WOOD B., 1992b – Early hominid species and speciation. *Journal of human Evolution*, 22: 351-365.
WOOD B., 1994 – Taxonomy and evolutionary relationships of *Homo erectus*. *Courier Forschungsinstitut Senckenberg*, 171: 159-165.
WOODWARD A.S., 1921 – A new cave man from Rhodesia, South Africa. *Nature,* 108: 371-372.
WU R., 1983 – Hominid fossils from China and their bearing on human evolution. *Canadian Journal of Anthropology/Revue Canadienne d'Anthropologie*, 3: 207-214.
WU R., DONG X., 1983 – Des fossiles d'*Homo erectus* découverts en Chine. *L'Anthropologie*, 87: 177-183.
WU R., WU X., 1982 – Comparison of Tautavel man with *Homo erectus* and early *Homo sapiens* in China. *In* Lumley de H. (Ed.), l'*Homo erectus* et la place de l'Homme de Tautavel parmi les Hominidés fossiles. *Ier Congrès international de Paléontologie Humaine*, prétirage, t.2, pp. 605-616.
WU X., 1991 – Continuité évolutive des Hommes fossiles chinois. *In* Hublin J.J. and Tillier A.M. (Eds.), Aux origines d'*Homo sapiens*. Paris: Presses Universitaires de France, pp.158-179.
ZEITOUN V., 1996a – Le tubercule préglénoïde: implication phylogénétique. *Comptes Rendus de l'Académie des Sciences*, 321: 997-1003.
ZEITOUN V., 1996b – Cladistique et paléoanthropologie: le cas *Homo erectus* (Dubois, 1894). Ph.D. dissertation, Université de Bordeaux I.
ZEITOUN V., 2000 – Adéquation entre changements environnementaux et spéciations humaines au Plio-Pléistocène. *C. R. Acad. Sci.* 330: 161-166.
ZEITOUN V., 2001 – The taxinomical position of the skull of Zuttiyeh. *C. R. Acad. Sci.* 332: 521-525.
ZEITOUN V., WIDIANTO H., DJUBIANTONO T., 2007 – The phylogeny of the Flores Man: The cladistic answer. *In* Indriati E. (Ed.), Recent advances on southeast asian paleoanthropology and archaeology. Proceedings international seminar on southeast asian paleoanthropology, Yogjakarta, pp. 54-60
ZHOU M., LI Y., WANG L., 1982 – Chronology of the chinese fossil hominids. *In* Lumley de H. (Ed.), l'*Homo erectus* et la place de l'Homme de Tautavel parmi les Hominidés fossiles. 1er Congrès international de Paléontologie humaine, prétirage, t.2, pp. 593-604.
ZIELINSKI G, MAYEWSKI P, MEEKER L., WHITELOW S., TWICKLER M., 1996 – Potential impact of the Toba megaeruption ~ 71 000 years ago. Geophys. *Res. Lett.* 23: 837-840.

LIST OF THE FIGURES

www.ingramcontent.com/pod-product-compliance
Lightning Source LLC
LaVergne TN
LVHW071211210926
848482LV00015B/776